Springer Texts in Statistics

For further volumes:
http://www.springer.com/series/417

Vladimir Spokoiny • Thorsten Dickhaus

Basics of Modern Mathematical Statistics

Springer

Vladimir Spokoiny
Weierstrass Institute (WIAS)
Berlin, Germany

Thorsten Dickhaus
Weierstrass Institute for Applied Analysis
and Stochastics (WIAS)
Mohrenstr. 39
D-10117 Berlin, Germany

ISSN 1431-875X ISSN 2197-4136 (electronic)
ISBN 978-3-662-51348-4 ISBN 978-3-642-39909-1 (eBook)
DOI 10.1007/978-3-642-39909-1
Springer Heidelberg New York Dordrecht London

Mathematics Subject Classification (2010): 62Fxx, 62Jxx, 62Hxx

Printed on acid-free paper

Springer is part of Springer Science+Business Media (www.springer.com)

To my children Seva, Irina, Daniel,
Alexander, Michael, and Maria

To my family

Preface

Preface of the First Author

This book was written on the basis of a graduate course on mathematical statistics given at the mathematical faculty of the Humboldt-University Berlin.

The classical theory of parametric estimation, since the seminal works by Fisher, Wald, and Le Cam, among many others, has now reached maturity and an elegant form. It can be considered as more or less complete, at least for the so-called regular case. The question of the optimality and efficiency of the classical methods has been rigorously studied and typical results state the asymptotic normality and efficiency of the maximum likelihood and/or Bayes estimates; see an excellent monograph by Ibragimov and Khas'minskij (1981) for a comprehensive study.

In the time around 1984 when I started my own Ph.D. at the Lomonosoff University, a popular joke in our statistical community in Moscow was that all the problems in the parametric statistical theory have been solved and described in a complete way in Ibragimov and Khas'minskij (1981), there is nothing to do any more for mathematical statisticians. If at all, only few nonparametric problems remain open. After finishing my Ph.D. I also moved to nonparametric statistics for a while with the focus on local adaptive estimation. In the year 2005 I started to write a monograph on nonparametric estimation using local parametric methods which was supposed to systemize my previous experience in this area. The very first draft of this book was available already in the autumn 2005, and it only included few sections about basics of parametric estimation. However, attempts to prepare a more systematic and more general presentation of the nonparametric theory led me back to the very basic parametric concepts. In 2007 I extended significantly the part about parametric methods. In the spring 2009 I taught a graduate course on parametric statistics at the mathematical faculty of the Humboldt-University Berlin. My intention was to present a "modern" version of the classical theory which in particular addresses the following questions:

what do you need to know from parametric statistics to work on modern parametric and nonparametric methods?

how to identify the borderline between the classical parametric and the modern nonparametric statistics?

The basic assumptions of the classical parametric theory are that the parametric specification is exact and the sample size is large relative to the dimension of the parameter space. Unfortunately, this viewpoint limits applicability of the classical theory: it is usually unrealistic to assume that the parametric specification is fulfilled exactly. So, the modern version of the parametric theory has to include a possible model misspecification. The issue of large samples is even more critical. Many modern applications face a situation when the number of parameters p is not only comparable with the sample size n, it can be even much larger than n. It is probably the main challenge of the modern parametric theory to include in a rigorous way the case of "large p small n." One can say that the parametric theory that is able to systematically treat the issues of model misspecification and of small fixed samples already includes the nonparametric statistics. The present study aims at reconsidering the basics of the parametric theory in this sense. The "modern parametric" view can be stressed as follows:

- any model is parametric;
- any parametric model is wrong;
- even a wrong model can be useful.

The *model* mentioned in the first item can be understood as a set of assumptions describing the unknown distribution of the underlying data. This description is usually given in terms of some parameters. The parameter space can be large or infinite dimensional, however, the model is uniquely specified by the parameter value. In this sense "any model is parametric."

The second statement "any parametric model is wrong" means that any imaginary model is only an idealization (approximation) of reality. It is unrealistic to assume that the data exactly follow the parametric model, even if this model is flexible and involves a lot of parameters. Model misspecification naturally leads to the notion of the *modeling bias* measuring the distance between the underlying model and the selected parametric family. It also separates parametric and nonparametric viewpoint. The parametric approach focuses on "estimation within the model" ignoring the modeling bias. The nonparametric approach attempts to account for the modeling bias and to optimize the joint impact of two kinds of errors: estimation error within the model and the modeling bias. This volume is limited to parametric estimation and testing for some special models like exponential families or linear models. However, it prepares some important tools for doing the general parametric theory presented in the second volume.

The last statement "even a wrong model can be useful" introduces the notion of a "useful" parametric specification. In some sense it indicates a change of a paradigm in the parametric statistics. Trying to find the true model is hopeless anyway. Instead, one aims at taking a potentially wrong parametric model which, however, possesses some useful properties. Among others, one can figure out the following "useful" features:

- a nice geometric structure of the likelihood leading to a numerically efficient estimation procedure;
- parameter identifiability.

Lack of identifiability in the considered model is just an indication that the considered parametric model is poorly selected. A proper parametrization should involve a reasonable regularization ensuring both features: numerical efficiency/stability and a proper parameter identification. The present volume presents some examples of "useful models" like linear or exponential families. The second volume will extend such models to a quite general *regular* case involving some smoothness and moment conditions on the log-likelihood process of the considered parametric family.

This book does not pretend to systematically cover the scope of the classical parametric theory. Some very important and even fundamental issues are not considered at all in this book. One characteristic example is given by the notion of sufficiency, which can be hardly combined with model misspecification. At the same time, much more attention is paid to the questions of nonasymptotic inference under model misspecification including concentration and confidence sets in dependence of the sample size and dimensionality of the parameter space. In the first volume we especially focus on linear models. This can be explained by their role for the general theory in which a linear model naturally arises from local approximation of a general regular model.

This volume can be used as textbook for a graduate course in mathematical statistics. It assumes that the reader is familiar with the basic notions of the probability theory including the Lebesgue measure, Radon–Nycodim derivative, etc. Knowledge of basic statistics is not required. I tried to be as self-contained as possible; the most of the presented results are proved in a rigorous way. Sometimes the details are left to the reader as exercises, in those cases some hints are given.

Preface of the Second Author

It was in early 2012 when Prof. Spokoiny approached me with the idea of a joint lecture on Mathematical Statistics at Humboldt-University Berlin, where I was a junior professor at that time. Up to then, my own education in statistical inference had been based on the German textbooks by Witting (1985) and Witting and Müller-Funk (1995), and for teaching in English I had always used the books by Lehmann and Casella (1998), Lehmann and Romano (2005), and Lehmann (1999). However, I was aware of Prof. Spokoiny's own textbook project and so the question was which text to use as the basis for the lecture. Finally, the first part of the lecture (estimation theory) was given by Prof. Spokoiny based on the by then already substantiated Chaps. 1–5 of the present work, while I gave the second part on test theory based on my own teaching material which was mainly based on Lehmann and Romano (2005).

This joint teaching activity turned out to be the starting point of a collaboration between Prof. Spokoiny and myself, and I was invited to join him as a coauthor of the present work for the Chaps. 6–8 on test theory, matching my own research interests. By the summer term of 2013, the book manuscript had substantially been extended, and I used it as the sole basis for the Mathematical Statistics lecture. During the course of this 2013 lecture, I received many constructive comments and suggestions from students and teaching assistants, which led to a further improvement of the text.

Berlin, Germany Vladimir Spokoiny
Berlin, Germany Thorsten Dickhaus

Acknowledgments

Many people made useful and extremely helpful suggestions which allowed to improve the composition and presentation and to clean up the mistakes and typos. The authors especially thank Wolfgang Härdle and Vladimir Panov who assisted us on the long way of writing the book.

Contents

Chapter 1
Basic Notions

The starting point of any statistical analysis is *data*, also called *observations* or a *sample*. A statistical model is used to explain the nature of the data. A standard approach assumes that the data is random and utilizes some probabilistic framework. On the contrary to probability theory, the distribution of the data is not known precisely and the goal of the analysis is to infer on this unknown distribution.

The *parametric* approach assumes that the distribution of the data is known up to the value of a parameter $\boldsymbol{\theta}$ from some subset Θ of a finite-dimensional space $\mathbb{R}^p$. In this case the statistical analysis is naturally reduced to the estimation of the parameter $\boldsymbol{\theta}$: as soon as $\boldsymbol{\theta}$ is known, we know the whole distribution of the data. Before introducing the general notion of a statistical model, we discuss some popular examples.

1.1 Example of a Bernoulli Experiment

Let $\boldsymbol{Y} = (Y_1, \ldots, Y_n)^\top$ be a sequence of binary digits zero or one. We distinguish between deterministic and random sequences. Deterministic sequences appear, e.g., from the binary representation of a real number, or from digitally coded images, etc. Random binary sequences appear, e.g., from coin throw, games, etc. In many situations incomplete information can be treated as random data: the classification of healthy and sick patients, individual vote results, the bankruptcy of a firm or credit default, etc.

Basic assumptions behind a Bernoulli experiment are:

- the observed data Y_i are independent and identically distributed.
- each Y_i assumes the value one with probability $\theta \in [0, 1]$.

The parameter θ completely identifies the distribution of the data $\boldsymbol{Y}$. Indeed, for every $i \le n$ and $y \in \{0, 1\}$,

V. Spokoiny and T. Dickhaus, *Basics of Modern Mathematical Statistics*,
Springer Texts in Statistics, DOI 10.1007/978-3-642-39909-1_1,

$$\mathbb{P}(Y_i = y) = \theta^y (1-\theta)^{1-y},$$

and the independence of the Y_i's implies for every sequence $\boldsymbol{y} = (y_1, \ldots, y_n)$ that

$$\mathbb{P}\bigl(\boldsymbol{Y} = \boldsymbol{y}\bigr) = \prod_{i=1}^{n} \theta^{y_i} (1-\theta)^{1-y_i}. \tag{1.1}$$

To indicate this fact, we write $\mathbb{P}_\theta$ in place of $\mathbb{P}$.

Equation (1.1) can be rewritten as

$$\mathbb{P}_\theta\bigl(\boldsymbol{Y} = \boldsymbol{y}\bigr) = \theta^{s_n} (1-\theta)^{n-s_n},$$

where

$$s_n = \sum_{i=1}^{n} y_i .$$

The value s_n is often interpreted as the number of successes in the sequence $\boldsymbol{y}$.

Probabilistic theory focuses on the probabilistic properties of the data $\boldsymbol{Y}$ under the given measure $\mathbb{P}_\theta$. The aim of the statistical analysis is to infer on the measure $\mathbb{P}_\theta$ for an unknown θ based on the available data $\boldsymbol{Y}$. Typical examples of statistical problems are:

1. *Estimate* the parameter θ, i.e. build a function $\tilde{\theta}$ of the data $\boldsymbol{Y}$ into $[0, 1]$ which approximates the unknown value θ as well as possible;
2. Build a *confidence set* for θ, i.e. a random (data-based) set (usually an interval) containing θ with a prescribed probability;
3. *Testing a simple hypothesis* that θ coincides with a prescribed value θ_0, e.g. $\theta_0 = 1/2$;
4. *Testing a composite hypothesis* that θ belongs to a prescribed subset Θ_0 of the interval $[0, 1]$.

Usually any statistical method is based on a preliminary probabilistic analysis of the model under the given θ.

Theorem 1.1.1. *Let $\boldsymbol{Y}$ be i.i.d. Bernoulli with the parameter θ. Then the mean and the variance of the sum $S_n = Y_1 + \ldots + Y_n$ satisfy*

$$\begin{aligned} \mathbb{E}_\theta S_n &= n\theta, \\ \operatorname{Var}_\theta S_n &\stackrel{\text{def}}{=} \mathbb{E}_\theta\bigl(S_n - \mathbb{E}_\theta S_n\bigr)^2 = n\theta(1-\theta). \end{aligned}$$

Exercise 1.1.1. Prove this theorem.

This result suggests that the empirical mean $\tilde{\theta} = S_n/n$ is a reasonable estimate of θ. Indeed, the result of the theorem implies

$$\mathbb{E}_\theta \tilde{\theta} = \theta, \qquad \mathbb{E}_\theta \bigl(\tilde{\theta} - \theta\bigr)^2 = \theta(1-\theta)/n.$$

The first equation means that $\tilde{\theta}$ is an *unbiased* estimate of θ, that is, $\mathbb{E}_\theta \tilde{\theta} = \theta$ for all θ. The second equation yields a kind of concentration (consistency) property of $\tilde{\theta}$: with n growing, the estimate $\tilde{\theta}$ concentrates in a small neighborhood of the point θ. By the Chebyshev inequality

$$\mathbb{P}_\theta\bigl(|\tilde{\theta} - \theta| > \delta\bigr) \le \theta(1-\theta)/(n\delta^2).$$

This result is refined by the famous de Moivre–Laplace theorem.

Theorem 1.1.2. *Let Y be i.i.d. Bernoulli with the parameter θ. Then for every $k \le n$*

$$\mathbb{P}_\theta\bigl(S_n = k\bigr) = \binom{n}{k} \theta^k (1-\theta)^{n-k}$$
$$\approx \frac{1}{\sqrt{2\pi n\theta(1-\theta)}} \exp\biggl\{-\frac{(k-n\theta)^2}{2n\theta(1-\theta)}\biggr\},$$

where $a_n \approx b_n$ means $a_n/b_n \to 1$ as $n \to \infty$. Moreover, for any fixed $z > 0$,

$$\mathbb{P}_\theta\Bigl(\Bigl|\frac{S_n}{n} - \theta\Bigr| > z\sqrt{\theta(1-\theta)/n}\Bigr) \approx \frac{2}{\sqrt{2\pi}} \int_z^\infty e^{-t^2/2} dt.$$

This *concentration* result yields that the estimate $\tilde{\theta}$ deviates from a root-n neighborhood $A(z,\theta) \stackrel{\text{def}}{=} \{u : |\theta - u| \le z\sqrt{\theta(1-\theta)/n}\}$ with probability of order $e^{-z^2/2}$.

This result bounding the difference $|\tilde{\theta} - \theta|$ can also be used to build random *confidence intervals* around the point $\tilde{\theta}$. Indeed, by the result of the theorem, the random interval $E^*(z) = \{u : |\tilde{\theta} - u| \le z\sqrt{\theta(1-\theta)/n}\}$ fails to cover the true point θ with approximately the same probability:

$$\mathbb{P}_\theta\bigl(E^*(z) \not\ni \theta\bigr) \approx \frac{2}{\sqrt{2\pi}} \int_z^\infty e^{-t^2/2} dt. \tag{1.2}$$

Unfortunately, the construction of this interval $E^*(z)$ is not entirely data-based. Its width involves the true unknown value θ. A data-based confidence set can be obtained by replacing the population variance $\sigma^2 \stackrel{\text{def}}{=} \mathbb{E}_\theta\bigl(Y_1 - \theta\bigr)^2 = \theta(1-\theta)$ with its empirical counterpart

$$\tilde{\sigma}^2 \stackrel{\text{def}}{=} \frac{1}{n} \sum_{i=1}^n (Y_i - \tilde{\theta})^2$$

The resulting confidence set $E(z)$ reads as

$$E(z) \stackrel{\text{def}}{=} \{u : |\tilde{\theta} - u| \le z\sqrt{n^{-1}\tilde{\sigma}^2}\}.$$

It possesses the same asymptotic properties as $E^*(z)$ including (1.2).

The *hypothesis* that the value θ is equal to a prescribed value θ_0, e.g. $\theta_0 = 1/2$, can be checked by examining the difference $|\tilde{\theta} - 1/2|$. If this value is too large compared to $\sigma n^{-1/2}$ or with $\tilde{\sigma} n^{-1/2}$, then the hypothesis is wrong with high probability. Similarly one can consider a *composite hypothesis* that θ belongs to some interval $[\theta_1, \theta_2] \subset [0, 1]$. If $\tilde{\theta}$ deviates from this interval at least by the value $z\tilde{\sigma} n^{-1/2}$ with a large z, then the data significantly contradict this hypothesis.

1.2 Least Squares Estimation in a Linear Model

A linear model assumes a linear systematic dependence between the *output* (also called *response* or *explained variable*) Y from the *input* (also called *regressor* or *explanatory variable*) Ψ which in general can be multidimensional. The linear model is usually written in the form

$$\mathbb{E}(Y) = \Psi^\top \boldsymbol{\theta}^*$$

with an unknown vector of coefficients $\boldsymbol{\theta}^* = (\theta_1^*, \ldots, \theta_p^*)^\top$. Equivalently one writes

$$\boldsymbol{Y} = \Psi^\top \boldsymbol{\theta}^* + \boldsymbol{\varepsilon} \tag{1.3}$$

where $\boldsymbol{\varepsilon}$ stands for the individual error with zero mean: $\mathbb{E}\boldsymbol{\varepsilon} = 0$. Such a linear model is often used to describe the influence of the response on the regressor Ψ from the collection of data in the form of a sample (Y_i, Ψ_i) for $i = 1, \ldots, n$.

Let $\boldsymbol{\theta}$ be a vector of coefficients considered as a candidate for $\boldsymbol{\theta}^*$. Then each observation Y_i is approximated by $\Psi_i^\top \boldsymbol{\theta}$. One often measures the quality of approximation by the sum of quadratic errors $|Y_i - \Psi_i^\top \boldsymbol{\theta}|^2$. Under the model assumption (1.3), the expected value of this sum is

$$\mathbb{E}\sum |Y_i - \Psi_i^\top \boldsymbol{\theta}|^2 = \mathbb{E}\sum \bigl|\Psi_i^\top(\boldsymbol{\theta}^* - \boldsymbol{\theta}) + \varepsilon_i\bigr|^2 = \sum \bigl|\Psi_i^\top(\boldsymbol{\theta}^* - \boldsymbol{\theta})\bigr|^2 + \sum \mathbb{E}\varepsilon_i^2.$$

The cross term cancels in view of $\mathbb{E}\varepsilon_i = 0$. Note that minimizing this expression w.r.t. $\boldsymbol{\theta}$ is equivalent to minimizing the first sum because the second sum does not depend on $\boldsymbol{\theta}$. Therefore,

$$\operatorname*{argmin}_{\boldsymbol{\theta}} \mathbb{E}\sum |Y_i - \Psi_i^\top \boldsymbol{\theta}|^2 = \operatorname*{argmin}_{\boldsymbol{\theta}} \sum \bigl|\Psi_i^\top(\boldsymbol{\theta}^* - \boldsymbol{\theta})\bigr|^2 = \boldsymbol{\theta}^*.$$

In other words, the true parameter vector $\boldsymbol{\theta}^*$ minimizes the expected quadratic error of fitting the data with a linear combinations of the Ψ_i's. The least squares estimate of the parameter vector $\boldsymbol{\theta}^*$ is defined by minimizing in $\boldsymbol{\theta}$ its empirical counterpart, that is, the sum of the squared errors $\left|Y_i - \Psi_i^\top \boldsymbol{\theta}\right|^2$ over all i:

$$\tilde{\boldsymbol{\theta}} \stackrel{\text{def}}{=} \operatorname*{argmin}_{\boldsymbol{\theta}} \sum_{i=1}^{n} \left|Y_i - \Psi_i^\top \boldsymbol{\theta}\right|^2.$$

This equation can be solved explicitly under some condition on the Ψ_i's. Define the $p \times n$ design matrix $\Psi = (\Psi_1, \ldots, \Psi_n)$. The aforementioned condition means that this matrix is of rank p.

Theorem 1.2.1. *Let $Y_i = \Psi_i^\top \boldsymbol{\theta}^* + \varepsilon_i$ for $i = 1, \ldots, n$, where ε_i are independent and satisfy $\mathbb{E}\varepsilon_i = 0$, $\mathbb{E}\varepsilon_i^2 = \sigma^2$. Suppose that the matrix Ψ is of rank p. Then*

$$\tilde{\boldsymbol{\theta}} = \left(\Psi\Psi^\top\right)^{-1}\Psi \boldsymbol{Y},$$

where $\boldsymbol{Y} = (Y_1, \ldots, Y_n)^\top$. Moreover, $\tilde{\boldsymbol{\theta}}$ is unbiased in the sense that

$$\mathbb{E}_{\boldsymbol{\theta}^*}\tilde{\boldsymbol{\theta}} = \boldsymbol{\theta}^*$$

and its variance satisfies $\operatorname{Var}(\tilde{\boldsymbol{\theta}}) = \sigma^2\left(\Psi\Psi^\top\right)^{-1}$.

For each vector $\boldsymbol{h} \in \mathbb{R}^p$, the random value $\tilde{a} = \langle \boldsymbol{h}, \tilde{\boldsymbol{\theta}} \rangle = \boldsymbol{h}^\top \tilde{\boldsymbol{\theta}}$ is an unbiased estimate of $a^* = \boldsymbol{h}^\top \boldsymbol{\theta}^*$:

$$\mathbb{E}_{\boldsymbol{\theta}^*}(\tilde{a}) = a^* \tag{1.4}$$

with the variance

$$\operatorname{Var}(\tilde{a}) = \sigma^2 \boldsymbol{h}^\top \left(\Psi\Psi^\top\right)^{-1} \boldsymbol{h}.$$

Proof. Define

$$Q(\boldsymbol{\theta}) \stackrel{\text{def}}{=} \sum_{i=1}^{n} \left|Y_i - \Psi_i^\top \boldsymbol{\theta}\right|^2 = \|\boldsymbol{Y} - \Psi^\top \boldsymbol{\theta}\|^2,$$

where $\|\boldsymbol{y}\|^2 \stackrel{\text{def}}{=} \sum_i y_i^2$. The normal equation $dQ(\boldsymbol{\theta})/d\boldsymbol{\theta} = 0$ can be written as $\Psi\Psi^\top\boldsymbol{\theta} = \Psi\boldsymbol{Y}$ yielding the representation of $\tilde{\boldsymbol{\theta}}$. Now the model equation yields $\mathbb{E}_{\boldsymbol{\theta}}\boldsymbol{Y} = \Psi^\top\boldsymbol{\theta}^*$ and thus

$$\mathbb{E}_{\boldsymbol{\theta}^*}\tilde{\boldsymbol{\theta}} = \left(\Psi\Psi^\top\right)^{-1}\Psi\mathbb{E}_{\boldsymbol{\theta}^*}\boldsymbol{Y} = \left(\Psi\Psi^\top\right)^{-1}\Psi\Psi^\top\boldsymbol{\theta}^* = \boldsymbol{\theta}^*$$

as required.

Exercise 1.2.1. Check that $\operatorname{Var}(\tilde{\boldsymbol{\theta}}) = \sigma^2 \bigl(\Psi\Psi^\top\bigr)^{-1}$.

Similarly one obtains $\mathbb{E}_{\boldsymbol{\theta}^*}(\tilde{a}) = \mathbb{E}_{\boldsymbol{\theta}^*}\bigl(\boldsymbol{h}^\top\tilde{\boldsymbol{\theta}}\bigr) = \boldsymbol{h}^\top\boldsymbol{\theta}^* = a^*$, that is, $\tilde{a}$ is an unbiased estimate of a^*. Also

$$\operatorname{Var}(\tilde{a}) = \operatorname{Var}\bigl(\boldsymbol{h}^\top\tilde{\boldsymbol{\theta}}\bigr) = \boldsymbol{h}^\top \operatorname{Var}(\tilde{\boldsymbol{\theta}})\boldsymbol{h} = \sigma^2\boldsymbol{h}^\top\bigl(\Psi\Psi^\top\bigr)^{-1}\boldsymbol{h}.$$

which completes the proof.

The next result states that the proposed estimate $\tilde{a}$ is in some sense the best possible one. Namely, we consider the class of *all linear unbiased* estimates $\tilde{a}$ satisfying the identity (1.4). It appears that the variance $\sigma^2\boldsymbol{h}^\top\bigl(\Psi\Psi^\top\bigr)^{-1}\boldsymbol{h}$ of $\tilde{a}$ is the smallest possible in this class.

Theorem 1.2.2 (Gauss–Markov). *Let $Y_i = \Psi_i^\top\boldsymbol{\theta}^* + \varepsilon_i$ for $i = 1,\ldots,n$ with uncorrelated ε_i satisfying $\mathbb{E}\varepsilon_i = 0$ and $\mathbb{E}\varepsilon_i^2 = \sigma^2$. Let $\operatorname{rank}(\Psi) = p$. Suppose that the value $a^* \stackrel{\text{def}}{=} \langle\boldsymbol{h},\boldsymbol{\theta}^*\rangle = \boldsymbol{h}^\top\boldsymbol{\theta}^*$ is to be estimated for a given vector $\boldsymbol{h} \in \mathbb{R}^p$. Then $\tilde{a} = \langle\boldsymbol{h},\tilde{\boldsymbol{\theta}}\rangle = \boldsymbol{h}^\top\tilde{\boldsymbol{\theta}}$ is an unbiased estimate of a^*. Moreover, $\tilde{a}$ has the minimal possible variance over the class of all linear unbiased estimates of a^*.*

This result was historically one of the first optimality results in statistics. It presents a lower efficiency bound of any statistical procedure. Under the imposed restrictions it is impossible to do better than the LSE does. This and more general results will be proved later in Chap. 4.

Define also the vector of residuals

$$\hat{\boldsymbol{\varepsilon}} \stackrel{\text{def}}{=} \boldsymbol{Y} - \Psi^\top\tilde{\boldsymbol{\theta}}\,.$$

If $\tilde{\boldsymbol{\theta}}$ is a good estimate of the vector $\boldsymbol{\theta}^*$, then due to the model equation, $\hat{\boldsymbol{\varepsilon}}$ is a good estimate of the vector $\boldsymbol{\varepsilon}$ of individual errors. Many statistical procedures utilize this observation by checking the quality of estimation via the analysis of the estimated vector $\hat{\boldsymbol{\varepsilon}}$. In the case when this vector still shows a nonzero systematic component, there is evidence that the assumed linear model is incorrect. This vector can also be used to estimate the noise variance σ^2.

Theorem 1.2.3. *Consider the linear model $Y_i = \Psi_i^\top\boldsymbol{\theta}^* + \varepsilon_i$ with independent homogeneous errors ε_i. Then the variance $\sigma^2 = \mathbb{E}\varepsilon_i^2$ can be estimated by*

$$\hat{\sigma}^2 = \frac{\|\hat{\boldsymbol{\varepsilon}}\|^2}{n-p} = \frac{\|\boldsymbol{Y} - \Psi^\top\tilde{\boldsymbol{\theta}}\|^2}{n-p}$$

and $\hat{\sigma}^2$ is an unbiased estimate of σ^2, that is, $\mathbb{E}_{\boldsymbol{\theta}^*}\hat{\sigma}^2 = \sigma^2$ for all $\boldsymbol{\theta}^*$ and σ.

Theorems 1.2.2 and 1.2.3 can be used to describe the concentration properties of the estimate $\tilde{a}$ and to build confidence sets based on $\tilde{a}$ and $\hat{\sigma}$, especially if the errors ε_i are normally distributed.

Theorem 1.2.4. *Let* $Y_i = \Psi_i^\top \boldsymbol{\theta}^* + \varepsilon_i$ *for* $i = 1, \ldots, n$ *with* $\varepsilon_i \sim \mathcal{N}(0, \sigma^2)$. *Let* $\operatorname{rank}(\Psi) = p$. *Then it holds for the estimate* $\tilde{a} = \boldsymbol{h}^\top \tilde{\boldsymbol{\theta}}$ *of* $a^* = \boldsymbol{h}^\top \boldsymbol{\theta}^*$

$$\tilde{a} - a^* \sim \mathcal{N}(0, s^2)$$

with $s^2 = \sigma^2 \boldsymbol{h}^\top (\Psi \Psi^\top)^{-1} \boldsymbol{h}$.

Corollary 1.2.1 (Concentration). *If for some* $\alpha > 0$, z_α *is the* $1 - \alpha/2$*-quantile of the standard normal law (i.e.,* $\Phi(z_\alpha) = 1 - \alpha/2$*), then*

$$\mathbb{P}_{\boldsymbol{\theta}^*}\bigl(|\tilde{a} - a^*| > z_\alpha\, s\bigr) = \alpha$$

Exercise 1.2.2. Check Corollary 1.2.1.

The next result describes the confidence set for a^*. The unknown variance s^2 is replaced by its estimate

$$\hat{s}^2 \stackrel{\text{def}}{=} \hat{\sigma}^2 \boldsymbol{h}^\top (\Psi \Psi^\top)^{-1} \boldsymbol{h}$$

Corollary 1.2.2 (Confidence Set). *If* $E(z_\alpha) \stackrel{\text{def}}{=} \{a : |\tilde{a} - a| \le \hat{s}\, z_\alpha\}$, *then*

$$\mathbb{P}_{\boldsymbol{\theta}^*}\bigl(E(z_\alpha) \not\ni a^*\bigr) \approx \alpha.$$

1.3 General Parametric Model

Let $\boldsymbol{Y}$ denote the observed data with values in the observation space $\mathcal{Y}$. In most cases, $\boldsymbol{Y} \in \mathbb{R}^n$, that is, $\boldsymbol{Y} = (Y_1, \ldots, Y_n)^\top$. Here n denotes the sample size (number of observations). The basic assumption about these data is that the vector $\boldsymbol{Y}$ is a random variable on a probability space $(\mathcal{Y}, \mathcal{B}(\mathcal{Y}), \mathbb{P})$, where $\mathcal{B}(\mathcal{Y})$ is the Borel σ-algebra on $\mathcal{Y}$. The probabilistic approach assumes that the probability measure $\mathbb{P}$ is known and studies the distributional (population) properties of the vector $\boldsymbol{Y}$. On the contrary, the statistical approach assumes that the data $\boldsymbol{Y}$ are given and tries to recover the distribution $\mathbb{P}$ on the basis of the available data $\boldsymbol{Y}$. One can say that the statistical problem is inverse to the probabilistic one.

The statistical analysis is usually based on the notion of *statistical experiment*. This notion assumes that a family $\mathcal{P}$ of probability measures on $(\mathcal{Y}, \mathcal{B}(\mathcal{Y}))$ is fixed and the unknown underlying measure $\mathbb{P}$ belongs to this family. Often this family is parameterized by the value $\boldsymbol{\theta}$ from some parameter set Θ: $\mathcal{P} = (\mathbb{P}_{\boldsymbol{\theta}}, \boldsymbol{\theta} \in \Theta)$. The corresponding statistical experiment can be written as

$$\bigl(\mathcal{Y}, \mathcal{B}(\mathcal{Y}), (\mathbb{P}_{\boldsymbol{\theta}}, \boldsymbol{\theta} \in \Theta)\bigr).$$

The value $\boldsymbol{\theta}^*$ denotes the "true" parameter value, that is, $\mathbb{P} = \mathbb{P}_{\boldsymbol{\theta}^*}$.

The statistical experiment is *dominated* if there exists a *dominating* σ-finite measure $\boldsymbol{\mu}_0$ such that all the $\mathbb{P}_{\boldsymbol{\theta}}$ are absolutely continuous w.r.t. $\boldsymbol{\mu}_0$. In what follows we assume without further mention that the considered statistical models are dominated. Usually the choice of a dominating measure is unimportant and any one can be used.

The *parametric* approach assumes that Θ is a subset of a finite-dimensional Euclidean space $\mathbb{R}^p$. In this case, the unknown data distribution is specified by the value of a finite-dimensional parameter $\boldsymbol{\theta}$ from $\Theta \subseteq \mathbb{R}^p$. Since in this case the parameter $\boldsymbol{\theta}$ completely identifies the distribution of the observations $\boldsymbol{Y}$, the statistical estimation problem is reduced to recovering (estimating) this parameter from the data. The nice feature of the parametric theory is that the estimation problem can be solved in a rather general way.

1.4 Statistical decision problem. Loss and Risk

The statistical decision problem is usually formulated in terms of game theory, the statistician playing as it were against nature. Let $\mathcal{D}$ denote the *decision space* that is assumed to be a topological space. Next, let $\wp(\cdot,\cdot)$ be a *loss function* given on the product $\mathcal{D} \times \Theta$. The value $\wp(d, \boldsymbol{\theta})$ denotes the loss associated with the decision $d \in \mathcal{D}$ when the true parameter value is $\boldsymbol{\theta} \in \Theta$. The *statistical decision problem* is composed of a statistical experiment $(\mathcal{Y}, \mathcal{B}(\mathcal{Y}), \mathcal{P})$, a decision space $\mathcal{D}$ and a loss function $\wp(\cdot,\cdot)$.

A *statistical decision* $\rho = \rho(\boldsymbol{Y})$ is a measurable function of the observed data $\boldsymbol{Y}$ with values in the decision space $\mathcal{D}$. Clearly, $\rho(\boldsymbol{Y})$ can be considered as a random $\mathcal{D}$-valued element on the space $(\mathcal{Y}, \mathcal{B}(\mathcal{Y}))$. The corresponding *loss* under the true model $(\mathcal{Y}, \mathcal{B}(\mathcal{Y}), \mathbb{P}_{\boldsymbol{\theta}^*})$ reads as $\wp(\rho(\boldsymbol{Y}), \boldsymbol{\theta}^*)$. Finally, the *risk* is defined as the expected value of the loss:

$$\mathcal{R}(\rho, \boldsymbol{\theta}^*) \stackrel{\text{def}}{=} \mathbb{E}_{\boldsymbol{\theta}^*} \wp(\rho(\boldsymbol{Y}), \boldsymbol{\theta}^*).$$

Below we present a list of typical statistical decision problems.

Example 1.4.1 (Point Estimation Problem). Let the target of analysis be the true parameter $\boldsymbol{\theta}^*$ itself, that is, $\mathcal{D}$ coincides with Θ. Let $\wp(\cdot,\cdot)$ be a kind of distance on Θ, that is, $\wp(\boldsymbol{\theta}, \boldsymbol{\theta}^*)$ denotes the loss of estimation, when the selected value is $\boldsymbol{\theta}$ while the true parameter is $\boldsymbol{\theta}^*$. Typical examples of the loss function are quadratic loss $\wp(\boldsymbol{\theta}, \boldsymbol{\theta}^*) = \|\boldsymbol{\theta} - \boldsymbol{\theta}^*\|^2$, l_1-loss $\wp(\boldsymbol{\theta}, \boldsymbol{\theta}^*) = \|\boldsymbol{\theta} - \boldsymbol{\theta}^*\|_1$ or sup-loss $\wp(\boldsymbol{\theta}, \boldsymbol{\theta}^*) = \|\boldsymbol{\theta} - \boldsymbol{\theta}^*\|_\infty = \max_{j=1,\ldots,p} |\theta_j - \theta_j^*|$.

If $\tilde{\boldsymbol{\theta}}$ is an estimate of $\boldsymbol{\theta}^*$, that is, $\tilde{\boldsymbol{\theta}}$ is a Θ-valued function of the data $\boldsymbol{Y}$, then the corresponding risk is

$$\mathcal{R}(\rho, \boldsymbol{\theta}^*) \stackrel{\text{def}}{=} \mathbb{E}_{\boldsymbol{\theta}^*} \wp(\tilde{\boldsymbol{\theta}}, \boldsymbol{\theta}^*).$$

Particularly, the quadratic risk reads as $\mathbb{E}_{\boldsymbol{\theta}^*} \|\tilde{\boldsymbol{\theta}} - \boldsymbol{\theta}^*\|^2$.

Example 1.4.2 (Testing Problem). Let Θ_0 and Θ_1 be two complementary subsets of Θ, that is, $\Theta_0 \cap \Theta_1 = \emptyset$, $\Theta_0 \cup \Theta_1 = \Theta$. Our target is to check whether the true parameter $\boldsymbol{\theta}^*$ belongs to the subset Θ_0. The decision space consists of two points $\{0, 1\}$ for which $d = 0$ means the acceptance of the *hypothesis* $H_0 : \boldsymbol{\theta}^* \in \Theta_0$ while $d = 1$ rejects H_0 in favor of the *alternative* $H_1 : \boldsymbol{\theta}^* \in \Theta_1$. Define the loss

$$\wp(d, \boldsymbol{\theta}) = \mathbf{1}(d = 1, \boldsymbol{\theta} \in \Theta_0) + \mathbf{1}(d = 0, \boldsymbol{\theta} \in \Theta_1).$$

A test ϕ is a binary valued function of the data, $\phi = \phi(\boldsymbol{Y}) \in \{0, 1\}$. The corresponding risk $\mathcal{R}(\phi, \boldsymbol{\theta}^*) = \mathbb{E}_{\boldsymbol{\theta}^*}\phi(\boldsymbol{Y})$ can be interpreted as the probability of selecting the wrong subset.

Example 1.4.3 (Confidence Estimation). Let the target of analysis again be the parameter $\boldsymbol{\theta}^*$. However, we aim to identify a subset A of Θ, as small as possible, that covers with a prescribed probability the true value $\boldsymbol{\theta}^*$. Our decision space $\mathcal{D}$ is now the set of all measurable subsets in Θ. For any $A \in \mathcal{D}$, the loss function is defined as $\wp(A, \boldsymbol{\theta}^*) = \mathbf{1}(A \not\ni \boldsymbol{\theta}^*)$. A confidence set is a random set $\mathcal{E}$ selected from the data $\boldsymbol{Y}$, $\mathcal{E} = \mathcal{E}(\boldsymbol{Y})$. The corresponding risk $\mathcal{R}(\mathcal{E}, \boldsymbol{\theta}^*) = \mathbb{E}_{\boldsymbol{\theta}^*}\wp(\mathcal{E}, \boldsymbol{\theta}^*)$ is just the probability that $\mathcal{E}$ does not cover $\boldsymbol{\theta}^*$.

Example 1.4.4 (Estimation of a Functional). Let the target of estimation be a given function $f(\boldsymbol{\theta}^*)$ of the parameter $\boldsymbol{\theta}^*$ with values in another space F. A typical example is given by a single component of the vector $\boldsymbol{\theta}^*$. An estimate ρ of $f(\boldsymbol{\theta}^*)$ is a function of the data $\boldsymbol{Y}$ into F: $\rho = \rho(\boldsymbol{Y}) \in F$. The loss function $\wp$ is defined on the product $F \times F$, yielding the loss $\wp(\rho(\boldsymbol{Y}), f(\boldsymbol{\theta}^*))$ and the risk $\mathcal{R}(\rho(\boldsymbol{Y}), f(\boldsymbol{\theta}^*)) = \mathbb{E}_{\boldsymbol{\theta}^*}\wp(\rho(\boldsymbol{Y}), f(\boldsymbol{\theta}^*))$.

Exercise 1.4.1. Define the statistical decision problem for testing a simple hypothesis $\boldsymbol{\theta}^* = \boldsymbol{\theta}_0$ for a given point $\boldsymbol{\theta}_0$.

1.5 Efficiency

After the statistical decision problem is stated, one can ask for its optimal solution. Equivalently one can say that the aim of statistical analysis is to build a decision with the minimal possible risk. However, a comparison of any two decisions on the basis of risk can be a nontrivial problem. Indeed, the risk $\mathcal{R}(\rho, \boldsymbol{\theta}^*)$ of a decision ρ depends on the true parameter value $\boldsymbol{\theta}^*$. It may happen that one decision performs better for some points $\boldsymbol{\theta}^* \in \Theta$ but worse at other points $\boldsymbol{\theta}^*$. An extreme example of such an estimate is the trivial deterministic decision $\tilde{\boldsymbol{\theta}} = \boldsymbol{\theta}_0$ which sets the estimate equal to the value $\boldsymbol{\theta}_0$ whatever the data is. This is, of course, a very strange and poor estimate, but it clearly outperforms all other methods if the true parameter $\boldsymbol{\theta}^*$ is indeed $\boldsymbol{\theta}_0$.

Two approaches are typically used to compare different statistical decisions: the *minimax* approach considers the maximum $\mathcal{R}(\rho)$ of the risks $\mathcal{R}(\rho, \boldsymbol{\theta})$ over the parameter set Θ while the *Bayes* approach is based on the weighted sum (integral) $\mathcal{R}_\pi(\rho)$ of such risks with respect to some measure π on the parameter set Θ which is called the *prior* distribution:

$$\mathcal{R}(\rho) = \sup_{\theta \in \Theta} \mathcal{R}(\rho, \boldsymbol{\theta}),$$

$$\mathcal{R}_\pi(\rho) = \int \mathcal{R}(\rho, \boldsymbol{\theta}) \pi(d\boldsymbol{\theta}).$$

The decision ρ^* is called *minimax* if

$$\mathcal{R}(\rho^*) = \inf_\rho \mathcal{R}(\rho) = \inf_\rho \sup_{\theta \in \Theta} \mathcal{R}(\rho, \boldsymbol{\theta}),$$

where the infimum is taken over the set of all possible decisions ρ. The value $\mathcal{R}^* = \mathcal{R}(\rho^*)$ is called the *minimax risk*.

Similarly, the decision ρ_π is called *Bayes* for the prior π if

$$\mathcal{R}_\pi(\rho_\pi) = \inf_\rho \mathcal{R}_\pi(\rho).$$

The corresponding value $\mathcal{R}_\pi(\rho_\pi)$ is called the *Bayes risk*.

Exercise 1.5.1. Show that the minimax risk is greater than or equal to the Bayes risk whatever the prior measure π is.

Hint: show that for any decision ρ, it holds $\mathcal{R}(\rho) \geq \mathcal{R}_\pi(\rho)$.

Usually the problem of finding a minimax or Bayes estimate is quite hard and a closed form solution is available only in very few special cases. A standard way out of this problem is to switch to an asymptotic setup in which the sample size grows to infinity.

Chapter 2
Parameter Estimation for an i.i.d. Model

This chapter is very important for understanding the whole book. It starts with very classical stuff: Glivenko–Cantelli results for the empirical measure that motivate the famous substitution principle. Then the method of moments is studied in more detail including the risk analysis and asymptotic properties. Some other classical estimation procedures are briefly discussed including the methods of minimum distance, M-estimates, and its special cases: least squares, least absolute deviations, and maximum likelihood estimates (MLEs). The concept of efficiency is discussed in context of the Cramér–Rao risk bound which is given in univariate and multivariate case. The last sections of Chap. 2 start a kind of smooth transition from classical to "modern" parametric statistics and they reveal the approach of the book. The presentation is focused on the (quasi) likelihood-based concentration and confidence sets. The basic concentration result is first introduced for the simplest Gaussian shift model and then extended to the case of a univariate exponential family in Sect. 2.11.

Below in this chapter we consider the estimation problem for a sample of independent identically distributed (i.i.d.) observations. Throughout the chapter the data $\boldsymbol{Y}$ are assumed to be given in the form of a sample $(Y_1, \ldots, Y_n)^\top$. We assume that the observations $Y_1, \ldots, Y_n$ are i.i.d.; each Y_i is from an unknown distribution P, also called a marginal measure. The joint data distribution $\mathbb{P}$ is the n-fold product of P: $\mathbb{P} = P^{\otimes n}$. Thus, the measure $\mathbb{P}$ is uniquely identified by P and the statistical problem can be reduced to recovering P.

The further step in model specification is based on a parametric assumption (PA): the measure P belongs to a given parametric family.

2.1 Empirical Distribution: Glivenko–Cantelli Theorem

Let $\boldsymbol{Y} = (Y_1, \ldots, Y_n)^\top$ be an i.i.d. sample. For simplicity we assume that the Y_i's are univariate with values in $\mathbb{R}$. Let P denote the distribution of each Y_i:

V. Spokoiny and T. Dickhaus, *Basics of Modern Mathematical Statistics*, Springer Texts in Statistics, DOI 10.1007/978-3-642-39909-1_2,

$$P(B) = \mathbb{P}(Y_i \in B), \qquad B \in \mathcal{B}(\mathbb{R}).$$

One often says that Y is an i.i.d. sample from P. Let also F be the corresponding distribution function (cdf):

$$F(y) = \mathbb{P}(Y_1 \le y) = P((-\infty, y]).$$

The assumption that the Y_i's are i.i.d. implies that the joint distribution $\mathbb{P}$ of the data Y is given by the n-fold product of the marginal measure P:

$$\mathbb{P} = P^{\otimes n}.$$

Let also P_n (resp. F_n) be the empirical measure (resp. empirical distribution function (edf))

$$P_n(B) = \frac{1}{n}\sum \mathbf{1}(Y_i \in B), \qquad F_n(y) = \frac{1}{n}\sum \mathbf{1}(Y_i \le y).$$

Here and everywhere in this chapter the symbol $\sum$ stands for $\sum_{i=1}^n$. One can consider F_n as the distribution function of the empirical measure P_n defined as the atomic measure at the Y_i's:

$$P_n(A) \stackrel{\text{def}}{=} \frac{1}{n}\sum_{i=1}^{n} \mathbf{1}(Y_i \in A).$$

So, $P_n(A)$ is the empirical frequency of the event A, that is, the fraction of observations Y_i belonging to A. By the law of large numbers one can expect that this empirical frequency is close to the true probability $P(A)$ if the number of observations is sufficiently large.

An equivalent definition of the empirical measure and edf can be given in terms of the empirical mean $\mathbb{E}_n g$ for a measurable function g:

$$\mathbb{E}_n g \stackrel{\text{def}}{=} \int_{-\infty}^{\infty} g(y)\mathbb{P}_n(dy) = \int_{-\infty}^{\infty} g(y)\, dF_n(y) = \frac{1}{n}\sum_{i=1}^{n} g(Y_i).$$

The first results claims that indeed, for every Borel set B on the real line, the empirical mass $P_n(B)$ (which is random) is close in probability to the population counterpart $P(B)$.

Theorem 2.1.1. *For any Borel set B, it holds*

1. $\mathbb{E}P_n(B) = P(B)$.
2. $\operatorname{Var}\{P_n(B)\} = n^{-1}\sigma_B^2$ *with* $\sigma_B^2 = P(B)\{1 - P(B)\}$.
3. $P_n(B) \to P(B)$ *in probability as* $n \to \infty$.
4. $\sqrt{n}\{P_n(B) - P(B)\} \xrightarrow{w} \mathcal{N}(0, \sigma_B^2)$.

Proof. Denote $\xi_i = \mathbf{1}(Y_i \in B)$. This is a Bernoulli r.v. with parameter $P(B) = \mathbb{E}\xi_i$. The first statement holds by definition of $P_n(B) = n^{-1}\sum_i \xi_i$. Next, for each $i \le n$,

$$\operatorname{Var}\xi_i \stackrel{\text{def}}{=} \mathbb{E}\xi_i^2 - (\mathbb{E}\xi_i)^2 = P(B)\{1 - P(B)\}$$

in view of $\xi_i^2 = \xi_i$. Independence of the ξ_i's yields

$$\operatorname{Var}\{P_n(B)\} = \operatorname{Var}\left(n^{-1}\sum_{i=1}^n \xi_i\right) = n^{-2}\sum_{i=1}^n \operatorname{Var}\xi_i = n^{-1}\sigma_B^2.$$

The third statement follows by the law of large numbers for the i.i.d. r.v. ξ_i:

$$\frac{1}{n}\sum_{i=1}^n \xi_i \xrightarrow{\mathbb{P}} \mathbb{E}\xi_1 .$$

Finally, the last statement follows by the Central Limit Theorem for the ξ_i:

$$\frac{1}{\sqrt{n}}\sum_{i=1}^n (\xi_i - \mathbb{E}\xi_i) \xrightarrow{w} \mathcal{N}(0, \sigma_B^2).$$

The next important result shows that the edf F_n is a good approximation of the cdf F in the uniform norm.

Theorem 2.1.2 (Glivenko–Cantelli). *It holds*

$$\sup_y \left| F_n(y) - F(y) \right| \to 0, \qquad n \to \infty$$

Proof. Consider first the case when the function F is continuous in y. Fix any integer N and define with $\varepsilon = 1/N$ the points $t_1 < t_2 < \ldots < t_N = +\infty$ such that $F(t_j) - F(t_{j-1}) = \varepsilon$ for $j = 2, \ldots, N$. For every j, by (3) of Theorem 2.1.1, it holds $F_n(t_j) \to F(t_j)$. This implies that for some $n(N)$, it holds for all $n \ge n(N)$ with a probability at least $1 - \varepsilon$

$$\left| F_n(t_j) - F(t_j) \right| \le \varepsilon, \qquad j = 1, \ldots, N. \tag{2.1}$$

Now for every $t \in [t_{j-1}, t_j]$, it holds by definition

$$F(t_{j-1}) \le F(t) \le F(t_j), \qquad F_n(t_{j-1}) \le F_n(t) \le F_n(t_j).$$

This together with (2.1) implies

$$\mathbb{P}\left(\left| F_n(t) - F(t) \right| > 2\varepsilon \right) \le \varepsilon.$$

If the function $F(\cdot)$ is not continuous, then for every positive ε, there exists a finite set $\mathcal{S}_\varepsilon$ of points of discontinuity s_m with $F(s_m) - F(s_m - 0) \ge \varepsilon$. One can proceed as in the continuous case by adding the points from $\mathcal{S}_\varepsilon$ to the discrete set $\{t_j\}$.

Exercise 2.1.1. Check the details of the proof of Theorem 2.1.2.

The results of Theorems 2.1.1 and 2.1.2 can be extended to certain functionals of the distribution P. Let $g(y)$ be a function on the real line. Consider its expectation

$$s_0 \stackrel{\text{def}}{=} \mathbb{E}g(Y_1) = \int_{-\infty}^{\infty} g(y)\, dF(y).$$

Its empirical counterpart is defined by

$$S_n \stackrel{\text{def}}{=} \int_{-\infty}^{\infty} g(y)\, dF_n(y) = \frac{1}{n}\sum_{i=1}^{n} g(Y_i).$$

It appears that S_n indeed well estimates s_0, at least for large n.

Theorem 2.1.3. *Let $g(y)$ be a function on the real line such that*

$$\int_{-\infty}^{\infty} g^2(y)\, dF(y) < \infty$$

Then

$$S_n \xrightarrow{\mathbb{P}} s_0, \qquad \sqrt{n}(S_n - s_0) \xrightarrow{w} \mathcal{N}(0, \sigma_g^2), \qquad n \to \infty,$$

where

$$\sigma_g^2 \stackrel{\text{def}}{=} \int_{-\infty}^{\infty} g^2(y)\, dF(y) - s_0^2 = \int_{-\infty}^{\infty} \bigl[g(y) - s_0\bigr]^2\, dF(y).$$

Moreover, if $h(z)$ is a twice continuously differentiable function on the real line, and $h'(s_0) \ne 0$, then

$$h(S_n) \xrightarrow{\mathbb{P}} h(s_0), \qquad \sqrt{n}\{h(S_n) - h(s_0)\} \xrightarrow{w} \mathcal{N}(0, \sigma_h^2), \qquad n \to \infty,$$

where $\sigma_h^2 \stackrel{\text{def}}{=} |h'(s_0)|^2 \sigma_g^2$.

Proof. The first statement is again the CLT for the i.i.d. random variables $\xi_i = g(Y_i)$ having mean value s_0 and variance σ_g^2.

It also implies the second statement in view of the Taylor expansion $h(S_n) - h(s_0) \approx h'(s_0)\,(S_n - s_0)$.

Exercise 2.1.2. Complete the proof.

Hint: use the first result to show that S_n belongs with high probability to a small neighborhood U of the point s_0.

Then apply the Taylor expansion of second order to $h(S_n) - h(s_0) = h\bigl(s_0 + n^{-1/2}\xi_n\bigr) - h(s_0)$ with $\xi_n = \sqrt{n}(S_n - s_0)$:

$$\bigl|n^{1/2}[h(S_n) - h(s_0)] - h'(s_0)\,\xi_n\bigr| \le n^{-1/2} H^* \xi_n^2/2,$$

where $H^* = \max_U |h''(y)|$. Show that $n^{-1/2}\xi_n^2 \xrightarrow{\mathbb{P}} 0$ because ξ_n is stochastically bounded by the first statement of the theorem.

The results of Theorems 2.1.2 and 2.1.3 can be extended to the case of a vectorial function $\boldsymbol{g}(\cdot)\colon \mathbb{R}^1 \to \mathbb{R}^m$, that is, $\boldsymbol{g}(y) = \bigl(g_1(y), \dots, g_m(y)\bigr)^\top$ for $y \in \mathbb{R}^1$. Then $\boldsymbol{s}_0 = (s_{0,1}, \dots, s_{0,m})^\top$ and its empirical counterpart $\boldsymbol{S}_n = (S_{n,1}, \dots, S_{n,m})^\top$ are vectors in $\mathbb{R}^m$ as well:

$$s_{0,j} \stackrel{\text{def}}{=} \int_{-\infty}^{\infty} g_j(y)\,dF(y), \qquad S_{n,j} \stackrel{\text{def}}{=} \int_{-\infty}^{\infty} g_j(y)\,dF_n(y), \qquad j = 1, \dots, m.$$

Theorem 2.1.4. *Let $\boldsymbol{g}(y)$ be an $\mathbb{R}^m$-valued function on the real line with a bounded covariance matrix $\Sigma = (\Sigma_{jk})_{j,k=1,\dots,m}$:*

$$\Sigma_{jk} \stackrel{\text{def}}{=} \int_{-\infty}^{\infty} \bigl[g_j(y) - s_{0,j}\bigr]\bigl[g_k(y) - s_{0,k}\bigr]\,dF(y) < \infty, \qquad j,k \le m$$

Then

$$\boldsymbol{S}_n \xrightarrow{\mathbb{P}} \boldsymbol{s}_0, \qquad \sqrt{n}\bigl(\boldsymbol{S}_n - \boldsymbol{s}_0\bigr) \xrightarrow{w} \mathcal{N}(0, \Sigma), \qquad n \to \infty.$$

Moreover, if $H(z)$ is a twice continuously differentiable function on $\mathbb{R}^m$ and $\Sigma\, H'(s_0) \ne 0$ where $H'(z)$ stands for the gradient of H at z, then

$$H(\boldsymbol{S}_n) \xrightarrow{\mathbb{P}} H(\boldsymbol{s}_0), \qquad \sqrt{n}\{H(\boldsymbol{S}_n) - H(\boldsymbol{s}_0)\} \xrightarrow{w} \mathcal{N}(0, \sigma_H^2), \qquad n \to \infty,$$

where $\sigma_H^2 \stackrel{\text{def}}{=} H'(s_0)^\top\, \Sigma\, H'(\boldsymbol{s}_0)$.

Exercise 2.1.3. Prove Theorem 2.1.4.

Hint: consider for every $\boldsymbol{h} \in \mathbb{R}^m$ the scalar products $\boldsymbol{h}^\top g(y)$, $\boldsymbol{h}^\top s_0$, $\boldsymbol{h}^\top S_n$. For the first statement, it suffices to show that

$$\boldsymbol{h}^\top \boldsymbol{S}_n \xrightarrow{\mathbb{P}} \boldsymbol{h}^\top \boldsymbol{s}_0, \qquad \sqrt{n}\boldsymbol{h}^\top\bigl(\boldsymbol{S}_n - \boldsymbol{s}_0\bigr) \xrightarrow{w} \mathcal{N}(0, \boldsymbol{h}^\top \Sigma \boldsymbol{h}), \qquad n \to \infty.$$

For the second statement, consider the expansion

$$\bigl|n^{1/2}[H(\boldsymbol{S}_n) - H(\boldsymbol{s}_0)] - \boldsymbol{\xi}_n^\top H'(\boldsymbol{s}_0)\bigr| \le n^{-1/2} H^*\, \|\boldsymbol{\xi}_n\|^2/2 \xrightarrow{\mathbb{P}} 0,$$

with $\xi_n = n^{1/2}(S_n - s_0)$ and $H^* = \max_{y \in U} \|H''(y)\|$ for a neighborhood U of s_0. Here $\|A\|$ means the maximal eigenvalue of a symmetric matrix A.

2.2 Substitution Principle: Method of Moments

By the Glivenko–Cantelli theorem the empirical measure P_n (resp. edf F_n) is a good approximation of the true measure P (resp. pdf F), at least, if n is sufficiently large. This leads to the important substitution method of statistical estimation: represent the target of estimation as a function of the distribution P, then replace P by P_n.

Suppose that there exists some functional $\boldsymbol{g}$ of a measure $P_{\boldsymbol{\theta}}$ from the family $\mathcal{P} = (P_{\boldsymbol{\theta}}, \boldsymbol{\theta} \in \Theta)$ such that the following identity holds:

$$\boldsymbol{\theta} \equiv \boldsymbol{g}(P_{\boldsymbol{\theta}}), \qquad \boldsymbol{\theta} \in \Theta.$$

This particularly implies $\boldsymbol{\theta}^* = \boldsymbol{g}(P_{\boldsymbol{\theta}^*}) = \boldsymbol{g}(P)$. The *substitution* estimate is defined by substituting P_n for P:

$$\tilde{\boldsymbol{\theta}} = \boldsymbol{g}(P_n).$$

Sometimes the obtained value $\tilde{\boldsymbol{\theta}}$ can lie outside the parameter set Θ. Then one can redefine the estimate $\tilde{\boldsymbol{\theta}}$ as the value providing the best fit of $\boldsymbol{g}(P_n)$:

$$\tilde{\boldsymbol{\theta}} = \operatorname*{argmin}_{\boldsymbol{\theta}} \|\boldsymbol{g}(P_{\boldsymbol{\theta}}) - \boldsymbol{g}(P_n)\|.$$

Here $\|\cdot\|$ denotes some norm on the parameter set Θ, e.g. the Euclidean norm.

2.2.1 *Method of Moments: Univariate Parameter*

The method of moments is a special but at the same time the most frequently used case of the substitution method. For illustration, we start with the univariate case. Let $\Theta \subseteq \mathbb{R}$, that is, θ is a univariate parameter. Let $g(y)$ be a function on $\mathbb{R}$ such that the first moment

$$m(\theta) \stackrel{\text{def}}{=} E_\theta g(Y_1) = \int g(y)\, dP_\theta(y)$$

is continuous and monotonic. Then the parameter θ can be uniquely identified by the value $m(\theta)$, that is, there exists an inverse function m^{-1} satisfying

$$\theta = m^{-1}\Bigl(\int g(y)\, dP_\theta(y)\Bigr).$$

The substitution method leads to the estimate

$$\tilde{\theta} = m^{-1}\Bigl(\int g(y)\,dP_n(y)\Bigr) = m^{-1}\Bigl(\frac{1}{n}\sum g(Y_i)\Bigr).$$

Usually $g(x) = x$ or $g(x) = x^2$, which explains the name of the method. This method was proposed by Pearson and is historically the first regular method of constructing a statistical estimate.

2.2.2 *Method of Moments: Multivariate Parameter*

The method of moments can be easily extended to the multivariate case. Let $\Theta \subseteq \mathbb{R}^p$, and let $\boldsymbol{g}(y) = \bigl(g_1(y), \ldots, g_p(y)\bigr)^\top$ be a function with values in $\mathbb{R}^p$. Define the moments $\boldsymbol{m}(\boldsymbol{\theta}) = \bigl(m_1(\boldsymbol{\theta}), \ldots, m_p(\boldsymbol{\theta})\bigr)$ by

$$m_j(\boldsymbol{\theta}) = E_{\boldsymbol{\theta}} g_j(Y_1) = \int g_j(y)\,dP_{\boldsymbol{\theta}}(y).$$

The main requirement on the choice of the vector function $\boldsymbol{g}$ is that the function $\boldsymbol{m}$ is invertible, that is, the system of equations

$$m_j(\boldsymbol{\theta}) = t_j$$

has a unique solution for any $\boldsymbol{t} \in \mathbb{R}^p$. The empirical counterpart $\boldsymbol{M}_n$ of the true moments $\boldsymbol{m}(\boldsymbol{\theta}^*)$ is given by

$$\boldsymbol{M}_n \stackrel{\text{def}}{=} \int \boldsymbol{g}(y)\,dP_n(y) = \Bigl(\frac{1}{n}\sum g_1(Y_i), \ldots, \frac{1}{n}\sum g_p(Y_i)\Bigr)^\top.$$

Then the estimate $\tilde{\boldsymbol{\theta}}$ can be defined as

$$\tilde{\boldsymbol{\theta}} \stackrel{\text{def}}{=} \boldsymbol{m}^{-1}(\boldsymbol{M}_n) = \boldsymbol{m}^{-1}\Bigl(\frac{1}{n}\sum g_1(Y_i), \ldots, \frac{1}{n}\sum g_p(Y_i)\Bigr).$$

2.2.3 *Method of Moments: Examples*

This section lists some widely used parametric families and discusses the problem of constructing the parameter estimates by different methods. In all the examples we assume that an i.i.d. sample from a distribution P is observed, and this measure P belongs to a given parametric family $(P_{\boldsymbol{\theta}}, \boldsymbol{\theta} \in \Theta)$, that is, $P = P_{\boldsymbol{\theta}^*}$ for $\boldsymbol{\theta}^* \in \Theta$.

2.2.3.1 Gaussian Shift

Let P_θ be the normal distribution on the real line with mean θ and the known variance σ^2. The corresponding density w.r.t. the Lebesgue measure reads as

$$p(y,\theta) = \frac{1}{\sqrt{2\pi\sigma^2}} \exp\Bigl\{-\frac{(y-\theta)^2}{2\sigma^2}\Bigr\}.$$

It holds $\mathbb{E}_\theta Y_1 = \theta$ and $\operatorname{Var}_\theta(Y_1) = \sigma^2$ leading to the moment estimate

$$\tilde{\theta} = \int y \, dP_n(y) = \frac{1}{n} \sum Y_i$$

with mean $\mathbb{E}_\theta \tilde{\theta} = \theta$ and variance

$$\operatorname{Var}_\theta(\tilde{\theta}) = \sigma^2/n.$$

2.2.3.2 Univariate Normal Distribution

Let $Y_i \sim \mathcal{N}(\alpha, \sigma^2)$ as in the previous example but both mean α and the variance σ^2 are unknown. This leads to the problem of estimating the vector $\boldsymbol{\theta} = (\theta_1, \theta_2) = (\alpha, \sigma^2)$ from the i.i.d. sample $\boldsymbol{Y}$.

The method of moments suggests to estimate the parameters from the first two empirical moments of the Y_i's using the equations $m_1(\boldsymbol{\theta}) = \mathbb{E}_{\boldsymbol{\theta}} Y_1 = \alpha$, $m_2(\boldsymbol{\theta}) = \mathbb{E}_{\boldsymbol{\theta}} Y_1^2 = \alpha^2 + \sigma^2$. Inverting these equalities leads to

$$\alpha = m_1(\boldsymbol{\theta}), \qquad \sigma^2 = m_2(\boldsymbol{\theta}) - m_1^2(\boldsymbol{\theta}).$$

Substituting the empirical measure P_n yields the expressions for $\tilde{\boldsymbol{\theta}}$:

$$\tilde{\alpha} = \frac{1}{n} \sum Y_i, \qquad \tilde{\sigma}^2 = \frac{1}{n} \sum Y_i^2 - \Bigl(\frac{1}{n} \sum Y_i\Bigr)^2 = \frac{1}{n} \sum (Y_i - \tilde{\alpha})^2. \tag{2.2}$$

As previously for the case of a known variance, it holds under $\mathbb{P} = \mathbb{P}_{\boldsymbol{\theta}}$:

$$\mathbb{E}\tilde{\alpha} = \alpha, \qquad \operatorname{Var}_{\boldsymbol{\theta}}(\tilde{\alpha}) = \sigma^2/n.$$

However, for the estimate $\tilde{\sigma}^2$ of σ^2, the result is slightly different and it is described in the next theorem.

Theorem 2.2.1. *It holds*

$$\mathbb{E}_{\boldsymbol{\theta}} \tilde{\sigma}^2 = \frac{n-1}{n} \sigma^2, \qquad \operatorname{Var}_{\boldsymbol{\theta}}(\tilde{\sigma}^2) = \frac{2(n-1)}{n^2} \sigma^4.$$

Proof. We use vector notation. Consider the unit vector $\boldsymbol{e} = n^{-1/2}(1,\ldots,1)^\top \in \mathbb{R}^n$ and denote by Π_1 the projector on $\boldsymbol{e}$:

$$\Pi_1 \boldsymbol{h} = (\boldsymbol{e}^\top \boldsymbol{h})\boldsymbol{e}.$$

Then by definition $\tilde{\alpha} = n^{-1/2}\boldsymbol{e}^\top \Pi_1 \boldsymbol{Y}$ and $\tilde{\sigma}^2 = n^{-1}\|\boldsymbol{Y} - \Pi_1 \boldsymbol{Y}\|^2$. Moreover, the model equation $\boldsymbol{Y} = n^{1/2}\alpha \boldsymbol{e} + \boldsymbol{\varepsilon}$ implies in view of $\Pi_1 \boldsymbol{e} = \boldsymbol{e}$ that

$$\Pi_1 \boldsymbol{Y} = (n^{1/2}\alpha \boldsymbol{e} + \Pi_1 \boldsymbol{\varepsilon}).$$

Now

$$n\tilde{\sigma}^2 = \|\boldsymbol{Y} - \Pi_1 \boldsymbol{Y}\|^2 = \|\boldsymbol{\varepsilon} - \Pi_1 \boldsymbol{\varepsilon}\|^2 = \|(I_n - \Pi_1)\boldsymbol{\varepsilon}\|^2$$

where I_n is the identity operator in $\mathbb{R}^n$ and $I_n - \Pi_1$ is the projector on the hyperplane in $\mathbb{R}^n$ orthogonal to the vector $\boldsymbol{e}$. Obviously $(I_n - \Pi_1)\boldsymbol{\varepsilon}$ is a Gaussian vector with zero mean and the covariance matrix V defined by

$$\begin{aligned} V &= \mathbb{E}\bigl[(I_n - \Pi_1)\boldsymbol{\varepsilon}\boldsymbol{\varepsilon}^\top (I_n - \Pi_1)\bigr] = (I_n - \Pi_1)\mathbb{E}(\boldsymbol{\varepsilon}\boldsymbol{\varepsilon}^\top)(I_n - \Pi_1) \\ &= \sigma^2 (I_n - \Pi_1)^2 = \sigma^2 (I_n - \Pi_1). \end{aligned}$$

It remains to note that for any Gaussian vector $\boldsymbol{\xi} \sim \mathcal{N}(0, V)$ it holds

$$\mathbb{E}\|\boldsymbol{\xi}\|^2 = \operatorname{tr} V, \qquad \operatorname{Var}\bigl(\|\boldsymbol{\xi}\|^2\bigr) = 2\operatorname{tr}(V^2).$$

Exercise 2.2.1. Check the details of the proof.
Hint: reduce to the case of diagonal V.

Exercise 2.2.2. Compute the covariance $\mathbb{E}(\tilde{\alpha} - \alpha)(\tilde{\sigma}^2 - \sigma^2)$. Show that $\tilde{\alpha}$ and $\tilde{\sigma}^2$ are independent.
Hint: represent $\tilde{\alpha} - \alpha = n^{-1/2}\boldsymbol{e}^\top \Pi_1 \boldsymbol{\varepsilon}$ and $\tilde{\sigma}^2 = n^{-1}\|(I_n - \Pi_1)\boldsymbol{\varepsilon}\|^2$. Use that $\Pi_1 \boldsymbol{\varepsilon}$ and $(I_n - \Pi_1)\boldsymbol{\varepsilon}$ are independent if Π_1 is a projector and $\boldsymbol{\varepsilon}$ is a Gaussian vector.

2.2.3.3 Uniform Distribution on $[0, \theta]$

Let Y_i be uniformly distributed on the interval $[0, \theta]$ of the real line where the right end point θ is unknown. The density $p(y, \theta)$ of P_θ w.r.t. the Lebesgue measure is $\theta^{-1}\mathbf{1}(y \le \theta)$. It is easy to compute that for an integer k

$$\mathbb{E}_\theta(Y_1^k) = \theta^{-1}\int_0^\theta y^k \, dy = \theta^k/(k+1),$$

or $\theta = \{(k+1)\mathbb{E}_\theta(Y_1^k)\}^{1/k}$. This leads to the family of estimates

$$\tilde{\theta}_k = \left(\frac{k+1}{n}\sum Y_i^k\right)^{1/(k+1)}.$$

Letting k to infinity leads to the estimate

$$\tilde{\theta}_\infty = \max\{Y_1, \dots, Y_n\}.$$

This estimate is quite natural in the context of the univariate distribution. Later it will appear once again as the MLE. However, it is not a moment estimate.

2.2.3.4 Bernoulli or Binomial Model

Let P_θ be a Bernoulli law for $\theta \in [0,1]$. Then every Y_i is binary with

$$\mathbb{E}_\theta Y_i = \theta.$$

This leads to the moment estimate

$$\tilde{\theta} = \int y \, dP_n(y) = \frac{1}{n}\sum Y_i.$$

Exercise 2.2.3. Compute the moment estimate for $g(y) = y^k, k \geq 1$.

2.2.3.5 Multinomial Model

The multinomial distribution B_θ^m describes the number of successes in m experiments when each success has the probability $\theta \in [0,1]$. This distribution can be viewed as the sum of m binomials with the same parameter θ. Observed is the sample $\boldsymbol{Y}$ where each Y_i is the number of successes in the ith experiment. One has

$$P_\theta(Y_1 = k) = \binom{m}{k}\theta^k(1-\theta)^{m-k}, \qquad k = 0, \dots, m.$$

Exercise 2.2.4. Check that method of moments with $g(x) = x$ leads to the estimate

$$\tilde{\theta} = \frac{1}{mn}\sum Y_i.$$

Compute $\operatorname{Var}_\theta(\tilde{\theta})$.

Hint: Reduce the multinomial model to the sum of m Bernoulli.

2.2.3.6 Exponential Model

Let P_θ be an exponential distribution on the positive semiaxis with the parameter θ. This means

$$\mathbb{P}_\theta(Y_1 > y) = e^{-y/\theta}.$$

Exercise 2.2.5. Check that method of moments with $g(x) = x$ leads to the estimate

$$\tilde{\theta} = \frac{1}{n} \sum Y_i .$$

Compute $\mathrm{Var}_\theta(\tilde{\theta})$.

2.2.3.7 Poisson Model

Let P_θ be the Poisson distribution with the parameter θ. The Poisson random variable Y_1 is integer-valued with

$$P_\theta(Y_1 = k) = \frac{\theta^k}{k!} e^{-\theta}.$$

Exercise 2.2.6. Check that method of moments with $g(x) = x$ leads to the estimate

$$\tilde{\theta} = \frac{1}{n} \sum Y_i .$$

Compute $\mathrm{Var}_\theta(\tilde{\theta})$.

2.2.3.8 Shift of a Laplace (Double Exponential) Law

Let P_0 be a symmetric distribution defined by the equations

$$P_0(|Y_1| > y) = e^{-y/\sigma}, \qquad y \ge 0,$$

for some given $\sigma > 0$. Equivalently one can say that the absolute value of Y_1 is exponential with parameter σ under P_0. Now define P_θ by shifting P_0 by the value θ. This means that

$$P_\theta(|Y_1 - \theta| > y) = e^{-y/\sigma}, \qquad y \ge 0.$$

It is obvious that $\mathbb{E}_0 Y_1 = 0$ and $\mathbb{E}_\theta Y_1 = \theta$.

Exercise 2.2.7. Check that method of moments leads to the estimate

$$\tilde{\theta} = \frac{1}{n} \sum Y_i \,.$$

Compute $\operatorname{Var}_\theta(\tilde{\theta})$.

2.2.3.9 Shift of a Symmetric Density

Let the observations Y_i be defined by the equation

$$Y_i = \theta^* + \varepsilon_i$$

where θ^* is an unknown parameter and the errors ε_i are i.i.d. with a density symmetric around zero and finite second moment $\sigma^2 = \mathbb{E}\varepsilon_1^2$. This particularly yields that $\mathbb{E}\varepsilon_i = 0$ and $\mathbb{E}Y_i = \theta^*$. The method of moments immediately yields the empirical mean estimate

$$\tilde{\theta} = \frac{1}{n} \sum Y_i$$

with $\operatorname{Var}_\theta(\tilde{\theta}) = \sigma^2/n$.

2.3 Unbiased Estimates, Bias, and Quadratic Risk

Consider a parametric i.i.d. experiment corresponding to a sample $\boldsymbol{Y} = (Y_1, \ldots, Y_n)^\top$ from a distribution $P_{\boldsymbol{\theta}^*} \in (P_{\boldsymbol{\theta}}, \boldsymbol{\theta} \in \Theta \subseteq \mathbb{R}^p)$. By $\boldsymbol{\theta}^*$ we denote the true parameter from Θ. Let $\tilde{\boldsymbol{\theta}}$ be an estimate of $\boldsymbol{\theta}^*$, that is, a function of the available data $\boldsymbol{Y}$ with values in Θ: $\tilde{\boldsymbol{\theta}} = \tilde{\boldsymbol{\theta}}(\boldsymbol{Y})$.

An estimate $\tilde{\boldsymbol{\theta}}$ of the parameter $\boldsymbol{\theta}^*$ is called unbiased if

$$\mathbb{E}_{\boldsymbol{\theta}^*}\tilde{\boldsymbol{\theta}} = \boldsymbol{\theta}^*.$$

This property seems to be rather natural and desirable. However, it is often just matter of parametrization. Indeed, if $g : \Theta \to \Theta$ is a linear transformation of the parameter set Θ, that is, $g(\boldsymbol{\theta}) = A\boldsymbol{\theta} + b$, then the estimate $\tilde{\vartheta} \stackrel{\text{def}}{=} A\tilde{\boldsymbol{\theta}} + b$ of the new parameter $\vartheta = A\boldsymbol{\theta} + b$ is again unbiased. However, if $m(\cdot)$ is a nonlinear transformation, then the identity $\mathbb{E}_{\boldsymbol{\theta}^*} m(\tilde{\boldsymbol{\theta}}) = m(\boldsymbol{\theta}^*)$ is not preserved.

Example 2.3.1. Consider the Gaussian shift experiments for Y_i i.i.d. $\mathcal{N}(\theta^*, \sigma^2)$ with known variance σ^2 but the shift parameter θ^* is unknown. Then $\tilde{\theta} = n^{-1}(Y_1 + \ldots + Y_n)$ is an unbiased estimate of θ^*. However, for $m(\theta) = \theta^2$, it holds

$$\mathbb{E}_{\theta^*}|\tilde{\theta}|^2 = |\theta^*|^2 + \sigma^2/n,$$

that is, the estimate $|\tilde{\theta}|^2$ of $|\theta^*|^2$ is slightly biased.

The property of "no bias" is especially important in connection with the quadratic risk of the estimate $\tilde{\boldsymbol{\theta}}$. To illustrate this point, we first consider the case of a univariate parameter.

2.3.1 Univariate Parameter

Let $\theta \in \Theta \subseteq \mathbb{R}^1$. Denote by $\mathrm{Var}(\tilde{\theta})$ the *variance* of the estimate $\tilde{\theta}$:

$$\mathrm{Var}_{\theta^*}(\tilde{\theta}) = \mathbb{E}_{\theta^*}(\tilde{\theta} - \mathbb{E}_{\theta^*}\tilde{\theta})^2.$$

The quadratic risk of $\tilde{\theta}$ is defined by

$$\mathcal{R}(\tilde{\theta}, \theta^*) \stackrel{\text{def}}{=} \mathbb{E}_{\theta^*}|\tilde{\theta} - \theta^*|^2.$$

It is obvious that $\mathcal{R}(\tilde{\theta}, \theta^*) = \mathrm{Var}_{\theta^*}(\tilde{\theta})$ if $\tilde{\theta}$ is unbiased. It turns out that the quadratic risk of $\tilde{\theta}$ is larger than the variance when this property is not fulfilled. Define the *bias* of $\tilde{\theta}$ as

$$b(\tilde{\theta}, \theta^*) \stackrel{\text{def}}{=} \mathbb{E}_{\theta^*}\tilde{\theta} - \theta^*.$$

Theorem 2.3.1. *It holds for any estimate $\tilde{\theta}$ of the univariate parameter θ^*:*

$$\mathcal{R}(\tilde{\theta}, \theta^*) = \mathrm{Var}_{\theta^*}(\tilde{\theta}) + b^2(\tilde{\theta}, \theta^*).$$

Due to this result, the bias $b(\tilde{\theta}, \theta^*)$ contributes the value $b^2(\tilde{\theta}, \theta^*)$ in the quadratic risk. This particularly explains why one is interested in considering unbiased or at least nearly unbiased estimates.

2.3.2 Multivariate Case

Now we extend the result to the multivariate case with $\boldsymbol{\theta} \in \Theta \subseteq \mathbb{R}^p$. Then $\tilde{\boldsymbol{\theta}}$ is a vector in $\mathbb{R}^p$. The corresponding *variance–covariance matrix* $\mathrm{Var}_{\boldsymbol{\theta}^*}(\tilde{\boldsymbol{\theta}})$ is defined as

$$\mathrm{Var}_{\boldsymbol{\theta}^*}(\tilde{\boldsymbol{\theta}}) \stackrel{\text{def}}{=} \mathbb{E}_{\boldsymbol{\theta}^*}\bigl[(\tilde{\boldsymbol{\theta}} - \mathbb{E}_{\boldsymbol{\theta}^*}\tilde{\boldsymbol{\theta}})(\tilde{\boldsymbol{\theta}} - \mathbb{E}_{\boldsymbol{\theta}^*}\tilde{\boldsymbol{\theta}})^\top\bigr].$$

As previously, $\tilde{\boldsymbol{\theta}}$ is unbiased if $\mathbb{E}_{\boldsymbol{\theta}^*}\tilde{\boldsymbol{\theta}} = \boldsymbol{\theta}^*$, and the bias of $\tilde{\boldsymbol{\theta}}$ is $b(\tilde{\boldsymbol{\theta}}, \boldsymbol{\theta}^*) \stackrel{\text{def}}{=} \mathbb{E}_{\boldsymbol{\theta}^*}\tilde{\boldsymbol{\theta}} - \boldsymbol{\theta}^*$.

The quadratic risk of the estimate $\tilde{\boldsymbol{\theta}}$ in the multivariate case is usually defined via the Euclidean norm of the difference $\tilde{\boldsymbol{\theta}} - \boldsymbol{\theta}^*$:

$$\mathcal{R}(\tilde{\boldsymbol{\theta}}, \boldsymbol{\theta}^*) \stackrel{\text{def}}{=} \mathbb{E}_{\boldsymbol{\theta}^*} \big\| \tilde{\boldsymbol{\theta}} - \boldsymbol{\theta}^* \big\|^2.$$

Theorem 2.3.2. *It holds*

$$\mathcal{R}(\tilde{\boldsymbol{\theta}}, \boldsymbol{\theta}^*) = \operatorname{tr}\big[\operatorname{Var}_{\boldsymbol{\theta}^*}(\tilde{\boldsymbol{\theta}})\big] + \big\| b(\tilde{\boldsymbol{\theta}}, \boldsymbol{\theta}^*) \big\|^2$$

Proof. The result follows similarly to the univariate case using the identity $\|v\|^2 = \operatorname{tr}(vv^\top)$ for any vector $v \in \mathbb{R}^p$.

Exercise 2.3.1. Complete the proof of Theorem 2.3.2.

2.4 Asymptotic Properties

The properties of the previously introduced estimate $\tilde{\boldsymbol{\theta}}$ heavily depend on the *sample size* n. We therefore use the notation $\tilde{\boldsymbol{\theta}}_n$ to highlight this dependence. A natural extension of the condition that $\tilde{\boldsymbol{\theta}}$ is unbiased is the requirement that the bias $b(\tilde{\boldsymbol{\theta}}, \boldsymbol{\theta}^*)$ becomes negligible as the sample size n increases. This leads to the notion of *consistency*.

Definition 2.4.1. A sequence of estimates $\tilde{\boldsymbol{\theta}}_n$ is consistent if

$$\tilde{\boldsymbol{\theta}}_n \stackrel{\mathbb{P}}{\longrightarrow} \boldsymbol{\theta}^* \qquad n \to \infty.$$

$\tilde{\boldsymbol{\theta}}_n$ is mean consistent if

$$\mathbb{E}_{\boldsymbol{\theta}^*} \|\tilde{\boldsymbol{\theta}}_n - \boldsymbol{\theta}^*\| \to 0, \qquad n \to \infty.$$

Clearly mean consistency implies consistency and also *asymptotic unbiasedness*:

$$b(\tilde{\boldsymbol{\theta}}_n, \boldsymbol{\theta}^*) = \mathbb{E}\tilde{\boldsymbol{\theta}}_n - \boldsymbol{\theta}^* \stackrel{\mathbb{P}}{\longrightarrow} 0, \qquad n \to \infty.$$

The property of consistency means that the difference $\tilde{\boldsymbol{\theta}} - \boldsymbol{\theta}^*$ is small for n large. The next natural question to address is how fast this difference tends to zero with n. The Glivenko–Cantelli result suggests that $\sqrt{n}\big(\tilde{\boldsymbol{\theta}}_n - \boldsymbol{\theta}^*\big)$ is asymptotically normal.

Definition 2.4.2. A sequence of estimates $\tilde{\boldsymbol{\theta}}_n$ is root-n normal if

$$\sqrt{n}\big(\tilde{\boldsymbol{\theta}}_n - \boldsymbol{\theta}^*\big) \stackrel{w}{\longrightarrow} \mathcal{N}(0, V^2)$$

for some fixed matrix V^2.

We aim to show that the moment estimates are consistent and asymptotically root-n normal under very general conditions. We start again with the univariate case.

2.4.1 Root-n Normality: Univariate Parameter

Our first result describes the simplest situation when the parameter of interest θ^* can be represented as an integral $\int g(y)dP_{\theta^*}(y)$ for some function $g(\cdot)$.

Theorem 2.4.1. *Suppose that $\Theta \subseteq \mathbb{R}$ and a function $g(\cdot) : \mathbb{R} \to \mathbb{R}$ satisfies for every $\theta \in \Theta$*

$$\int g(y)\,dP_\theta(y) = \theta,$$

$$\int \left[g(y) - \theta\right]^2 dP_\theta(y) = \sigma^2(\theta) < \infty.$$

Then the moment estimates $\tilde{\theta}_n = n^{-1}\sum g(Y_i)$ satisfy the following conditions:

1. each $\tilde{\theta}_n$ is unbiased, that is, $\mathbb{E}_{\theta^*}\tilde{\theta}_n = \theta^*$.
2. the normalized quadratic risk $n\mathbb{E}_{\theta^*}\left(\tilde{\theta}_n - \theta^*\right)^2$ fulfills

$$n\mathbb{E}_{\theta^*}\left(\tilde{\theta}_n - \theta^*\right)^2 = \sigma^2(\theta^*).$$

3. $\tilde{\theta}_n$ is asymptotically root-n normal:

$$\sqrt{n}\left(\tilde{\theta}_n - \theta^*\right) \xrightarrow{w} \mathcal{N}(0, \sigma^2(\theta^*)).$$

This result has already been proved, see Theorem 2.1.3. Next we extend this result to the more general situation when θ^* is defined implicitly via the moment $s_0(\theta^*) = \int g(y)\,dP_{\theta^*}(y)$. This means that there exists another function $m(\theta^*)$ such that $m(\theta^*) = \int g(y)\,dP_{\theta^*}(y)$.

Theorem 2.4.2. *Suppose that $\Theta \subseteq \mathbb{R}$ and a functions $g(y) : \mathbb{R} \to \mathbb{R}$ and $m(\theta) : \Theta \to \mathbb{R}$ satisfy*

$$\int g(y)\,dP_\theta(y) \equiv m(\theta),$$

$$\int \left\{g(y) - m(\theta)\right\}^2 dP_\theta(y) \equiv \sigma_g^2(\theta) < \infty.$$

We also assume that $m(\cdot)$ is monotonic and twice continuously differentiable with $m'\left(m(\theta^*)\right) \neq 0$. Then the moment estimates $\tilde{\theta}_n = m^{-1}\left(n^{-1}\sum g(Y_i)\right)$ satisfy the following conditions:

1. $\tilde{\theta}_n$ *is consistent, that is,* $\tilde{\theta}_n \xrightarrow{\mathbb{P}} \theta^*$.
2. $\tilde{\theta}_n$ *is asymptotically root-n normal:*

$$\sqrt{n}(\tilde{\theta}_n - \theta^*) \xrightarrow{w} \mathcal{N}(0, \sigma^2(\theta^*)), \tag{2.3}$$

where $\sigma^2(\theta^*) = |m'(m(\theta^*))|^{-2}\sigma_g^2(\theta^*)$.

This result also follows directly from Theorem 2.1.3 with $h(s) = m^{-1}(s)$.

The property of asymptotic normality allows us to study the *asymptotic concentration* of $\tilde{\theta}_n$ and to build *asymptotic confidence sets.*

Corollary 2.4.1. *Let* $\tilde{\theta}_n$ *be asymptotically root-n normal: see* (2.3)*. Then for any* $z > 0$

$$\lim_{n\to\infty} \mathbb{P}_{\theta^*}\bigl(\sqrt{n}\bigl|\tilde{\theta}_n - \theta^*\bigr| > z\sigma(\theta^*)\bigr) = 2\Phi(-z)$$

where $\Phi(z)$ *is the cdf of the standard normal law.*

In particular, this result implies that the estimate $\tilde{\theta}_n$ belongs to a small root-*n* neighborhood

$$A(z) \stackrel{\text{def}}{=} [\theta^* - n^{-1/2}\sigma(\theta^*)z, \theta^* + n^{-1/2}\sigma(\theta^*)z]$$

with the probability about $2\Phi(-z)$ which is small provided that z is sufficiently large.

Next we briefly discuss the problem of interval (or confidence) estimation of the parameter θ^*. This problem differs from the problem of point estimation: the target is to build an interval (a set) E_α on the basis of the observations $\boldsymbol{Y}$ such that $\mathbb{P}(E_\alpha \ni \theta^*) \approx 1 - \alpha$ for a given $\alpha \in (0, 1)$. This problem can be attacked similarly to the problem of concentration by considering the interval of width $2\sigma(\theta^*)z$ centered at the estimate $\tilde{\theta}$. However, the major difficulty is raised by the fact that this construction involves the true parameter value θ^* via the variance $\sigma^2(\theta^*)$. In some situations this variance does not depend on θ^*: $\sigma^2(\theta^*) \equiv \sigma^2$ with a known value σ^2. In this case the construction is immediate.

Corollary 2.4.2. *Let* $\tilde{\theta}_n$ *be asymptotically root-n normal: see* (2.3)*. Let also* $\sigma^2(\theta^*) \equiv \sigma^2$*. Then for any* $\alpha \in (0, 1)$*, the set*

$$E^\circ(z_\alpha) \stackrel{\text{def}}{=} [\tilde{\theta}_n - n^{-1/2}\sigma z_\alpha, \tilde{\theta}_n + n^{-1/2}\sigma z_\alpha],$$

where z_α *is defined by* $2\Phi(-z_\alpha) = \alpha$*, satisfies*

$$\lim_{n\to\infty} \mathbb{P}_{\theta^*}\bigl(E(z_\alpha) \ni \theta^*)\bigr) = 1 - \alpha. \tag{2.4}$$

Exercise 2.4.1. Check Corollaries 2.4.1 and 2.4.2.

Next we consider the case when the variance $\sigma^2(\theta^*)$ is unknown. Instead we assume that a consistent variance estimate $\tilde{\sigma}^2$ is available. Then we plug this estimate in the construction of the confidence set in place of the unknown true variance $\sigma^2(\theta^*)$ leading to the following confidence set:

$$E(z_\alpha) \stackrel{\text{def}}{=} [\tilde{\theta}_n - n^{-1/2}\tilde{\sigma} z_\alpha, \tilde{\theta}_n + n^{-1/2}\tilde{\sigma} z_\alpha]. \tag{2.5}$$

Theorem 2.4.3. *Let $\tilde{\theta}_n$ be asymptotically root-n normal: see* (2.3)*. Let $\sigma(\theta^*) > 0$ and $\tilde{\sigma}^2$ be a consistent estimate of $\sigma^2(\theta^*)$ in the sense that $\tilde{\sigma}^2 \xrightarrow{\mathbb{P}} \sigma^2(\theta^*)$. Then for any $\alpha \in (0,1)$, the set $E(z_\alpha)$ is asymptotically α-confident in the sense of* (2.4).

One natural estimate of the variance $\sigma(\theta^*)$ can be obtained by plugging in the estimate $\tilde{\theta}$ in place of θ^* leading to $\tilde{\sigma} = \sigma(\tilde{\theta})$. If $\sigma(\theta)$ is a continuous function of θ in a neighborhood of θ^*, then consistency of $\tilde{\theta}$ implies consistency of $\tilde{\sigma}$.

Corollary 2.4.3. *Let $\tilde{\theta}_n$ be asymptotically root-n normal and let the variance $\sigma^2(\theta)$ be a continuous function of θ at θ^*. Then $\tilde{\sigma} \stackrel{\text{def}}{=} \sigma(\tilde{\theta}_n)$ is a consistent estimate of $\sigma(\theta^*)$ and the set $E(z_\alpha)$ from* (2.5) *is asymptotically α-confident.*

2.4.2 Root-n Normality: Multivariate Parameter

Let now $\Theta \subseteq \mathbb{R}^p$ and $\boldsymbol{\theta}^*$ be the true parameter vector. The method of moments requires at least p different moment functions for identifying p parameters. Let $\boldsymbol{g}(y) : \mathbb{R} \to \mathbb{R}^p$ be a vector of moment functions, $\boldsymbol{g}(y) = \bigl(g_1(y), \ldots, g_p(y)\bigr)^\top$. Suppose first that the true parameter can be obtained just by integration: $\boldsymbol{\theta}^* = \int \boldsymbol{g}(y)\, dP_{\boldsymbol{\theta}^*}(y)$. This yields the moment estimate $\tilde{\boldsymbol{\theta}}_n = n^{-1} \sum \boldsymbol{g}(Y_i)$.

Theorem 2.4.4. *Suppose that a vector-function $\boldsymbol{g}(y) : \mathbb{R} \to \mathbb{R}^p$ satisfies the following conditions:*

$$\int \boldsymbol{g}(y)\, dP_{\boldsymbol{\theta}}(y) = \boldsymbol{\theta},$$

$$\int \{\boldsymbol{g}(y) - \boldsymbol{\theta}\}\{\boldsymbol{g}(y) - \boldsymbol{\theta}\}^\top dP_{\boldsymbol{\theta}}(y) = \Sigma(\boldsymbol{\theta}).$$

Then it holds for the moment estimate $\tilde{\boldsymbol{\theta}}_n = n^{-1} \sum \boldsymbol{g}(Y_i)$:

1. *$\tilde{\boldsymbol{\theta}}$ is unbiased, that is, $\mathbb{E}_{\boldsymbol{\theta}^*} \tilde{\boldsymbol{\theta}} = \boldsymbol{\theta}^*$.*
2. *$\tilde{\boldsymbol{\theta}}_n$ is asymptotically root-n normal:*

$$\sqrt{n}\bigl(\tilde{\boldsymbol{\theta}}_n - \boldsymbol{\theta}^*\bigr) \xrightarrow{w} \mathcal{N}(0, \Sigma(\boldsymbol{\theta}^*)). \tag{2.6}$$

3. *the normalized quadratic risk* $n\mathbb{E}_{\theta^*}\|\tilde{\theta}_n - \theta^*\|^2$ *fulfills*

$$n\mathbb{E}_{\theta^*}\|\tilde{\theta}_n - \theta^*\|^2 = \operatorname{tr}\Sigma(\theta^*).$$

Similarly to the univariate case, this result yields corollaries about *concentration* and *confidence sets* with intervals replaced by ellipsoids. Indeed, due to the second statement, the vector

$$\xi_n \stackrel{\text{def}}{=} \sqrt{n}\{\Sigma(\boldsymbol{\theta}^*)\}^{-1/2}(\boldsymbol{\theta} - \boldsymbol{\theta}^*)$$

is asymptotically standard normal: $\xi_n \xrightarrow{w} \xi \sim \mathcal{N}(0, I_p)$. This also implies that the squared norm of ξ_n is asymptotically χ_p^2-distributed where ξ_p^2 is the law of $\|\xi\|^2 = \xi_1^2 + \ldots + \xi_p^2$. Define the value z_α via the quantiles of χ_p^2 by the relation

$$\mathbb{P}\bigl(\|\xi\| > z_\alpha\bigr) = \alpha. \tag{2.7}$$

Corollary 2.4.4. *Suppose that $\tilde{\boldsymbol{\theta}}_n$ is root-n normal, see* (2.6). *Define for a given z the ellipsoid*

$$A(z) \stackrel{\text{def}}{=} \{\boldsymbol{\theta} : (\boldsymbol{\theta} - \boldsymbol{\theta}^*)^\top\{\Sigma(\boldsymbol{\theta}^*)\}^{-1}(\boldsymbol{\theta} - \boldsymbol{\theta}^*) \le z^2/n\}.$$

Then $A(z_\alpha)$ is asymptotically $(1-\alpha)$*-concentration set for $\tilde{\boldsymbol{\theta}}_n$ in the sense that*

$$\lim_{n\to\infty} \mathbb{P}\bigl(\tilde{\boldsymbol{\theta}} \notin A(z_\alpha)\bigr) = \alpha.$$

The weak convergence $\xi_n \xrightarrow{w} \xi$ suggests to build confidence sets also in form of ellipsoids with the axis defined by the covariance matrix $\Sigma(\boldsymbol{\theta}^*)$. Define for $\alpha > 0$

$$E^\circ(z_\alpha) \stackrel{\text{def}}{=} \bigl\{\boldsymbol{\theta} : \sqrt{n}\bigl\|\{\Sigma(\boldsymbol{\theta}^*)\}^{-1/2}(\tilde{\boldsymbol{\theta}} - \boldsymbol{\theta})\bigr\| \le z_\alpha\bigr\}.$$

The result of Theorem 2.4.4 implies that this set covers the true value $\boldsymbol{\theta}^*$ with probability approaching $1-\alpha$.

Unfortunately, in typical situations the matrix $\Sigma(\boldsymbol{\theta}^*)$ is unknown because it depends on the unknown parameter $\boldsymbol{\theta}^*$. It is natural to replace it with the matrix $\Sigma(\tilde{\boldsymbol{\theta}})$ replacing the true value $\boldsymbol{\theta}^*$ with its consistent estimate $\tilde{\boldsymbol{\theta}}$. If $\Sigma(\boldsymbol{\theta})$ is a continuous function of $\boldsymbol{\theta}$, then $\Sigma(\tilde{\boldsymbol{\theta}})$ provides a consistent estimate of $\Sigma(\boldsymbol{\theta}^*)$. This leads to the data-driven confidence set:

$$E(z_\alpha) \stackrel{\text{def}}{=} \bigl\{\boldsymbol{\theta} : \sqrt{n}\bigl\|\{\Sigma(\tilde{\boldsymbol{\theta}})\}^{-1/2}(\tilde{\boldsymbol{\theta}} - \boldsymbol{\theta})\bigr\| \le z\bigr\}.$$

Corollary 2.4.5. *Suppose that* $\tilde{\boldsymbol{\theta}}_n$ *is root-n normal, see* (2.6)*, with a non-degenerate matrix* $\Sigma(\boldsymbol{\theta}^*)$*. Let the matrix function* $\Sigma(\boldsymbol{\theta})$ *be continuous at* $\boldsymbol{\theta}^*$*. Let* z_α *be defined by* (2.7)*. Then* $E^\circ(z_\alpha)$ *and* $E(z_\alpha)$ *are asymptotically* $(1-\alpha)$*-confidence sets for* $\boldsymbol{\theta}^*$*:*

$$\lim_{n\to\infty} \mathbb{P}\bigl(E^\circ(z_\alpha) \ni \boldsymbol{\theta}^*\bigr) = \lim_{n\to\infty} \mathbb{P}\bigl(E(z_\alpha) \ni \boldsymbol{\theta}^*\bigr) = 1 - \alpha.$$

Exercise 2.4.2. Check Corollaries 2.4.4 and 2.4.5 about the set $E^\circ(z_\alpha)$.

Exercise 2.4.3. Check Corollary 2.4.5 about the set $E(z_\alpha)$.

Hint: $\tilde{\boldsymbol{\theta}}$ is consistent and $\Sigma(\boldsymbol{\theta})$ is continuous and invertible at $\boldsymbol{\theta}^*$. This implies

$$\Sigma(\tilde{\boldsymbol{\theta}}) - \Sigma(\boldsymbol{\theta}^*) \xrightarrow{\mathbb{P}} 0, \qquad \{\Sigma(\tilde{\boldsymbol{\theta}})\}^{-1} - \{\Sigma(\boldsymbol{\theta}^*)\}^{-1} \xrightarrow{\mathbb{P}} 0,$$

and hence, the sets $E^\circ(z_\alpha)$ and $E(z_\alpha)$ are nearly the same.

Finally we discuss the general situation when the target parameter is a function of the moments. This means the relations

$$\boldsymbol{m}(\boldsymbol{\theta}) = \int \boldsymbol{g}(y)\, dP_{\boldsymbol{\theta}}(y), \qquad \boldsymbol{\theta} = \boldsymbol{m}^{-1}\bigl(\boldsymbol{m}(\boldsymbol{\theta})\bigr).$$

Of course, these relations assume that the vector function $m(\cdot)$ is invertible. The substitution principle leads to the estimate

$$\tilde{\boldsymbol{\theta}} \stackrel{\text{def}}{=} \boldsymbol{m}^{-1}(\boldsymbol{M}_n),$$

where $\boldsymbol{M}_n$ is the vector of empirical moments:

$$\boldsymbol{M}_n \stackrel{\text{def}}{=} \int \boldsymbol{g}(y)\, dP_n(y) = \frac{1}{n}\sum \boldsymbol{g}(Y_i).$$

The central limit theorem implies (see Theorem 2.1.4) that $\boldsymbol{M}_n$ is a consistent estimate of $\boldsymbol{m}(\boldsymbol{\theta}^*)$ and the vector $\sqrt{n}\bigl[\boldsymbol{M}_n - \boldsymbol{m}(\boldsymbol{\theta}^*)\bigr]$ is asymptotically normal with some covariance matrix $\Sigma_{\boldsymbol{g}}(\boldsymbol{\theta}^*)$. Moreover, if $\boldsymbol{m}^{-1}$ is differentiable at the point $\boldsymbol{m}(\boldsymbol{\theta}^*)$, then $\sqrt{n}(\tilde{\boldsymbol{\theta}} - \boldsymbol{\theta}^*)$ is asymptotically normal as well:

$$\sqrt{n}(\tilde{\boldsymbol{\theta}} - \boldsymbol{\theta}^*) \xrightarrow{w} \mathcal{N}(0, \Sigma(\boldsymbol{\theta}^*))$$

where $\Sigma(\boldsymbol{\theta}^*) = H^\top \Sigma_{\boldsymbol{g}}(\boldsymbol{\theta}^*) H$ and H is the $p \times p$-Jacobi matrix of $\boldsymbol{m}^{-1}$ at $\boldsymbol{m}(\boldsymbol{\theta}^*)$: $H \stackrel{\text{def}}{=} \frac{d}{d\boldsymbol{\theta}} \boldsymbol{m}^{-1}(\boldsymbol{m}(\boldsymbol{\theta}^*))$.

2.5 Some Geometric Properties of a Parametric Family

The parametric situation means that the true marginal distribution P belongs to some given parametric family $(P_{\boldsymbol{\theta}}, \boldsymbol{\theta} \in \Theta \subseteq \mathbb{R}^p)$. By $\boldsymbol{\theta}^*$ we denote the true value, that is, $P = P_{\boldsymbol{\theta}^*} \in (P_{\boldsymbol{\theta}})$. The natural target of estimation in this situation is the parameter $\boldsymbol{\theta}^*$ itself. Below we assume that the family $(P_{\boldsymbol{\theta}})$ is *dominated*, that is, there exists a dominating measure μ_0. The corresponding density is denoted by

$$p(y, \boldsymbol{\theta}) = \frac{dP_{\theta}}{d\mu_0}(y).$$

We also use the notation

$$\ell(y, \boldsymbol{\theta}) \stackrel{\text{def}}{=} \log p(y, \boldsymbol{\theta})$$

for the log-density.

The following two important characteristics of the parametric family $(P_{\boldsymbol{\theta}})$ will be frequently used in the sequel: *the Kullback–Leibler divergence* and *Fisher information*.

2.5.1 *Kullback–Leibler Divergence*

For any two parameters $\boldsymbol{\theta}, \boldsymbol{\theta}'$, the value

$$\mathcal{K}(P_{\boldsymbol{\theta}}, P_{\boldsymbol{\theta}'}) = \int \log \frac{p(y, \boldsymbol{\theta})}{p(y, \boldsymbol{\theta}')} p(y, \boldsymbol{\theta}) d\mu_0(y) = \int \bigl[\ell(y, \boldsymbol{\theta}) - \ell(y, \boldsymbol{\theta}')\bigr] p(y, \boldsymbol{\theta}) d\mu_0(y)$$

is called the *Kullback–Leibler divergence* (KL-divergence) between $P_{\boldsymbol{\theta}}$ and $P_{\boldsymbol{\theta}'}$. We also write $\mathcal{K}(\boldsymbol{\theta}, \boldsymbol{\theta}')$ instead of $\mathcal{K}(P_{\boldsymbol{\theta}}, P_{\boldsymbol{\theta}'})$ if there is no risk of confusion. Equivalently one can represent the KL-divergence as

$$\mathcal{K}(\boldsymbol{\theta}, \boldsymbol{\theta}') = E_{\boldsymbol{\theta}} \log \frac{p(Y, \boldsymbol{\theta})}{p(Y, \boldsymbol{\theta}')} = E_{\boldsymbol{\theta}} \bigl[\ell(Y, \boldsymbol{\theta}) - \ell(Y, \boldsymbol{\theta}')\bigr],$$

where $Y \sim P_{\boldsymbol{\theta}}$. An important feature of the Kullback–Leibler divergence is that it is always non-negative and it is equal to zero iff the measures $P_{\boldsymbol{\theta}}$ and $P_{\boldsymbol{\theta}'}$ coincide.

Lemma 2.5.1. *For any $\boldsymbol{\theta}, \boldsymbol{\theta}'$, it holds*

$$\mathcal{K}(\boldsymbol{\theta}, \boldsymbol{\theta}') \geq 0.$$

Moreover, $\mathcal{K}(\boldsymbol{\theta}, \boldsymbol{\theta}') = 0$ implies that the densities $p(y, \boldsymbol{\theta})$ and $p(y, \boldsymbol{\theta}')$ coincide μ_0-a.s.

Proof. Define $Z(y) = p(y, \boldsymbol{\theta}')/p(y, \boldsymbol{\theta})$. Then

$$\int Z(y) p(y, \boldsymbol{\theta})\, d\mu_0(y) = \int p(y, \boldsymbol{\theta}')\, d\mu_0(y) = 1$$

because $p(y, \boldsymbol{\theta}')$ is the density of $P_{\boldsymbol{\theta}'}$ w.r.t. μ_0. Next, $\frac{d^2}{dt^2} \log(t) = -t^{-2} < 0$, thus, the log-function is strictly concave. The Jensen inequality implies

$$\begin{aligned}\mathcal{K}(\boldsymbol{\theta}, \boldsymbol{\theta}') &= -\int \log(Z(y)) p(y, \boldsymbol{\theta})\, d\mu_0(y) \geq -\log\Bigl(\int Z(y) p(y, \boldsymbol{\theta})\, d\mu_0(y)\Bigr) \\ &= -\log(1) = 0.\end{aligned}$$

Moreover, the strict concavity of the log-function implies that the equality in this relation is only possible if $Z(y) \equiv 1$ $P_{\boldsymbol{\theta}}$-a.s. This implies the last statement of the lemma.

The two mentioned features of the Kullback–Leibler divergence suggest to consider it as a kind of distance on the parameter space. In some sense, it measures how far $P_{\boldsymbol{\theta}'}$ is from $P_{\boldsymbol{\theta}}$. Unfortunately, it is not a metric because it is not symmetric:

$$\mathcal{K}(\boldsymbol{\theta}, \boldsymbol{\theta}') \neq \mathcal{K}(\boldsymbol{\theta}', \boldsymbol{\theta})$$

with very few exceptions for some special situations.

Exercise 2.5.1. Compute KL-divergence for the Gaussian shift, Bernoulli, Poisson, volatility, and exponential families. Check in which cases it is symmetric.

Exercise 2.5.2. Consider the shift experiment given by the equation $Y = \theta + \varepsilon$ where ε is an error with the given density function $p(\cdot)$ on $\mathbb{R}$. Compute the KL-divergence and check for symmetry.

One more important feature of the KL-divergence is its additivity.

Lemma 2.5.2. *Let* $(P_{\boldsymbol{\theta}}^{(1)}, \boldsymbol{\theta} \in \Theta)$ *and* $(P_{\boldsymbol{\theta}}^{(2)}, \boldsymbol{\theta} \in \Theta)$ *be two parametric families with the same parameter set* Θ, *and let* $(P_{\boldsymbol{\theta}} = P_{\boldsymbol{\theta}}^{(1)} \times P_{\boldsymbol{\theta}}^{(2)}, \boldsymbol{\theta} \in \Theta)$ *be the product family. Then for any* $\boldsymbol{\theta}, \boldsymbol{\theta}' \in \Theta$

$$\mathcal{K}(P_{\boldsymbol{\theta}}, P_{\boldsymbol{\theta}'}) = \mathcal{K}(P_{\boldsymbol{\theta}}^{(1)}, P_{\boldsymbol{\theta}'}^{(1)}) + \mathcal{K}(P_{\boldsymbol{\theta}}^{(2)}, P_{\boldsymbol{\theta}'}^{(2)})$$

Exercise 2.5.3. Prove Lemma 2.5.2. Extend the result to the case of the m-fold product of measures.

Hint: use that the log-density $\ell(y_1, y_2, \boldsymbol{\theta})$ of the product measure $P_{\boldsymbol{\theta}}$ fulfills $\ell(y_1, y_2, \boldsymbol{\theta}) = \ell^{(1)}(y_1, \boldsymbol{\theta}) + \ell^{(2)}(y_2, \boldsymbol{\theta})$.

The additivity of the KL-divergence helps to easily compute the KL quantity for two measures $\mathbb{P}_{\boldsymbol{\theta}}$ and $\mathbb{P}_{\boldsymbol{\theta}'}$ describing the i.i.d. sample $\boldsymbol{Y} = (Y_1, \ldots, Y_n)^{\top}$. The log-density of the measure $\mathbb{P}_{\boldsymbol{\theta}}$ w.r.t. $\boldsymbol{\mu}_0 = \mu_0^{\otimes n}$ at the point $\boldsymbol{y} = (y_1, \ldots, y_n)^\top$ is given by

$$L(\boldsymbol{y}, \boldsymbol{\theta}) = \sum \ell(y_i, \boldsymbol{\theta}).$$

An extension of the result of Lemma 2.5.2 yields

$$\mathcal{K}(\mathbb{P}_{\boldsymbol{\theta}}, \mathbb{P}_{\boldsymbol{\theta}'}) \stackrel{\text{def}}{=} \mathbb{E}_{\boldsymbol{\theta}}\bigl\{L(\boldsymbol{Y}, \boldsymbol{\theta}) - L(\boldsymbol{Y}, \boldsymbol{\theta}')\bigr\} = n\mathcal{K}(\boldsymbol{\theta}, \boldsymbol{\theta}').$$

2.5.2 *Hellinger Distance*

Another useful characteristic of a parametric family $(P_{\boldsymbol{\theta}})$ is the so-called *Hellinger distance*. For a fixed $\mu \in [0, 1]$ and any $\boldsymbol{\theta}, \boldsymbol{\theta}' \in \Theta$, define

$$\begin{aligned} h(\mu, P_{\boldsymbol{\theta}}, P_{\boldsymbol{\theta}'}) &= E_{\boldsymbol{\theta}}\left(\frac{dP_{\boldsymbol{\theta}'}}{dP_{\boldsymbol{\theta}}}(Y)\right)^{\mu} \\ &= \int \left(\frac{p(y, \boldsymbol{\theta}')}{p(y, \boldsymbol{\theta})}\right)^{\mu} dP_{\boldsymbol{\theta}}(y) \\ &= \int p^{\mu}(y, \boldsymbol{\theta}') p^{1-\mu}(y, \boldsymbol{\theta})\, d\mu_0(y). \end{aligned}$$

Note that this function can be represented as an exponential moment of the log-likelihood ratio $\ell(Y, \boldsymbol{\theta}, \boldsymbol{\theta}') = \ell(Y, \boldsymbol{\theta}) - \ell(Y, \boldsymbol{\theta}')$:

$$h(\mu, P_{\boldsymbol{\theta}}, P_{\boldsymbol{\theta}'}) = E_{\boldsymbol{\theta}} \exp\bigl\{\mu \ell(Y, \boldsymbol{\theta}', \boldsymbol{\theta})\bigr\} = E_{\boldsymbol{\theta}}\left(\frac{dP_{\boldsymbol{\theta}'}}{dP_{\boldsymbol{\theta}}}(Y)\right)^{\mu}.$$

It is obvious that $h(\mu, P_{\boldsymbol{\theta}}, P_{\boldsymbol{\theta}'}) \geq 0$. Moreover, $h(\mu, P_{\boldsymbol{\theta}}, P_{\boldsymbol{\theta}'}) \leq 1$. Indeed, the function x^{μ} for $\mu \in [0, 1]$ is concave and by the Jensen inequality:

$$E_{\boldsymbol{\theta}}\left(\frac{dP_{\boldsymbol{\theta}'}}{dP_{\boldsymbol{\theta}}}(Y)\right)^{\mu} \leq \left(\mathbb{E}_{\boldsymbol{\theta}} \frac{dP_{\boldsymbol{\theta}'}}{dP_{\boldsymbol{\theta}}}(Y)\right)^{\mu} = 1.$$

Similarly to the Kullback–Leibler, we often write $h(\mu, \boldsymbol{\theta}, \boldsymbol{\theta}')$ in place of $h(\mu, P_{\boldsymbol{\theta}}, P_{\boldsymbol{\theta}'})$.

Typically the Hellinger distance is considered for $\mu = 1/2$. Then

$$h(1/2, \boldsymbol{\theta}, \boldsymbol{\theta}') = \int p^{1/2}(y, \boldsymbol{\theta}') p^{1/2}(y, \boldsymbol{\theta}) d\mu_0(y).$$

In contrast to the Kullback–Leibler divergence, this quantity is symmetric and can be used to define a metric on the parameter set Θ.

Introduce

$$\mathfrak{m}(\mu,\boldsymbol{\theta},\boldsymbol{\theta}') = -\log h(\mu,\boldsymbol{\theta},\boldsymbol{\theta}') = -\log E_{\boldsymbol{\theta}} \exp\{\mu \ell(Y,\boldsymbol{\theta}',\boldsymbol{\theta})\}.$$

The property $h(\mu,\boldsymbol{\theta},\boldsymbol{\theta}') \le 1$ implies $\mathfrak{m}(\mu,\boldsymbol{\theta},\boldsymbol{\theta}') \ge 0$.

The rate function, similarly to the KL-divergence, is additive.

Lemma 2.5.3. *Let $(P_{\boldsymbol{\theta}}^{(1)}, \boldsymbol{\theta} \in \Theta)$ and $(P_{\boldsymbol{\theta}}^{(2)}, \boldsymbol{\theta} \in \Theta)$ be two parametric families with the same parameter set Θ, and let $(P_{\boldsymbol{\theta}} = P_{\boldsymbol{\theta}}^{(1)} \times P_{\boldsymbol{\theta}}^{(2)}, \boldsymbol{\theta} \in \Theta)$ be the product family. Then for any $\boldsymbol{\theta}, \boldsymbol{\theta}' \in \Theta$ and any $\mu \in [0,1]$*

$$\mathfrak{m}(\mu, P_{\boldsymbol{\theta}}, P_{\boldsymbol{\theta}'}) = \mathfrak{m}(P_{\boldsymbol{\theta}}^{(1)}, P_{\boldsymbol{\theta}'}^{(1)}) + \mathfrak{m}(P_{\boldsymbol{\theta}}^{(2)}, P_{\boldsymbol{\theta}'}^{(2)}).$$

Exercise 2.5.4. Prove Lemma 2.5.3. Extend the result to the case of an m-fold product of measures.

Hint: use that the log-density $\ell(y_1, y_2, \boldsymbol{\theta})$ of the product measure $P_{\boldsymbol{\theta}}$ fulfills $\ell(y_1, y_2, \boldsymbol{\theta}) = \ell^{(1)}(y_1, \boldsymbol{\theta}) + \ell^{(2)}(y_2, \boldsymbol{\theta})$.

Application of this lemma to the i.i.d. product family yields

$$\mathfrak{M}(\mu,\boldsymbol{\theta}',\boldsymbol{\theta}) \stackrel{\text{def}}{=} -\log \mathbb{E}_{\boldsymbol{\theta}} \exp\{\mu L(\boldsymbol{Y},\boldsymbol{\theta},\boldsymbol{\theta}')\} = n\,\mathfrak{m}(\mu,\boldsymbol{\theta}',\boldsymbol{\theta}).$$

2.5.3 *Regularity and the Fisher Information: Univariate Parameter*

An important assumption on the considered parametric family $(P_{\boldsymbol{\theta}})$ is that the corresponding density function $p(y,\boldsymbol{\theta})$ is absolutely continuous w.r.t. the parameter $\boldsymbol{\theta}$ for almost all y. Then the log-density $\ell(y,\boldsymbol{\theta})$ is differentiable as well with

$$\nabla \ell(y,\boldsymbol{\theta}) \stackrel{\text{def}}{=} \frac{\partial \ell(y,\boldsymbol{\theta})}{\partial \boldsymbol{\theta}} = \frac{1}{p(y,\boldsymbol{\theta})} \frac{\partial p(y,\boldsymbol{\theta})}{\partial \boldsymbol{\theta}}$$

with the convention $\frac{1}{0}\log(0) = 0$. In the case of a univariate parameter $\theta \in \mathbb{R}$, we also write $\ell'(y,\theta)$ instead of $\nabla \ell(y,\theta)$.

Moreover, we usually assume some regularity conditions on the density $p(y,\boldsymbol{\theta})$. The next definition presents one possible set of such conditions for the case of a univariate parameter θ.

Definition 2.5.1. The family $(P_\theta,\ \theta \in \Theta \subset \mathbb{R})$ is *regular* if the following conditions are fulfilled:

1. The sets $A(\theta) \stackrel{\text{def}}{=} \{y\colon p(y,\theta) = 0\}$ are the same for all $\theta \in \Theta$.

2. *Differentiability under the integration sign:* for any function $s(y)$ satisfying

$$\int s^2(y)p(y,\theta)d\mu_0(y) \le C, \qquad \theta \in \Theta$$

it holds

$$\frac{\partial}{\partial\theta}\int s(y)\,dP_\theta(y) = \frac{\partial}{\partial\theta}\int s(y)p(y,\theta)d\mu_0(y) = \int s(y)\frac{\partial p(y,\theta)}{\partial\theta}d\mu_0(y).$$

3. *Finite Fisher information:* the log-density function $\ell(y,\theta)$ is differentiable in θ and its derivative is square integrable w.r.t. P_θ:

$$\int \left|\ell'(y,\theta)\right|^2 dP_\theta(y) = \int \frac{|p'(y,\theta)|^2}{p(y,\theta)}d\mu_0(y). \tag{2.8}$$

The quantity in the condition (2.8) plays an important role in asymptotic statistics. It is usually referred to as the *Fisher information*.

Definition 2.5.2. Let $(P_\theta, \theta \in \Theta \subset \mathbb{R})$ be a regular parametric family with the univariate parameter. Then the quantity

$$\mathbb{F}(\theta) \stackrel{\text{def}}{=} \int \left|\ell'(y,\theta)\right|^2 p(y,\theta)d\mu_0(y) = \int \frac{|p'(y,\theta)|^2}{p(y,\theta)}d\mu_0(y)$$

is called the *Fisher information* of (P_θ) at $\theta \in \Theta$.

The definition of $\mathbb{F}(\theta)$ can be written as

$$\mathbb{F}(\theta) = \mathbb{E}_\theta\left|\ell'(Y,\theta)\right|^2$$

with $Y \sim P_\theta$.

A simple sufficient condition for regularity of a family (P_θ) is given by the next lemma.

Lemma 2.5.4. *Let the log-density $\ell(y,\theta) = \log p(y,\theta)$ of a dominated family (P_θ) be differentiable in θ and let the Fisher information $\mathbb{F}(\theta)$ be a continuous function on Θ. Then (P_θ) is regular.*

The proof is technical and can be found, e.g., in Borokov (1998). Some useful properties of the regular families are listed in the next lemma.

Lemma 2.5.5. *Let (P_θ) be a regular family. Then for any $\theta \in \Theta$ and $Y \sim P_\theta$*

1. $E_\theta\ell'(Y,\theta) = \int \ell'(y,\theta)\,p(y,\theta)\,d\mu_0(y) = 0$ *and* $\mathbb{F}(\theta) = \operatorname{Var}_\theta\left[\ell'(Y,\theta)\right]$.
2. $\mathbb{F}(\theta) = -E_\theta\ell''(Y,\theta) = -\int \ell''(y,\theta)p(y,\theta)d\mu_0(y)$.

Proof. Differentiating the identity $\int p(y,\theta)d\mu_0(y) = \int \exp\{\ell(y,\theta)\}d\mu_0(y) \equiv 1$ implies under the regularity conditions the first statement of the lemma. Differentiating once more yields the second statement with another representation of the Fisher information.

Like the KL-divergence, the Fisher information possesses the important additivity property.

Lemma 2.5.6. *Let $(P_\theta^{(1)}, \theta \in \Theta)$ and $(P_\theta^{(2)}, \theta \in \Theta)$ be two parametric families with the same parameter set Θ, and let $(P_\theta = P_\theta^{(1)} \times P_\theta^{(2)}, \theta \in \Theta)$ be the product family. Then for any $\theta \in \Theta$, the Fisher information $\mathbb{F}(\theta)$ satisfies*

$$\mathbb{F}(\theta) = \mathbb{F}^{(1)}(\theta) + \mathbb{F}^{(2)}(\theta)$$

where $\mathbb{F}^{(1)}(\theta)$ (resp. $\mathbb{F}^{(2)}(\theta)$) is the Fisher information for $(P_\theta^{(1)})$ (resp. for $(P_\theta^{(2)})$).

Exercise 2.5.5. Prove Lemma 2.5.6.

Hint: use that the log-density of the product experiment can be represented as $\ell(y_1, y_2, \theta) = \ell_1(y_1, \theta) + \ell_2(y_2, \theta)$. The independence of Y_1 and Y_2 implies

$$\begin{aligned} \mathbb{F}(\theta) &= \operatorname{Var}_\theta\bigl[\ell'(Y_1, Y_2, \theta)\bigr] = \operatorname{Var}_\theta\bigl[\ell_1'(Y_1, \theta) + \ell_2'(Y_2, \theta)\bigr] \\ &= \operatorname{Var}_\theta\bigl[\ell_1'(Y_1, \theta)\bigr] + \operatorname{Var}_\theta\bigl[\ell_2'(Y_2, \theta)\bigr]. \end{aligned}$$

Exercise 2.5.6. Compute the Fisher information for the Gaussian shift, Bernoulli, Poisson, volatility, and exponential families. Check in which cases it is constant.

Exercise 2.5.7. Consider the shift experiment given by the equation $Y = \theta + \varepsilon$ where ε is an error with the given density function $p(\cdot)$ on $\mathbb{R}$. Compute the Fisher information and check whether it is constant.

Exercise 2.5.8. Check that the i.i.d. experiment from the uniform distribution on the interval $[0, \theta]$ with unknown θ is not regular.

Now we consider the properties of the i.i.d. experiment from a given regular family (P_θ). The distribution of the whole i.i.d. sample $\boldsymbol{Y}$ is described by the product measure $\mathbb{P}_\theta = P_\theta^{\otimes n}$ which is dominated by the measure $\boldsymbol{\mu}_0 = \mu_0^{\otimes n}$. The corresponding log-density $L(\boldsymbol{y}, \boldsymbol{\theta})$ is given by

$$L(\boldsymbol{y}, \theta) \stackrel{\text{def}}{=} \log \frac{d\mathbb{P}_\theta}{d\boldsymbol{\mu}_0}(\boldsymbol{y}) = \sum \ell(y_i, \theta).$$

The function $\exp L(\boldsymbol{y}, \theta)$ is the density of $\mathbb{P}_\theta$ w.r.t. $\boldsymbol{\mu}_0$ and hence, for any r.v. ξ

$$\mathbb{E}_\theta \xi = \mathbb{E}_0\bigl[\xi \exp L(\boldsymbol{Y}, \theta)\bigr].$$

In particular, for $\xi \equiv 1$, this formula leads to the identity

$$\mathbb{E}_0\bigl[\exp L(\boldsymbol{Y}, \theta)\bigr] = \int \exp\{L(\boldsymbol{y}, \theta)\} \boldsymbol{\mu}_0(d\boldsymbol{y}) \equiv 1. \tag{2.9}$$

The next lemma claims that the product family $(\mathbb{P}_\theta)$ for an i.i.d. sample from a regular family is also regular.

Lemma 2.5.7. *Let (P_θ) be a regular family and $\mathbb{P}_\theta = P_\theta^{\otimes n}$. Then*

1. *The set $A_n \stackrel{\text{def}}{=} \{y = (y_1, \ldots, y_n)^\top : \prod p(y_i, \theta) = 0\}$ is the same for all $\theta \in \Theta$.*
2. *For any r.v. $S = S(Y)$ with $\mathbb{E}_\theta S^2 \le C$, $\theta \in \Theta$, it holds*

$$\frac{\partial}{\partial\theta}\mathbb{E}_\theta S = \frac{\partial}{\partial\theta}\mathbb{E}_0\bigl[S \exp L(Y,\theta)\bigr] = \mathbb{E}_0\bigl[SL'(Y,\theta)\exp L(Y,\theta)\bigr],$$

where $L'(Y,\theta) \stackrel{\text{def}}{=} \frac{\partial}{\partial\theta}L(Y,\theta)$.

3. *The derivative $L'(Y,\theta)$ is square integrable and*

$$\mathbb{E}_\theta\bigl|L'(Y,\theta)\bigr|^2 = n\mathbb{F}(\theta).$$

2.5.4 *Local Properties of the Kullback–Leibler Divergence and Hellinger Distance*

Here we show that the quantities introduced so far are closely related to each other. We start with the Kullback–Leibler divergence.

Lemma 2.5.8. *Let (P_θ) be a regular family. Then the KL-divergence $\mathcal{K}(\theta,\theta')$ satisfies:*

$$\mathcal{K}(\theta,\theta')\Big|_{\theta'=\theta} = 0,$$

$$\frac{d}{d\theta'}\mathcal{K}(\theta,\theta')\Big|_{\theta'=\theta} = 0,$$

$$\frac{d^2}{d\theta'^2}\mathcal{K}(\theta,\theta')\Big|_{\theta'=\theta} = \mathbb{F}(\theta).$$

In a small neighborhood of θ, the KL-divergence can be approximated by

$$\mathcal{K}(\theta,\theta') \approx \mathbb{F}(\theta)|\theta' - \theta|^2/2.$$

Similar properties can be established for the rate function $\mathfrak{m}(\mu,\theta,\theta')$.

Lemma 2.5.9. *Let (P_θ) be a regular family. Then the rate function $\mathfrak{m}(\mu,\theta,\theta')$ satisfies:*

$$\mathfrak{m}(\mu,\theta,\theta')\Big|_{\theta'=\theta} = 0,$$

$$\frac{d}{d\theta'}\mathfrak{m}(\mu,\theta,\theta')\Big|_{\theta'=\theta} = 0,$$

$$\frac{d^2}{d\theta'^2}\mathfrak{m}(\mu,\theta,\theta')\Big|_{\theta'=\theta} = \mu(1-\mu)\mathbb{F}(\theta).$$

In a small neighborhood of θ, the rate function $\mathfrak{m}(\mu,\theta,\theta')$ can be approximated by

$$\mathfrak{m}(\mu,\theta,\theta') \approx \mu(1-\mu)\mathbb{F}(\theta)|\theta'-\theta|^2/2.$$

Moreover, for any $\theta,\theta' \in \Theta$

$$\begin{aligned}
\mathfrak{m}(\mu,\theta,\theta')\Big|_{\mu=0} &= 0,\\
\frac{d}{d\mu}\mathfrak{m}(\mu,\theta,\theta')\Big|_{\mu=0} &= E_\theta \ell(Y,\theta,\theta') = \mathcal{K}(\theta,\theta'),\\
\frac{d^2}{d\mu^2}\mathfrak{m}(\mu,\theta,\theta')\Big|_{\mu=0} &= -\operatorname{Var}_\theta\bigl[\ell(Y,\theta,\theta')\bigr].
\end{aligned}$$

This implies an approximation for μ small

$$\mathfrak{m}(\mu,\theta,\theta') \approx \mu\mathcal{K}(\theta,\theta') - \frac{\mu^2}{2}\operatorname{Var}_\theta\bigl[\ell(Y,\theta,\theta')\bigr].$$

Exercise 2.5.9. Check the statements of Lemmas 2.5.8 and 2.5.9.

2.6 Cramér–Rao Inequality

Let $\tilde{\boldsymbol{\theta}}$ be an estimate of the parameter $\boldsymbol{\theta}^*$. We are interested in establishing a lower bound for the risk of this estimate. This bound indicates that under some conditions the quadratic risk of this estimate can never be below a specific value.

2.6.1 Univariate Parameter

We again start with the univariate case and consider the case of an unbiased estimate $\tilde{\theta}$. Suppose that the family $(P_\theta, \theta \in \Theta)$ is dominated by a σ-finite measure μ_0 on the real line and denote by $p(y,\theta)$ the density of P_θ w.r.t. μ_0:

$$p(y,\theta) \stackrel{\text{def}}{=} \frac{dP_\theta}{d\mu_0}(y).$$

Theorem 2.6.1 (Cramér–Rao Inequality). *Let $\tilde{\theta} = \tilde{\theta}(Y)$ be an unbiased estimate of θ for an i.i.d. sample from a regular family (P_θ). Then*

$$\mathbb{E}_\theta|\tilde{\theta}-\theta|^2 = \operatorname{Var}_\theta(\tilde{\theta}) \ge \frac{1}{n\mathbb{F}(\theta)}$$

with the equality iff $\tilde{\theta}-\theta = \{n\mathbb{F}(\theta)\}^{-1}L'(\boldsymbol{Y},\theta)$ *almost surely. Moreover, if* $\tilde{\theta}$ *is not unbiased and* $\tau(\theta) = \mathbb{E}_\theta\tilde{\theta}$, *then with* $\tau'(\theta) \stackrel{\text{def}}{=} \frac{d}{d\theta}\tau(\theta)$, *it holds*

$$\operatorname{Var}_\theta(\tilde{\theta}) \ge \frac{|\tau'(\theta)|^2}{n\mathbb{F}(\theta)}$$

and

$$\mathbb{E}_\theta|\tilde{\theta}-\theta|^2 = \operatorname{Var}_\theta(\tilde{\theta}) + |\tau(\theta)-\theta|^2 \ge \frac{|\tau'(\theta)|^2}{n\mathbb{F}(\theta)} + |\tau(\theta)-\theta|^2.$$

Proof. Consider first the case of an unbiased estimate $\tilde{\theta}$ with $\mathbb{E}_\theta\tilde{\theta} \equiv \theta$. Differentiating the identity (2.9) $\mathbb{E}_\theta \exp L(\boldsymbol{Y},\theta) \equiv 1$ w.r.t. θ yields

$$0 \equiv \int \bigl[L'(y,\theta)\exp\{L(y,\theta)\}\bigr]\mu_0(dy) = \mathbb{E}_\theta L'(\boldsymbol{Y},\theta). \tag{2.10}$$

Similarly, the identity $\mathbb{E}_\theta\tilde{\theta} = \theta$ implies

$$1 \equiv \int \bigl[\tilde{\theta}L'(\boldsymbol{Y},\theta)\exp\{L(\boldsymbol{Y},\theta)\}\bigr]\mu_0(dy) = \mathbb{E}_\theta\bigl[\tilde{\theta}L'(\boldsymbol{Y},\theta)\bigr].$$

Together with (2.10), this gives

$$\mathbb{E}_\theta\bigl[(\tilde{\theta}-\theta)L'(\boldsymbol{Y},\theta)\bigr] \equiv 1. \tag{2.11}$$

Define $h = \{n\mathbb{F}(\theta)\}^{-1}L'(\boldsymbol{Y},\theta)$. Then $\mathbb{E}\bigl\{hL'(\boldsymbol{Y},\theta)\bigr\} = 1$ and (2.11) yields

$$\mathbb{E}_\theta\bigl[(\tilde{\theta}-\theta-h)h\bigr] \equiv 0.$$

Now

$$\begin{aligned}\mathbb{E}_\theta(\tilde{\theta}-\theta)^2 &= \mathbb{E}_\theta(\tilde{\theta}-\theta-h+h)^2 = \mathbb{E}_\theta(\tilde{\theta}-\theta-h)^2 + \mathbb{E}_\theta h^2 \\ &= \mathbb{E}_\theta(\tilde{\theta}-\theta-h)^2 + \bigl\{n\mathbb{F}(\theta)\bigr\}^{-1} \ge \bigl\{n\mathbb{F}(\theta)\bigr\}^{-1}\end{aligned}$$

with the equality iff $\tilde{\theta}-\theta = \{n\mathbb{F}(\theta)\}^{-1}L'(\boldsymbol{Y},\theta)$ almost surely. This implies the first assertion.

Now we consider the general case. The proof is similar. The property (2.10) continues to hold. Next, the identity $\mathbb{E}_\theta\tilde{\theta} = \theta$ is replaced with $\mathbb{E}_\theta\tilde{\theta} = \tau(\theta)$ yielding

$$\mathbb{E}_\theta\bigl[\tilde{\theta}L'(\boldsymbol{Y},\theta)\bigr] \equiv \tau'(\theta)$$

and

$$\mathbb{E}_\theta\big[\{\tilde{\theta} - \tau(\theta)\}L'(\boldsymbol{Y},\theta)\big] \equiv \tau'(\theta).$$

Again by the Cauchy–Schwartz inequality

$$\begin{aligned}\left|\tau'(\theta)\right|^2 &= \mathbb{E}_\theta^2\big[\{\tilde{\theta} - \tau(\theta)\}L'(\boldsymbol{Y},\theta)\big]\\ &\le \mathbb{E}_\theta\{\tilde{\theta} - \tau(\theta)\}^2\, \mathbb{E}_\theta|L'(\boldsymbol{Y},\theta)|^2\\ &= \operatorname{Var}_\theta(\tilde{\theta})\; n\mathbb{F}(\theta)\end{aligned}$$

and the second assertion follows. The last statement is the usual decomposition of the quadratic risk into the squared bias and the variance of the estimate.

2.6.2 *Exponential Families and R-Efficiency*

An interesting question is how good (precise) the Cramér–Rao lower bound is. In particular, when it is an equality. Indeed, if we restrict ourselves to unbiased estimates, no estimate can have quadratic risk smaller than $[n\mathbb{F}(\theta)]^{-1}$. If an estimate has exactly the risk $[n\mathbb{F}(\theta)]^{-1}$, then this estimate is automatically efficient in the sense that it is the best in the class in terms of the quadratic risk.

Definition 2.6.1. An unbiased estimate $\tilde{\theta}$ is *R-efficient* if

$$\operatorname{Var}_\theta(\tilde{\theta}) = [n\mathbb{F}(\theta)]^{-1}.$$

Theorem 2.6.2. *An unbiased estimate $\tilde{\theta}$ is R-efficient if and only if*

$$\tilde{\theta} = n^{-1}\sum U(Y_i),$$

where the function $U(\cdot)$ on $\mathbb{R}$ satisfies $\int U(y)\,dP_\theta(y) \equiv \theta$ and the log-density $\ell(y,\theta)$ of P_θ can be represented as

$$\ell(y,\theta) = C(\theta)U(y) - B(\theta) + \ell(y), \tag{2.12}$$

for some functions $C(\cdot)$ and $B(\cdot)$ on Θ and a function $\ell(\cdot)$ on $\mathbb{R}$.

Proof. Suppose first that the representation (2.12) for the log-density is correct. Then $\ell'(y,\theta) = C'(\theta)U(y) - B'(\theta)$ and the identity $E_\theta \ell'(y,\theta) = 0$ implies the relation between the functions $B(\cdot)$ and $C(\cdot)$:

$$\theta C'(\theta) = B'(\theta). \tag{2.13}$$

Next, differentiating the equality

$$0 \equiv \int \{U(y) - \theta\} \, dP_\theta(y) = \int \{U(y) - \theta\} e^{L(y,\theta)} d\mu_0(y)$$

w.r.t. θ implies in view of (2.13)

$$1 \equiv \mathbb{E}_\theta\bigl[\{U(Y) - \theta\} \times \{C'(\theta)U(Y) - B'(\theta)\}\bigr] = C'(\theta)\mathbb{E}_\theta\{U(Y) - \theta\}^2.$$

This yields $\operatorname{Var}_\theta\{U(Y)\} = 1/C'(\theta)$. This leads to the following representation for the Fisher information:

$$\begin{aligned}\mathbb{F}(\theta) &= \operatorname{Var}_\theta\{\ell'(Y,\theta)\} \\ &= \operatorname{Var}_\theta\{C'(\theta)U(Y) - B'(\theta)\} \\ &= \{C'(\theta)\}^2 \operatorname{Var}_\theta\{U(Y)\} = C'(\theta).\end{aligned}$$

The estimate $\tilde{\theta} = n^{-1} \sum U(Y_i)$ satisfies

$$\mathbb{E}_\theta \tilde{\theta} = \theta,$$

that is, it is unbiased. Moreover,

$$\operatorname{Var}_\theta(\tilde{\theta}) = \operatorname{Var}_\theta\Bigl\{\frac{1}{n}\sum U(Y_i)\Bigr\} = \frac{1}{n^2}\sum \operatorname{Var}\{U(Y_i)\} = \frac{1}{nC'(\theta)} = \frac{1}{n\mathbb{F}(\theta)}$$

and $\tilde{\theta}$ is R-efficient.

Now we show a reverse statement. Due to the proof of the Cramér–Rao inequality, the only possibility of getting the equality in this inequality is if

$$L'(\boldsymbol{Y},\theta) = n\mathbb{F}(\theta)\,(\tilde{\theta} - \theta).$$

This implies for some fixed θ_0 and any θ°

$$\begin{aligned}L(\boldsymbol{Y},\theta^\circ) - L(\boldsymbol{Y},\theta_0) &= \int_{\theta_0}^{\theta^\circ} L'(\boldsymbol{Y},\theta)d\theta \\ &= \int_{\theta_0}^{\theta} n\mathbb{F}(\theta)(\tilde{\theta} - \theta)d\theta = n\{\tilde{\theta}C(\theta) - B(\theta)\}\end{aligned}$$

with $C(\theta) = \int_{\theta_0}^{\theta} \mathbb{F}(\theta)d\theta$ and $B(\theta) = \int_{\theta_0}^{\theta} \theta\mathbb{F}(\theta)d\theta$. Applying this equality to a sample with $n = 1$ yields $U(Y_1) = \tilde{\theta}(Y_1)$, and

$$\ell(Y_1,\theta) = \ell(Y_1,\theta_0) + C(\theta)U(Y_1) - B(\theta).$$

The desired representation follows.

Exercise 2.6.1. Apply the Cramér–Rao inequality and check R-efficiency to the empirical mean estimate $\tilde{\theta} = n^{-1} \sum Y_i$ for the Gaussian shift, Bernoulli, Poisson, exponential, and volatility families.

2.7 Cramér–Rao Inequality: Multivariate Parameter

This section extends the notions and results of the previous sections from the case of a univariate parameter to the case of a multivariate parameter with $\boldsymbol{\theta} \in \Theta \subset \mathbb{R}^p$.

2.7.1 *Regularity and Fisher Information: Multivariate Parameter*

The definition of regularity naturally extends to the case of a multivariate parameter $\boldsymbol{\theta} = (\theta_1, \ldots, \theta_p)^\top$. It suffices to check the same conditions as in the univariate case for every partial derivative $\partial p(y, \boldsymbol{\theta})/\partial \theta_j$ of the density $p(y, \boldsymbol{\theta})$ for $j = 1, \ldots, p$.

Definition 2.7.1. The family $(P_{\boldsymbol{\theta}}, \boldsymbol{\theta} \in \Theta \subset \mathbb{R}^p)$ is *regular* if the following conditions are fulfilled:

1. The sets $A(\boldsymbol{\theta}) \stackrel{\text{def}}{=} \{y \colon p(y, \theta) = 0\}$ are the same for all $\boldsymbol{\theta} \in \Theta$.
2. *Differentiability under the integration sign:* for any function $s(y)$ satisfying

$$\int s^2(y) p(y, \boldsymbol{\theta}) d\mu_0(y) \le C, \qquad \boldsymbol{\theta} \in \Theta$$

it holds

$$\frac{\partial}{\partial \boldsymbol{\theta}} \int s(y)\, dP_{\boldsymbol{\theta}}(y) = \frac{\partial}{\partial \boldsymbol{\theta}} \int s(y) p(y, \boldsymbol{\theta}) d\mu_0(y) = \int s(y) \frac{\partial p(y, \boldsymbol{\theta})}{\partial \boldsymbol{\theta}} d\mu_0(y).$$

3. *Finite Fisher information:* the log-density function $\ell(y, \boldsymbol{\theta})$ is differentiable in $\boldsymbol{\theta}$ and its derivative $\nabla \ell(y, \boldsymbol{\theta}) = \partial \ell(y, \boldsymbol{\theta})/\partial \boldsymbol{\theta}$ is square integrable w.r.t. $P_{\boldsymbol{\theta}}$:

$$\int \bigl|\nabla \ell(y, \boldsymbol{\theta})\bigr|^2 dP_\theta(y) = \int \frac{|\nabla p(y, \theta)|^2}{p(y, \theta)} d\mu_0(y) < \infty.$$

In the case of a multivariate parameter, the notion of the Fisher information leads to the *Fisher information matrix*.

Definition 2.7.2. Let $(P_{\boldsymbol{\theta}}, \boldsymbol{\theta} \in \Theta \subset \mathbb{R}^p)$ be a parametric family. The matrix

$$\mathbb{F}(\boldsymbol{\theta}) \stackrel{\text{def}}{=} \int \nabla \ell(y, \boldsymbol{\theta}) \nabla^\top \ell(y, \boldsymbol{\theta}) p(y, \boldsymbol{\theta}) d\mu_0(y)$$
$$= \int \nabla p(y, \boldsymbol{\theta}) \nabla^\top p(y, \boldsymbol{\theta}) \frac{1}{p(y, \boldsymbol{\theta})} d\mu_0(y)$$

is called the *Fisher information matrix* of $(P_{\boldsymbol{\theta}})$ at $\boldsymbol{\theta} \in \Theta$.

This definition can be rewritten as

$$\mathbb{F}(\boldsymbol{\theta}) = \mathbb{E}_{\boldsymbol{\theta}}\bigl[\nabla \ell(Y_1, \theta)\{\nabla \ell(Y_1, \theta)\}^\top\bigr].$$

The additivity property of the Fisher information extends to the multivariate case as well.

Lemma 2.7.1. *Let $(P_{\boldsymbol{\theta}}, \boldsymbol{\theta} \in \Theta)$ be a regular family. Then the n-fold product family $(\mathbb{P}_{\boldsymbol{\theta}})$ with $\mathbb{P}_{\boldsymbol{\theta}} = P_{\boldsymbol{\theta}}^{\otimes n}$ is also regular. The Fisher information matrix $\mathbb{F}(\boldsymbol{\theta})$ satisfies*

$$\mathbb{E}_{\boldsymbol{\theta}}\bigl[\nabla L(\boldsymbol{Y}, \boldsymbol{\theta})\{\nabla L(\boldsymbol{Y}, \boldsymbol{\theta})\}^\top\bigr] = n\mathbb{F}(\boldsymbol{\theta}). \tag{2.14}$$

Exercise 2.7.1. Compute the Fisher information matrix for the i.i.d. experiment $Y_i = \theta + \sigma\varepsilon_i$ with unknown θ and σ and ε_i i.i.d. standard normal.

2.7.2 *Local Properties of the Kullback–Leibler Divergence and Hellinger Distance*

The local relations between the Kullback–Leibler divergence, rate function, and Fisher information naturally extend to the case of a multivariate parameter. We start with the Kullback–Leibler divergence.

Lemma 2.7.2. *Let $(P_{\boldsymbol{\theta}})$ be a regular family. Then the KL-divergence $\mathcal{K}(\boldsymbol{\theta}, \boldsymbol{\theta}')$ satisfies:*

$$\mathcal{K}(\boldsymbol{\theta}, \boldsymbol{\theta}')\Big|_{\boldsymbol{\theta}'=\boldsymbol{\theta}} = 0,$$
$$\frac{d}{d\boldsymbol{\theta}'}\mathcal{K}(\boldsymbol{\theta}, \boldsymbol{\theta}')\Big|_{\boldsymbol{\theta}'=\boldsymbol{\theta}} = 0,$$
$$\frac{d^2}{d\boldsymbol{\theta}'^2}\mathcal{K}(\boldsymbol{\theta}, \boldsymbol{\theta}')\Big|_{\boldsymbol{\theta}'=\boldsymbol{\theta}} = \mathbb{F}(\boldsymbol{\theta}).$$

In a small neighborhood of $\boldsymbol{\theta}$, the KL-divergence can be approximated by

$$\mathcal{K}(\boldsymbol{\theta}, \boldsymbol{\theta}') \approx (\boldsymbol{\theta}' - \boldsymbol{\theta})^\top \mathbb{F}(\boldsymbol{\theta})\,(\boldsymbol{\theta}' - \boldsymbol{\theta})/2.$$

Similar properties can be established for the rate function $\mathfrak{m}(\mu, \boldsymbol{\theta}, \boldsymbol{\theta}')$.

Lemma 2.7.3. *Let $(P_{\boldsymbol{\theta}})$ be a regular family. Then the rate function $\mathfrak{m}(\mu, \boldsymbol{\theta}, \boldsymbol{\theta}')$ satisfies:*

$$\begin{aligned} \mathfrak{m}(\mu, \boldsymbol{\theta}, \boldsymbol{\theta}')\Big|_{\boldsymbol{\theta}'=\boldsymbol{\theta}} &= 0, \\ \frac{d}{d\boldsymbol{\theta}'}\mathfrak{m}(\mu, \boldsymbol{\theta}, \boldsymbol{\theta}')\Big|_{\boldsymbol{\theta}'=\boldsymbol{\theta}} &= 0, \\ \frac{d^2}{d\boldsymbol{\theta}'^2}\mathfrak{m}(\mu, \boldsymbol{\theta}, \boldsymbol{\theta}')\Big|_{\boldsymbol{\theta}'=\boldsymbol{\theta}} &= \mu(1-\mu)\mathbb{F}(\boldsymbol{\theta}). \end{aligned}$$

In a small neighborhood of $\boldsymbol{\theta}$, the rate function can be approximated by

$$\mathfrak{m}(\mu, \boldsymbol{\theta}, \boldsymbol{\theta}') \approx \mu(1-\mu)(\boldsymbol{\theta}' - \boldsymbol{\theta})^\top \mathbb{F}(\boldsymbol{\theta})\,(\boldsymbol{\theta}' - \boldsymbol{\theta})/2.$$

Moreover, for any $\boldsymbol{\theta}, \boldsymbol{\theta}' \in \Theta$

$$\begin{aligned} \mathfrak{m}(\mu, \boldsymbol{\theta}, \boldsymbol{\theta}')\Big|_{\mu=0} &= 0, \\ \frac{d}{d\mu}\mathfrak{m}(\mu, \boldsymbol{\theta}, \boldsymbol{\theta}')\Big|_{\mu=0} &= E_{\boldsymbol{\theta}}\ell(Y, \boldsymbol{\theta}, \boldsymbol{\theta}') = \mathcal{K}(\boldsymbol{\theta}, \boldsymbol{\theta}'), \\ \frac{d^2}{d\mu^2}\mathfrak{m}(\mu, \boldsymbol{\theta}, \boldsymbol{\theta}')\Big|_{\mu=0} &= -\operatorname{Var}_{\boldsymbol{\theta}}\bigl[\ell(Y, \boldsymbol{\theta}, \boldsymbol{\theta}')\bigr]. \end{aligned}$$

This implies an approximation for μ small:

$$\mathfrak{m}(\mu, \boldsymbol{\theta}, \boldsymbol{\theta}') \approx \mu\mathcal{K}(\boldsymbol{\theta}, \boldsymbol{\theta}') - \frac{\mu^2}{2}\operatorname{Var}_{\boldsymbol{\theta}}\bigl[\ell(Y, \boldsymbol{\theta}, \boldsymbol{\theta}')\bigr].$$

Exercise 2.7.2. Check the statements of Lemmas 2.7.2 and 2.7.3.

2.7.3 Multivariate Cramér–Rao Inequality

Let $\tilde{\boldsymbol{\theta}} = \tilde{\boldsymbol{\theta}}(Y)$ be an estimate of the unknown parameter vector. This estimate is called *unbiased* if

$$\mathbb{E}_{\boldsymbol{\theta}}\tilde{\boldsymbol{\theta}} \equiv \boldsymbol{\theta}.$$

Theorem 2.7.1 (Multivariate Cramér–Rao Inequality). *Let $\tilde{\boldsymbol{\theta}} = \tilde{\boldsymbol{\theta}}(Y)$ be an unbiased estimate of $\boldsymbol{\theta}$ for an i.i.d. sample from a regular family $(P_{\boldsymbol{\theta}})$. Then*

$$\operatorname{Var}_{\theta}(\tilde{\theta}) \geq \{n\mathbb{F}(\theta)\}^{-1},$$
$$\mathbb{E}_{\theta}\|\tilde{\theta}-\theta\|^2 = \operatorname{tr}\{\operatorname{Var}_{\theta}(\tilde{\theta})\} \geq \operatorname{tr}\bigl[\{n\mathbb{F}(\theta)\}^{-1}\bigr].$$

Moreover, if $\tilde{\theta}$ *is not unbiased and* $\tau(\theta) = \mathbb{E}_{\theta}\tilde{\theta}$, *then with* $\nabla\tau(\theta) \stackrel{\text{def}}{=} \frac{d}{d\theta}\tau(\theta)$, *it holds*

$$\operatorname{Var}_{\theta}(\tilde{\theta}) \geq \nabla\tau(\theta)\,\{n\mathbb{F}(\theta)\}^{-1}\{\nabla\tau(\theta)\}^{\top},$$

and

$$\begin{aligned}\mathbb{E}_{\theta}\|\tilde{\theta}-\theta\|^2 &= \operatorname{tr}\bigl[\operatorname{Var}_{\theta}(\tilde{\theta})\bigr] + \|\tau(\theta)-\theta\|^2 \\ &\geq \operatorname{tr}\bigl[\nabla\tau(\theta)\,\{n\mathbb{F}(\theta)\}^{-1}\{\nabla\tau(\theta)\}^{\top}\bigr] + \|\tau(\theta)-\theta\|^2.\end{aligned}$$

Proof. Consider first the case of an unbiased estimate $\tilde{\theta}$ with $\mathbb{E}_{\theta}\tilde{\theta} \equiv \theta$. Differentiating the identity (2.9) $\mathbb{E}_{\theta}\exp L(Y,\theta) \equiv 1$ w.r.t. θ yields

$$0 \equiv \int \nabla L(y,\theta)\exp\{L(y,\theta)\}\,\mu_0(dy) = \mathbb{E}_{\theta}\bigl[\nabla L(Y,\theta)\bigr] \equiv 0. \tag{2.15}$$

Similarly, the identity $\mathbb{E}_{\theta}\tilde{\theta} = \theta$ implies

$$I \equiv \int \tilde{\theta}(y)\,\{\nabla L(y,\theta)\}^{\top}\exp\{L(y,\theta)\}\mu_0(dy) = \mathbb{E}_{\theta}\bigl[\tilde{\theta}\,\{\nabla L(Y,\theta)\}^{\top}\bigr].$$

Together with (2.15), this gives

$$\mathbb{E}_{\theta}\bigl[(\tilde{\theta}-\theta)\,\{\nabla L(Y,\theta)\}^{\top}\bigr] \equiv I. \tag{2.16}$$

Consider the random vector

$$h \stackrel{\text{def}}{=} \{n\mathbb{F}(\theta)\}^{-1}\nabla L(Y,\theta).$$

By (2.15) $\mathbb{E}_{\theta}h = 0$ and by (2.14)

$$\begin{aligned}\operatorname{Var}_{\theta}(h) &= \mathbb{E}_{\theta}\bigl(hh^{\top}\bigr) = n^{-2}\mathbb{E}_{\theta}\bigl[I^{-1}(\theta)\nabla L(Y,\theta)\{I^{-1}(\theta)\nabla L(Y,\theta)\}^{\top}\bigr] \\ &= n^{-2}I^{-1}(\theta)\mathbb{E}_{\theta}\bigl[\nabla L(Y,\theta)\{\nabla L(Y,\theta)\}^{\top}\bigr]I^{-1}(\theta) = \{n\mathbb{F}(\theta)\}^{-1}..\end{aligned}$$

and the identities (2.15) and (2.16) imply that

$$\mathbb{E}_{\theta}\bigl[(\tilde{\theta}-\theta-h)h^{\top}\bigr] = 0. \tag{2.17}$$

The "no bias" property yields $\mathbb{E}_{\theta}\big(\tilde{\theta}-\theta\big)=0$ and $\mathbb{E}_{\theta}\big[(\tilde{\theta}-\theta)(\tilde{\theta}-\theta)^{\top}\big]=\operatorname{Var}_{\theta}(\tilde{\theta})$. Finally by the orthogonality (2.17) and

$$\begin{aligned}\operatorname{Var}_{\theta}(\tilde{\theta}) &= \operatorname{Var}_{\theta}(h)+\operatorname{Var}\big(\tilde{\theta}-\theta-h\big)\\ &= \big\{n\mathbb{F}(\theta)\big\}^{-1}+\operatorname{Var}_{\theta}\big(\tilde{\theta}-\theta-h\big)\end{aligned}$$

and the variance of $\tilde{\theta}$ is not smaller than $\big\{n\mathbb{F}(\theta)\big\}^{-1}$. Moreover, the equality is only possible if $\tilde{\theta}-\theta-h$ is equal to zero almost surely.

Now we consider the general case. The proof is similar. The property (2.15) continues to hold. Next, the identity $\mathbb{E}_{\theta}\tilde{\theta}=\theta$ is replaced with $\mathbb{E}_{\theta}\tilde{\theta}=\tau(\theta)$ yielding

$$\mathbb{E}_{\theta}\big[\tilde{\theta}\,\{\nabla L(Y,\theta)\}^{\top}\big]\equiv\nabla\tau(\theta)$$

and

$$\mathbb{E}_{\theta}\big[\{\tilde{\theta}-\tau(\theta)\}\,\{\nabla L(Y,\theta)\}^{\top}\big]\equiv\nabla\tau(\theta).$$

Define

$$h\stackrel{\text{def}}{=}\nabla\tau(\theta)\,\big\{n\mathbb{F}(\theta)\big\}^{-1}\nabla L(Y,\theta).$$

Then similarly to the above

$$\begin{aligned}\mathbb{E}_{\theta}\big[hh^{\top}\big] &= \nabla\tau(\theta)\,\big\{n\mathbb{F}(\theta)\big\}^{-1}\big\{\nabla\tau(\theta)\big\}^{\top},\\ \mathbb{E}_{\theta}\big[(\tilde{\theta}-\theta-h)h^{\top}\big] &= 0,\end{aligned}$$

and the second assertion follows. The statements about the quadratic risk follow from its usual decomposition into squared bias and the variance of the estimate.

2.7.4 *Exponential Families and R-Efficiency*

The notion of R-efficiency naturally extends to the case of a multivariate parameter.

Definition 2.7.3. An unbiased estimate $\tilde{\theta}$ is *R-efficient* if

$$\operatorname{Var}_{\theta}(\tilde{\theta})=\big\{n\mathbb{F}(\theta)\big\}^{-1}.$$

Theorem 2.7.2. *An unbiased estimate $\tilde{\theta}$ is R-efficient if and only if*

$$\tilde{\theta}=n^{-1}\sum U(Y_i),$$

where the vector function $\boldsymbol{U}(\cdot)$ *on* $\mathbb{R}$ *satisfies* $\int \boldsymbol{U}(y)dP_{\boldsymbol{\theta}}(y) \equiv \boldsymbol{\theta}$ *and the log-density* $\ell(y,\boldsymbol{\theta})$ *of* $P_{\boldsymbol{\theta}}$ *can be represented as*

$$\ell(y,\boldsymbol{\theta}) = \boldsymbol{C}(\boldsymbol{\theta})^{\top}\boldsymbol{U}(y) - B(\boldsymbol{\theta}) + \ell(y), \tag{2.18}$$

for some functions $\boldsymbol{C}(\cdot)$ *and* $B(\cdot)$ *on* Θ *and a function* $\ell(\cdot)$ *on* $\mathbb{R}$.

Proof. Suppose first that the representation (2.18) for the log-density is correct. Denote by $\boldsymbol{C}'(\boldsymbol{\theta})$ the $p \times p$ Jacobi matrix of the vector function $\boldsymbol{C}$: $\boldsymbol{C}'(\boldsymbol{\theta}) \stackrel{\text{def}}{=} \frac{d}{d\boldsymbol{\theta}}\boldsymbol{C}(\boldsymbol{\theta})$. Then $\nabla\ell(y,\boldsymbol{\theta}) = \boldsymbol{C}'(\boldsymbol{\theta})\boldsymbol{U}(y) - \nabla B(\boldsymbol{\theta})$ and the identity $E_{\boldsymbol{\theta}}\nabla\ell(y,\boldsymbol{\theta}) = 0$ implies the relation between the functions $B(\cdot)$ and $\boldsymbol{C}(\cdot)$:

$$\boldsymbol{C}'(\boldsymbol{\theta})\,\boldsymbol{\theta} = \nabla B(\boldsymbol{\theta}). \tag{2.19}$$

Next, differentiating the equality

$$0 \equiv \int \bigl[\boldsymbol{U}(y) - \boldsymbol{\theta}\bigr]\,dP_{\boldsymbol{\theta}}(y) = \int [\boldsymbol{U}(y) - \boldsymbol{\theta}]e^{L(y,\boldsymbol{\theta})}d\mu_0(y)$$

w.r.t. $\boldsymbol{\theta}$ implies in view of (2.19)

$$\begin{aligned} I &\equiv \mathbb{E}_{\boldsymbol{\theta}}\bigl[\{\boldsymbol{U}(Y) - \boldsymbol{\theta}\}\{\boldsymbol{C}'(\boldsymbol{\theta})\boldsymbol{U}(Y) - \nabla B(\boldsymbol{\theta})\}\bigr]^{\top} \\ &= \boldsymbol{C}'(\boldsymbol{\theta})\mathbb{E}_{\boldsymbol{\theta}}\bigl[\{\boldsymbol{U}(Y) - \boldsymbol{\theta}\}\{\boldsymbol{U}(Y) - \boldsymbol{\theta}\}^{\top}\bigr]. \end{aligned}$$

This yields $\operatorname{Var}_{\boldsymbol{\theta}}\bigl[\boldsymbol{U}(Y)\bigr] = [\boldsymbol{C}'(\boldsymbol{\theta})]^{-1}$. This leads to the following representation for the Fisher information matrix:

$$\begin{aligned} \mathbb{F}(\boldsymbol{\theta}) &= \operatorname{Var}_{\boldsymbol{\theta}}\bigl[\nabla\ell(Y,\boldsymbol{\theta})\bigr] = \operatorname{Var}_{\boldsymbol{\theta}}[\boldsymbol{C}'(\boldsymbol{\theta})\boldsymbol{U}(Y) - \nabla B(\boldsymbol{\theta})] \\ &= \bigl[\boldsymbol{C}'(\boldsymbol{\theta})\bigr]^2 \operatorname{Var}_{\boldsymbol{\theta}}\bigl[\boldsymbol{U}(Y)\bigr] = \boldsymbol{C}(\boldsymbol{\theta}). \end{aligned}$$

The estimate $\tilde{\boldsymbol{\theta}} = n^{-1}\sum \boldsymbol{U}(Y_i)$ satisfies

$$\mathbb{E}_{\boldsymbol{\theta}}\tilde{\boldsymbol{\theta}} = \boldsymbol{\theta},$$

that is, it is unbiased. Moreover,

$$\begin{aligned} \operatorname{Var}_{\boldsymbol{\theta}}\bigl(\tilde{\boldsymbol{\theta}}\bigr) &= \operatorname{Var}_{\boldsymbol{\theta}}\Bigl(\frac{1}{n}\sum \boldsymbol{U}(Y_i)\Bigr) \\ &= \frac{1}{n^2}\sum \operatorname{Var}[\boldsymbol{U}(Y_i)] = \frac{1}{n}\bigl[\boldsymbol{C}'(\boldsymbol{\theta})\bigr]^{-1} = \bigl\{n\mathbb{F}(\boldsymbol{\theta})\bigr\}^{-1} \end{aligned}$$

and $\tilde{\boldsymbol{\theta}}$ is R-efficient.

As in the univariate case, one can show that equality in the Cramér–Rao bound is only possible if $\nabla L(\boldsymbol{Y}, \boldsymbol{\theta})$ and $\tilde{\boldsymbol{\theta}} - \boldsymbol{\theta}$ are linearly dependent. This leads again to the exponential family structure of the likelihood function.

Exercise 2.7.3. Complete the proof of the Theorem 2.7.2.

2.8 Maximum Likelihood and Other Estimation Methods

This section presents some other popular methods of estimating the unknown parameter including minimum distance and M-estimation, maximum likelihood procedure, etc.

2.8.1 Minimum Distance Estimation

Let $\rho(P, P')$ denote some functional (distance) defined for measures P, P' on the real line. We assume that ρ satisfies the following conditions: $\rho(P_{\boldsymbol{\theta}_1}, P_{\boldsymbol{\theta}_2}) \geq 0$ and $\rho(P_{\boldsymbol{\theta}_1}, P_{\boldsymbol{\theta}_2}) = 0$ iff $\boldsymbol{\theta}_1 = \boldsymbol{\theta}_2$. This implies for every $\boldsymbol{\theta}^* \in \Theta$ that

$$\operatorname*{argmin}_{\boldsymbol{\theta} \in \Theta} \rho(P_{\boldsymbol{\theta}}, P_{\boldsymbol{\theta}^*}) = \boldsymbol{\theta}^*.$$

The Glivenko–Cantelli theorem states that P_n converges weakly to the true distribution $P_{\boldsymbol{\theta}^*}$. Therefore, it is natural to define an estimate $\tilde{\boldsymbol{\theta}}$ of $\boldsymbol{\theta}^*$ by replacing in this formula the true measure $P_{\boldsymbol{\theta}^*}$ by its empirical counterpart P_n, that is, by minimizing the distance ρ between the measures $P_{\boldsymbol{\theta}}$ and P_n over the set $(P_{\boldsymbol{\theta}})$. This leads to the *minimum distance estimate*

$$\tilde{\boldsymbol{\theta}} = \operatorname*{argmin}_{\boldsymbol{\theta} \in \Theta} \rho(P_{\boldsymbol{\theta}}, P_n).$$

2.8.2 M-Estimation and Maximum Likelihood Estimation

Another general method of building an estimate of $\boldsymbol{\theta}^*$, the so-called M-estimation is defined via a contrast function $\psi(y, \boldsymbol{\theta})$ given for every $y \in \mathbb{R}$ and $\boldsymbol{\theta} \in \Theta$. The principal condition on ψ is that the integral $\mathbb{E}_{\boldsymbol{\theta}} \psi(Y_1, \boldsymbol{\theta}')$ is minimized for $\boldsymbol{\theta} = \boldsymbol{\theta}'$:

$$\boldsymbol{\theta} = \operatorname*{argmin}_{\boldsymbol{\theta}'} \int \psi(y, \boldsymbol{\theta}')\, dP_{\boldsymbol{\theta}}(y), \qquad \boldsymbol{\theta} \in \Theta. \tag{2.20}$$

In particular,

$$\boldsymbol{\theta}^* = \operatorname*{argmin}_{\boldsymbol{\theta}\in\Theta} \int \psi(y,\boldsymbol{\theta})\, dP_{\boldsymbol{\theta}^*}(y),$$

and the M-estimate is again obtained by substitution, that is, by replacing the true measure $P_{\boldsymbol{\theta}^*}$ with its empirical counterpart P_n:

$$\tilde{\boldsymbol{\theta}} = \operatorname*{argmin}_{\boldsymbol{\theta}\in\Theta} \int \psi(y,\boldsymbol{\theta})\, dP_n(y) = \operatorname*{argmin}_{\boldsymbol{\theta}\in\Theta} \frac{1}{n} \sum \psi(Y_i,\boldsymbol{\theta}).$$

Exercise 2.8.1. Let $\boldsymbol{Y}$ be an i.i.d. sample from $P \in (P_\theta, \theta \in \Theta \subset \mathbb{R})$.

(i) Let also $g(y)$ satisfy $\int g(y)\, dP_\theta(y) \equiv \theta$, leading to the moment estimate

$$\tilde{\theta} = n^{-1} \sum g(Y_i).$$

Show that this estimate can be obtained as the M-estimate for a properly selected function $\psi(\cdot)$.

(ii) Let $\int g(y)\, dP_\theta(y) \equiv m(\theta)$ for the given functions $g(\cdot)$ and $m(\cdot)$ whereas $m(\cdot)$ is monotonous. Show that the moment estimate $\tilde{\theta} = m^{-1}(M_n)$ with $M_n = n^{-1} \sum g(Y_i)$ can be obtained as the M-estimate for a properly selected function $\psi(\cdot)$.

We mention three prominent examples of the contrast function ψ and the resulting estimates: least squares, least absolute deviation (LAD), and maximum likelihood.

2.8.2.1 Least Squares Estimation

The *least squares estimate* (LSE) corresponds to the quadratic contrast

$$\psi(y,\boldsymbol{\theta}) = \|\boldsymbol{g}(y) - \boldsymbol{\theta}\|^2,$$

where $\boldsymbol{g}(y)$ is a p-dimensional function of the observation y satisfying

$$\int \boldsymbol{g}(y)\, dP_{\boldsymbol{\theta}}(y) \equiv \boldsymbol{\theta}, \qquad \boldsymbol{\theta} \in \Theta.$$

Then the true parameter $\boldsymbol{\theta}^*$ fulfills the relation

$$\boldsymbol{\theta}^* = \operatorname*{argmin}_{\boldsymbol{\theta}\in\Theta} \int \|\boldsymbol{g}(y) - \boldsymbol{\theta}\|^2\, dP_{\boldsymbol{\theta}^*}(y)$$

because

$$\int \|g(y) - \theta\|^2 \, dP_{\theta^*}(y) = \|\theta^* - \theta\|^2 + \int \|g(y) - \theta^*\|^2 \, dP_{\theta^*}(y).$$

The substitution method leads to the estimate $\tilde{\theta}$ of θ^* defined by minimization of the empirical version of the integral $\int \|g(y) - \theta\|^2 \, dP_{\theta^*}(y)$:

$$\tilde{\theta} \stackrel{\text{def}}{=} \operatorname*{argmin}_{\theta \in \Theta} \int \|g(y) - \theta\|^2 \, dP_n(y) = \operatorname*{argmin}_{\theta \in \Theta} \sum \|g(Y_i) - \theta\|^2.$$

This is again a quadratic optimization problem having a closed form solution called least squares or ordinary LSE.

Lemma 2.8.1. *It holds*

$$\tilde{\theta} = \operatorname*{argmin}_{\theta \in \Theta} \sum \|g(Y_i) - \theta\|^2 = \frac{1}{n} \sum g(Y_i).$$

One can see that the LSE $\tilde{\theta}$ coincides with the moment estimate based on the function $g(\cdot)$. Indeed, the equality $\int g(y) \, dP_{\theta^*}(y) = \theta^*$ leads directly to the LSE $\tilde{\theta} = n^{-1} \sum g(Y_i)$.

2.8.2.2 LAD (Median) Estimation

The next example of an M-estimate is given by the absolute deviation contrast fit. For simplicity of presentation, we consider here only the case of a univariate parameter. The contrast function $\psi(y, \theta)$ is given by $\psi(y, \theta) \stackrel{\text{def}}{=} |y - \theta|$. The solution of the related optimization problem (2.20) is given by the *median* $\operatorname{med}(P_\theta)$ of the distribution P_θ.

Definition 2.8.1. The value t is called the *median* of a distribution function F if

$$F(t) \geq 1/2, \qquad F(t-) < 1/2.$$

If $F(\cdot)$ is a continuous function, then the median $t = \operatorname{med}(F)$ satisfies $F(t) = 1/2$.

Theorem 2.8.1. *For any cdf* F, *the median* $\operatorname{med}(F)$ *satisfies*

$$\inf_{\theta \in \mathbb{R}} \int |y - \theta| \, dF(y) = \int |y - \operatorname{med}(F)| \, dF(y).$$

Proof. Consider for simplicity the case of a continuous distribution function F. One has $|y - \theta| = (\theta - y)\mathbf{1}(y < \theta) + (y - \theta)\mathbf{1}(y \geq \theta)$. Differentiating w.r.t. θ yields the following equation for any extreme point of $\int |y - \theta| \, dF(y)$:

$$-\int_{-\infty}^{\theta} dF(y) + \int_{\theta}^{\infty} dF(y) = 0.$$

The median is the only solution of this equation.

Let the family (P_θ) be such that $\theta = \operatorname{med}(P_\theta)$ for all $\theta \in \mathbb{R}$. Then the M-estimation approach leads to the *LAD* estimate

$$\tilde{\theta} \stackrel{\text{def}}{=} \operatorname*{argmin}_{\theta \in \mathbb{R}} \int |y - \theta| \, dF_n(y) = \operatorname*{argmin}_{\theta \in \mathbb{R}} \sum |Y_i - \theta|.$$

Due to Theorem 2.8.1, the solution of this problem is given by the median of the edf F_n.

2.8.2.3 Maximum Likelihood Estimation

Let now $\psi(y, \boldsymbol{\theta}) = -\ell(y, \boldsymbol{\theta}) = -\log p(y, \boldsymbol{\theta})$ where $p(y, \boldsymbol{\theta})$ is the density of the measure $P_{\boldsymbol{\theta}}$ at y w.r.t. some dominating measure μ_0. This choice leads to the *MLE*:

$$\tilde{\boldsymbol{\theta}} = \operatorname*{argmax}_{\boldsymbol{\theta} \in \Theta} \frac{1}{n} \sum \log p(Y_i, \boldsymbol{\theta}).$$

The condition (2.20) is fulfilled because

$$\begin{aligned}
\operatorname*{argmin}_{\boldsymbol{\theta}'} \int \psi(y, \boldsymbol{\theta}') \, dP_{\boldsymbol{\theta}}(y) &= \operatorname*{argmin}_{\boldsymbol{\theta}'} \int \{\psi(y, \boldsymbol{\theta}') - \psi(y, \boldsymbol{\theta})\} \, dP_{\boldsymbol{\theta}}(y) \\
&= \operatorname*{argmin}_{\boldsymbol{\theta}'} \int \log \frac{p(y, \boldsymbol{\theta})}{p(y, \boldsymbol{\theta}')} \, dP_{\boldsymbol{\theta}}(y) \\
&= \operatorname*{argmin}_{\boldsymbol{\theta}'} \mathcal{K}(\boldsymbol{\theta}, \boldsymbol{\theta}') = \boldsymbol{\theta}.
\end{aligned}$$

Here we used that the Kullback–Leibler divergence $\mathcal{K}(\boldsymbol{\theta}, \boldsymbol{\theta}')$ attains its minimum equal to zero at the point $\boldsymbol{\theta}' = \boldsymbol{\theta}$ which in turn follows from the concavity of the log-function by the Jensen inequality.

Note that the definition of the MLE does not depend on the choice of the dominating measure μ_0.

Exercise 2.8.2. Show that the MLE $\tilde{\boldsymbol{\theta}}$ does not change if another dominating measure is used.

Computing an M-estimate or MLE leads to solving an optimization problem for the empirical quantity $\sum \psi(Y_i, \boldsymbol{\theta})$ w.r.t. the parameter $\boldsymbol{\theta}$. If the function ψ is differentiable w.r.t. $\boldsymbol{\theta}$, then the solution can be found from the *estimating equation*

$$\frac{\partial}{\partial \boldsymbol{\theta}} \sum \psi(Y_i, \boldsymbol{\theta}) = 0.$$

Exercise 2.8.3. Show that any M-estimate and particularly the MLE can be represented as minimum distance estimate with a properly defined distance ρ.

Hint: define $\rho(P_{\theta}, P_{\theta^*})$ as $\int [\psi(y,\theta) - \psi(y,\theta^*)] dP_{\theta^*}(y)$.

Recall that the MLE $\tilde{\theta}$ is defined by maximizing the expression $L(\theta) = \sum \ell(Y_i, \theta)$ w.r.t. θ. Below we use the notation $L(\theta, \theta') \stackrel{\text{def}}{=} L(\theta) - L(\theta')$, often called the *log-likelihood ratio*.

In our study we will focus on the value of the maximum $L(\tilde{\theta}) = \max_{\theta} L(\theta)$. Let $L(\theta) = \sum \ell(Y_i, \theta)$ be the likelihood function. The value

$$L(\tilde{\theta}) \stackrel{\text{def}}{=} \max_{\theta} L(\theta)$$

is called the *maximum log-likelihood* or *fitted log-likelihood*. The *excess* $L(\tilde{\theta}) - L(\theta^*)$ is the difference between the maximum of the likelihood function $L(\theta)$ over θ and its particular value at the true parameter θ^*:

$$L(\tilde{\theta}, \theta^*) \stackrel{\text{def}}{=} \max_{\theta} L(\theta) - L(\theta^*).$$

The next section collects some examples of computing the MLE $\tilde{\theta}$ and the corresponding maximum log-likelihood.

2.9 Maximum Likelihood for Some Parametric Families

The examples of this section focus on the structure of the log-likelihood and the corresponding MLE $\tilde{\theta}$ and the maximum log-likelihood $L(\tilde{\theta})$.

2.9.1 Gaussian Shift

Let P_θ be the normal distribution on the real line with mean θ and the known variance σ^2. The corresponding density w.r.t. the Lebesgue measure reads as

$$p(y, \theta) = \frac{1}{\sqrt{2\pi\sigma^2}} \exp\Bigl\{-\frac{(y-\theta)^2}{2\sigma^2}\Bigr\}.$$

The log-likelihood $L(\theta)$ is

$$L(\theta) = \sum \log p(Y_i, \theta) = -\frac{n}{2}\log(2\pi\sigma^2) - \frac{1}{2\sigma^2}\sum (Y_i - \theta)^2.$$

The corresponding normal equation $L'(\theta) = 0$ yields

$$-\frac{1}{\sigma^2}\sum(Y_i - \theta) = 0 \tag{2.21}$$

leading to the empirical mean solution $\tilde{\theta} = n^{-1}\sum Y_i$.

The computation of the fitted likelihood is a bit more involved.

Theorem 2.9.1. *Let $Y_i = \theta^* + \varepsilon_i$ with $\varepsilon_i \sim \mathcal{N}(0, \sigma^2)$. For any θ*

$$L(\tilde{\theta}, \theta) = n\sigma^{-2}(\tilde{\theta} - \theta)^2/2. \tag{2.22}$$

Moreover,

$$L(\tilde{\theta}, \theta^*) = n\sigma^{-2}(\tilde{\theta} - \theta^*)^2/2 = \xi^2/2$$

where ξ is a standard normal r.v. so that $2L(\tilde{\theta}, \theta^*)$ has the fixed χ_1^2 distribution with one degree of freedom. If $\mathfrak{z}_\alpha$ is the quantile of $\chi_1^2/2$ with $P(\xi^2/2 > \mathfrak{z}_\alpha) = \alpha$, then

$$\mathcal{E}(\mathfrak{z}_\alpha) = \{u\colon L(\tilde{\theta}, u) \le \mathfrak{z}_\alpha\} \tag{2.23}$$

is an α-confidence set: $\mathbb{P}_{\theta^*}(\mathcal{E}(\mathfrak{z}_\alpha) \not\ni \theta^*) = \alpha$.

For every $r > 0$,

$$\mathbb{E}_{\theta^*}\big|2L(\tilde{\theta}, \theta^*)\big|^r = \mathfrak{c}_r\,,$$

where $\mathfrak{c}_r = E|\xi|^{2r}$ with $\xi \sim \mathcal{N}(0, 1)$.

Proof (Proof 1). Consider $L(\tilde{\theta}, \theta) \stackrel{\text{def}}{=} L(\tilde{\theta}) - L(\theta)$ as a function of the parameter θ. Obviously

$$L(\tilde{\theta}, \theta) = -\frac{1}{2\sigma^2}\sum\big[(Y_i - \tilde{\theta})^2 - (Y_i - \theta)^2\big],$$

so that $L(\tilde{\theta}, \theta)$ is a quadratic function of θ. Next, it holds $L(\tilde{\theta}, \theta)\big|_{\theta=\tilde{\theta}} = 0$ and $\frac{d}{d\theta}L(\tilde{\theta}, \theta)\big|_{\theta=\tilde{\theta}} = -\frac{d}{d\theta}L(\theta)\big|_{\theta=\tilde{\theta}} = 0$ due to the normal equation (2.21). Finally,

$$\frac{d^2}{d\theta^2}L(\tilde{\theta}, \theta)\big|_{\theta=\tilde{\theta}} = -\frac{d^2}{d\theta^2}L(\theta)\big|_{\theta=\tilde{\theta}} = n/\sigma^2.$$

This implies by the Taylor expansion of a quadratic function $L(\tilde{\theta}, \theta)$ at $\theta = \tilde{\theta}$:

$$L(\tilde{\theta}, \theta) = \frac{n}{2\sigma^2}(\tilde{\theta} - \theta)^2.$$

Proof 2. First observe that for any two points θ', θ, the log-likelihood ratio $L(\theta', \theta) = \log(d\mathbb{P}_{\theta'}/d\mathbb{P}_\theta) = L(\theta') - L(\theta)$ can be represented in the form

$$L(\theta', \theta) = L(\theta') - L(\theta) = \sigma^{-2}(S - n\theta)(\theta' - \theta) - n\sigma^{-2}(\theta' - \theta)^2/2.$$

Substituting the MLE $\tilde{\theta} = S/n$ in place of θ' implies

$$L(\tilde{\theta}, \theta) = n\sigma^{-2}(\tilde{\theta} - \theta)^2/2.$$

Now we consider the second statement about the distribution of $L(\tilde{\theta}, \theta^*)$. The substitution $\theta = \theta^*$ in (2.22) and the model equation $Y_i = \theta^* + \varepsilon_i$ imply $\tilde{\theta} - \theta^* = n^{-1/2}\sigma\xi$, where

$$\xi \stackrel{\text{def}}{=} \frac{1}{\sigma\sqrt{n}} \sum \varepsilon_i$$

is standard normal. Therefore,

$$L(\tilde{\theta}, \theta^*) = \xi^2/2.$$

This easily implies the result of the theorem.

We see that under $\mathbb{P}_{\theta^*}$ the variable $2L(\tilde{\theta}, \theta^*)$ is χ_1^2 distributed with one degree of freedom, and this distribution does not depend on the sample size n and the scale parameter σ. This fact is known in a more general form as chi-squared theorem.

Exercise 2.9.1. Check that the confidence sets

$$E^{\circ}(z_\alpha) \stackrel{\text{def}}{=} [\tilde{\theta} - n^{-1/2}\sigma z_\alpha, \tilde{\theta} + n^{-1/2}\sigma z_\alpha],$$

where z_α is defined by $2\Phi(-z_\alpha) = \alpha$, and $\mathcal{E}(\mathfrak{z}_\alpha)$ from (2.23) coincide.

Exercise 2.9.2. Compute the constant $\mathfrak{c}_r$ from Theorem 2.9.1 for $r = 0.5, 1, 1.5, 2$.

Already now we point out an interesting feature of the fitted log-likelihood $L(\tilde{\theta}, \theta^*)$. It can be viewed as the normalized squared loss of the estimate $\tilde{\theta}$ because $L(\tilde{\theta}, \theta^*) = n\sigma^{-2}|\tilde{\theta} - \theta^*|^2$. The last statement of Theorem 2.9.1 yields that

$$\mathbb{E}_{\theta^*}|\tilde{\theta} - \theta^*|^{2r} = \mathfrak{c}_r \sigma^{2r} n^{-r}.$$

2.9.2 Variance Estimation for the Normal Law

Let Y_i be i.i.d. normal with mean zero and unknown variance θ^*:

$$Y_i \sim \mathcal{N}(0, \theta^*), \qquad \theta^* \in \mathbb{R}_+ .$$

The likelihood function reads as

$$L(\theta) = \sum \log p(Y_i, \theta) = -\frac{n}{2}\log(2\pi\theta) - \frac{1}{2\theta}\sum Y_i^2.$$

The normal equation $L'(\theta) = 0$ yields

$$L'(\theta) = -\frac{n}{2\theta} + \frac{1}{2\theta^2} \sum Y_i^2 = 0$$

leading to

$$\tilde{\theta} = \frac{1}{n} S_n$$

with $S_n = \sum Y_i^2$. Moreover, for any θ

$$L(\tilde{\theta}, \theta) = -\frac{n}{2} \log(\tilde{\theta}/\theta) - \frac{S_n}{2}(1/\tilde{\theta} - 1/\theta) = n\mathcal{K}(\tilde{\theta}, \theta)$$

where

$$\mathcal{K}(\theta, \theta') = -\frac{1}{2}\big[\log(\theta/\theta') + 1 - \theta/\theta'\big]$$

is the Kullback–Leibler divergence for two Gaussian measures $\mathcal{N}(0, \theta)$ and $\mathcal{N}(0, \theta')$.

2.9.3 Univariate Normal Distribution

Let Y_i be as in previous example $\mathcal{N}\{\alpha, \sigma^2\}$ but neither the mean α nor the variance σ^2 are known. This leads to estimating the vector $\boldsymbol{\theta} = (\theta_1, \theta_2) = (\alpha, \sigma^2)$ from the i.i.d. sample $\boldsymbol{Y}$.

The maximum likelihood approach leads to maximizing the log-likelihood w.r.t. the vector $\boldsymbol{\theta} = (\alpha, \sigma^2)^\top$:

$$L(\boldsymbol{\theta}) = \sum \log p(Y_i, \boldsymbol{\theta}) = -\frac{n}{2} \log(2\pi\theta_2) - \frac{1}{2\theta_2} \sum (Y_i - \theta_1)^2.$$

Exercise 2.9.3. Check that the ML approach leads to the same estimates (2.2) as the method of moments.

2.9.4 Uniform Distribution on $[0, \theta]$

Let Y_i be uniformly distributed on the interval $[0, \theta^*]$ of the real line where the right end point θ^* is unknown. The density $p(y, \theta)$ of P_θ w.r.t. the Lebesgue measure is $\theta^{-1}\mathbf{1}(y \le \theta)$. The likelihood reads as

$$Z(\theta) = \theta^{-n}\mathbf{1}(\max_i Y_i \le \theta).$$

This density is positive iff $\theta \ge \max_i Y_i$ and it is maximized exactly for $\tilde{\theta} = \max_i Y_i$. One can see that the MLE $\tilde{\theta}$ is the limiting case of the moment estimate $\tilde{\theta}_k$ as k grows to infinity.

2.9.5 Bernoulli or Binomial Model

Let P_θ be a Bernoulli law for $\theta \in [0, 1]$. The density of Y_i under P_θ can be written as

$$p(y, \theta) = \theta^y (1-\theta)^{1-y}.$$

The corresponding log-likelihood reads as

$$L(\theta) = \sum \{Y_i \log\theta + (1 - Y_i)\log(1-\theta)\} = S_n \log\frac{\theta}{1-\theta} + n\log(1-\theta)$$

with $S_n = \sum Y_i$. Maximizing this expression w.r.t. θ results again in the empirical mean

$$\tilde{\theta} = S_n/n.$$

This implies

$$L(\tilde{\theta}, \theta) = n\tilde{\theta}\log\frac{\tilde{\theta}}{\theta} + n(1-\tilde{\theta})\log\frac{1-\tilde{\theta}}{1-\theta} = n\mathcal{K}(\tilde{\theta}, \theta)$$

where $\mathcal{K}(\theta, \theta') = \theta\log(\theta/\theta') + (1-\theta)\log\{(1-\theta)/(1-\theta')$ is the Kullback–Leibler divergence for the Bernoulli law.

2.9.6 Multinomial Model

The multinomial distribution B_θ^m describes the number of successes in m experiments when one success has the probability $\theta \in [0, 1]$. This distribution can be viewed as the sum of m binomials with the same parameter θ.

One has

$$P_\theta(Y_1 = k) = \binom{m}{k}\theta^k(1-\theta)^{m-k}, \qquad k = 0, \dots, m.$$

Exercise 2.9.4. Check that the ML approach leads to the estimate

$$\tilde{\theta} = \frac{1}{mn} \sum Y_i .$$

Compute $L(\tilde{\theta}, \theta)$.

2.9.7 *Exponential Model*

Let $Y_1, \ldots, Y_n$ be i.i.d. exponential random variables with parameter $\theta^* > 0$. This means that Y_i are nonnegative and satisfy $\mathbb{P}(Y_i > t) = e^{-t/\theta^*}$. The density of the exponential law w.r.t. the Lebesgue measure is $p(y, \theta^*) = e^{-y/\theta^*}/\theta^*$. The corresponding log-likelihood can be written as

$$L(\theta) = -n \log \theta - \sum_{i=1}^{n} Y_i/\theta = -S/\theta - n \log \theta,$$

where $S = Y_1 + \ldots + Y_n$.

The ML estimating equation yields $S/\theta^2 = n/\theta$ or

$$\tilde{\theta} = S/n.$$

For the fitted log-likelihood $L(\tilde{\theta}, \theta)$ this gives

$$L(\tilde{\theta}, \theta) = -n(1 - \tilde{\theta}/\theta) - n \log(\tilde{\theta}/\theta) = n\mathcal{K}(\tilde{\theta}, \theta).$$

Here once again $\mathcal{K}(\theta, \theta') = \theta/\theta' - 1 - \log(\theta/\theta')$ is the Kullback–Leibler divergence for the exponential law.

2.9.8 *Poisson Model*

Let $Y_1, \ldots, Y_n$ be i.i.d. Poisson random variables satisfying $\mathbb{P}(Y_i = m) = |\theta^*|^m e^{-\theta^*}/m!$ for $m = 0, 1, 2, \ldots$. The corresponding log-likelihood can be written as

$$L(\theta) = \sum_{i=1}^{n} \log\bigl(\theta^{Y_i} e^{-\theta}/Y_i!\bigr) = \log \theta \sum_{i=1}^{n} Y_i - \theta - \log(Y_i!) = S \log \theta - n\theta + R,$$

where $S = Y_1 + \ldots + Y_n$ and $R = \sum_{i=1}^{n} \log(Y_i!)$. Here we leave out that $0! = 1$.

The ML estimating equation immediately yields $S/\theta = n$ or

$$\tilde{\theta} = S/n.$$

For the fitted log-likelihood $L(\tilde{\theta}, \theta)$ this gives

$$L(\tilde{\theta}, \theta) = n\tilde{\theta} \log(\tilde{\theta}/\theta) - n(\tilde{\theta} - \theta) = n\mathcal{K}(\tilde{\theta}, \theta).$$

Here again $\mathcal{K}(\theta, \theta') = \theta \log(\theta/\theta') - (\theta - \theta')$ is the Kullback–Leibler divergence for the Poisson law.

2.9.9 *Shift of a Laplace (Double Exponential) Law*

Let P_0 be the symmetric distribution defined by the equations

$$P_0(|Y_1| > y) = e^{-y/\sigma}, \qquad y \geq 0,$$

for some given $\sigma > 0$. Equivalently one can say that the absolute value of Y_1 is exponential with parameter σ under P_0. Now define P_θ by shifting P_0 by the value θ. This means that

$$P_\theta(|Y_1 - \theta| > y) = e^{-y/\sigma}, \qquad y \geq 0.$$

The density of $Y_1 - \theta$ under P_θ is $p(y) = (2\sigma)^{-1} e^{-|y|/\sigma}$. The maximum likelihood approach leads to maximizing the sum

$$L(\theta) = -n \log(2\sigma) - \sum |Y_i - \theta|/\sigma,$$

or equivalently to minimizing the sum $\sum |Y_i - \theta|$:

$$\tilde{\theta} = \operatorname*{argmin}_{\theta} \sum |Y_i - \theta|. \tag{2.24}$$

This is just the *LAD* estimate given by the *median* of the edf:

$$\tilde{\theta} = \operatorname{med}(F_n).$$

Exercise 2.9.5. Show that the median solves the problem (2.24).

Hint: suppose that n is odd. Consider the ordered observations $Y_{(1)} \leq Y_{(2)} \leq \ldots \leq Y_{(n)}$. Show that the median of P_n is given by $Y_{((n+1)/2)}$. Show that this point solves (2.24).

2.10 Quasi Maximum Likelihood Approach

Let $Y = (Y_1, \ldots, Y_n)^\top$ be a sample from a marginal distribution P. Let also $(P_{\boldsymbol{\theta}}, \boldsymbol{\theta} \in \Theta)$ be a given parametric family with the log-likelihood $\ell(y, \boldsymbol{\theta})$. The parametric approach is based on the assumption that the underlying distribution P belongs to this family. The *quasi maximum likelihood method* applies the maximum likelihood approach for family $(P_{\boldsymbol{\theta}})$ even if the underlying distribution P does not belong to this family. This leads again to the estimate $\tilde{\boldsymbol{\theta}}$ that maximizes the expression $L(\boldsymbol{\theta}) = \sum \ell(Y_i, \boldsymbol{\theta})$ and is called the *quasi MLE*. It might happen that the true distribution belongs to some other parametric family for which one also can construct the MLE. However, there could be serious reasons for applying the quasi maximum likelihood approach even in this misspecified case. One of them is that the properties of the estimate $\tilde{\boldsymbol{\theta}}$ are essentially determined by the geometrical structure of the log-likelihood. The use of a parametric family with a nice geometric structure (which are quadratic or convex functions of the parameter) can seriously simplify the algorithmic burdens and improve the behavior of the method.

2.10.1 LSE as Quasi Likelihood Estimation

Consider the model

$$Y_i = \theta^* + \varepsilon_i \tag{2.25}$$

where θ^* is the parameter of interest from $\mathbb{R}$ and ε_i are random errors satisfying $\mathbb{E}\varepsilon_i = 0$. The assumption that ε_i are i.i.d. normal $\mathcal{N}(0, \sigma^2)$ leads to the quasi log-likelihood

$$L(\theta) = -\frac{n}{2}\log(2\pi\sigma^2) - \frac{1}{2\sigma^2}\sum(Y_i - \theta)^2.$$

Maximizing the expression $L(\theta)$ leads to minimizing the sum of squared residuals $(Y_i - \theta)^2$:

$$\tilde{\theta} = \operatorname*{argmin}_{\theta} \sum(Y_i - \theta)^2 = \frac{1}{n}\sum Y_i\,.$$

This estimate is called a *LSE* or ordinary least squares estimate (oLSE).

Example 2.10.1. Consider the model (2.25) with heterogeneous errors, that is, ε_i are independent normal with zero mean and variances σ_i^2. The corresponding log-likelihood reads

$$L^\circ(\theta) = -\frac{1}{2}\sum\Bigl\{\log(2\pi\sigma_i^2) + \frac{(Y_i - \theta)^2}{\sigma_i^2}\Bigr\}.$$

The MLE $\tilde{\theta}^{\circ}$ is

$$\tilde{\theta}^{\circ} \stackrel{\text{def}}{=} \operatorname*{argmax}_{\theta} L^{\circ}(\theta) = N^{-1} \sum Y_i / \sigma_i^2, \qquad N = \sum \sigma_i^{-2}.$$

We now compare the estimates $\tilde{\theta}$ and $\tilde{\theta}^{\circ}$.

Lemma 2.10.1. *The following assertions hold for the estimate* $\tilde{\theta}$:

1. $\tilde{\theta}$ *is unbiased:* $\mathbb{E}_{\theta^*} \tilde{\theta} = \theta^*$.
2. *The quadratic risk of* $\tilde{\theta}$ *is equal to the variance* $\operatorname{Var}(\tilde{\theta})$ *given by*

$$\mathcal{R}(\tilde{\theta}, \theta^*) \stackrel{\text{def}}{=} \mathbb{E}_{\theta^*} |\tilde{\theta} - \theta^*|^2 = \operatorname{Var}(\tilde{\theta}) = n^{-2} \sum \sigma_i^2.$$

3. $\tilde{\theta}$ *is not R-efficient unless all* σ_i^2 *are equal.*

Now we consider the MLE $\tilde{\theta}^{\circ}$.

Lemma 2.10.2. *The following assertions hold for the estimate* $\tilde{\theta}^{\circ}$:

1. $\tilde{\theta}^{\circ}$ *is unbiased:* $\mathbb{E}_{\theta^*} \tilde{\theta}^{\circ} = \theta^*$.
2. *The quadratic risk of* $\tilde{\theta}^{\circ}$ *is equal to the variance* $\operatorname{Var}(\tilde{\theta}^{\circ})$ *given by*

$$\mathcal{R}(\tilde{\theta}^{\circ}, \theta^*) \stackrel{\text{def}}{=} \mathbb{E}_{\theta^*} |\tilde{\theta}^{\circ} - \theta^*|^2 = \operatorname{Var}(\tilde{\theta}^{\circ}) = N^{-2} \sum \sigma_i^{-2} = N^{-1}.$$

3. $\tilde{\theta}^{\circ}$ *is R-efficient.*

Exercise 2.10.1. Check the statements of Lemmas 2.10.1 and 2.10.2.

Hint: compute the Fisher information for the model (2.25) using the property of additivity:

$$\mathbb{F}(\theta) = \sum \mathbb{F}^{(i)}(\theta) = \sum \sigma_i^{-2} = N,$$

where $\mathbb{F}^{(i)}(\theta)$ is the Fisher information in the marginal model $Y_i = \theta + \varepsilon_i$ with just one observation Y_i. Apply the Cramér–Rao inequality for one observation of the vector $\boldsymbol{Y}$.

2.10.2 LAD and Robust Estimation as Quasi Likelihood Estimation

Consider again the model (2.25). The classical least squares approach faces serious problems if the available data $\boldsymbol{Y}$ are contaminated with outliers. The reasons for contamination could be missing data or typing errors, etc. Unfortunately, even a single outlier can significantly disturb the sum $L(\theta)$ and thus, the estimate $\tilde{\theta}$.

A typical approach proposed and developed by Huber is to apply another "influence function" $\psi(Y_i - \theta)$ in the sum $L(\theta)$ in place of the squared residual $|Y_i - \theta|^2$ leading to the M-estimate

$$\tilde{\theta} = \operatorname*{argmin}_{\theta} \sum \psi(Y_i - \theta). \tag{2.26}$$

A popular ψ-function for robust estimation is the absolute value $|Y_i - \theta|$. The resulting estimate

$$\tilde{\theta} = \operatorname*{argmin}_{\theta} \sum |Y_i - \theta|$$

is called *LAD* and the solution is the *median* of the empirical distribution P_n. Another proposal is called the Huber function: it is quadratic in a vicinity of zero and linear outside:

$$\psi(x) = \begin{cases} x^2 & \text{if } |x| \le t, \\ a|x| + b & \text{otherwise.} \end{cases}$$

Exercise 2.10.2. Show that for each $t > 0$, the coefficients $a = a(t)$ and $b = b(t)$ can be selected to provide that $\psi(x)$ and its derivatives are continuous.

A remarkable fact about this approach is that every such estimate can be viewed as a quasi MLE for the model (2.25). Indeed, for a given function ψ, define the measure P_θ with the log-density $\ell(y, \theta) = -\psi(y - \theta)$. Then the log-likelihood is $L(\theta) = -\sum \psi(Y_i - \theta)$ and the corresponding (quasi) MLE coincides with (2.26).

Exercise 2.10.3. Suggest a σ-finite measure μ such that $\exp\{-\psi(y - \theta)\}$ is the density of Y_i for the model (2.25) w.r.t. the measure μ.

Hint: suppose for simplicity that

$$C_\psi \stackrel{\text{def}}{=} \int \exp\{-\psi(x)\}\, dx < \infty.$$

Show that $C_\psi^{-1} \exp\{-\psi(y - \theta)\}$ is a density w.r.t. the Lebesgue measure for any θ.

Exercise 2.10.4. Show that the LAD $\tilde{\theta} = \operatorname{argmin}_\theta \sum |Y_i - \theta|$ is the quasi MLE for the model (2.25) when the errors ε_i are assumed Laplacian (double exponential) with density $p(x) = (1/2)e^{-|x|}$.

2.11 Univariate Exponential Families

Most parametric families considered in the previous sections are particular cases of exponential families (EF) distributions. This includes the Gaussian shift, Bernoulli, Poisson, exponential, volatility models. The notion of an EF already appeared in the context of the Cramér–Rao inequality. Now we study such families in further detail.

We say that $\mathcal{P}$ is an *exponential family* if all measures $P_\theta \in \mathcal{P}$ are dominated by a σ-finite measure μ_0 on $\mathcal{Y}$ and the density functions $p(y,\theta) = dP_\theta/d\mu_0(y)$ are of the form

$$p(y,\theta) \stackrel{\text{def}}{=} \frac{dP_\theta}{d\mu_0}(y) = p(y)e^{yC(\theta)-B(\theta)}.$$

Here $C(\theta)$ and $B(\theta)$ are some given nondecreasing functions on Θ and $p(y)$ is a nonnegative function on $\mathcal{Y}$.

Usually one assumes some regularity conditions on the family $\mathcal{P}$. One possibility was already given when we discussed the Cramér–Rao inequality; see Definition 2.5.1. Below we assume that condition is always fulfilled. It basically means that we can differentiate w.r.t. θ under the integral sign.

For an EF, the log-likelihood admits an especially simple representation, nearly linear in y:

$$\ell(y,\theta) \stackrel{\text{def}}{=} \log p(y,\theta) = yC(\theta) - B(\theta) + \log p(y)$$

so that the log-likelihood ratio for $\theta, \theta' \in \Theta$ reads as

$$\ell(y,\theta,\theta') \stackrel{\text{def}}{=} \ell(y,\theta) - \ell(y,\theta') = y\bigl[C(\theta) - C(\theta')\bigr] - \bigl[B(\theta) - B(\theta')\bigr].$$

2.11.1 Natural Parametrization

Let $\mathcal{P} = (P_\theta)$ be an EF. By Y we denote one observation from the distribution $P_\theta \in \mathcal{P}$. In addition to the regularity conditions, one often assumes the *natural* parametrization for the family $\mathcal{P}$ which means the relation $E_\theta Y = \theta$. Note that this relation is fulfilled for all the examples of EF's that we considered so far in the previous section. It is obvious that the natural parametrization is only possible if the following identifiability condition is fulfilled: for any two different measures from the considered parametric family, the corresponding mean values are different. Otherwise the natural parametrization is always possible: just define θ as the expectation of Y. Below we use the abbreviation EFn for an exponential family with natural parametrization.

2.11.1.1 Some Properties of an EFn

The natural parametrization implies an important property for the functions $B(\theta)$ and $C(\theta)$.

Lemma 2.11.1. *Let $\big(P_\theta\big)$ be a naturally parameterized EF. Then*

$$B'(\theta) = \theta C'(\theta).$$

Proof. Differentiating both sides of the equation $\int p(y,\theta)\mu_0(dy) = 1$ w.r.t. θ yields

$$\begin{aligned} 0 &= \int \big\{yC'(\theta) - B'(\theta)\big\} p(y,\theta)\mu_0(dy) \\ &= \int \big\{yC'(\theta) - B'(\theta)\big\} P_\theta(dy) \\ &= \theta C'(\theta) - B'(\theta) \end{aligned}$$

and the result follows.

The next lemma computes the important characteristics of a natural EF such as the Kullback–Leibler divergence $\mathcal{K}(\theta,\theta') = E_\theta \log\big(p(Y,\theta)/p(Y,\theta')\big)$, the Fisher information $\mathbb{F}(\theta) \stackrel{\text{def}}{=} E_\theta|\ell'(Y,\theta)|^2$, and the rate function $\mathfrak{m}(\mu,\theta,\theta') = -\log E_\theta \exp\big\{\mu\ell(Y,\theta,\theta')\big\}$.

Lemma 2.11.2. *Let (P_θ) be an EFn. Then with $\theta,\theta' \in \Theta$ fixed, it holds for*

- *the Kullback–Leibler divergence* $\mathcal{K}(\theta,\theta') = E_\theta \log\big(p(Y,\theta)/p(Y,\theta')\big)$:

$$\begin{aligned} \mathcal{K}(\theta,\theta') &= \int \log\frac{p(y,\theta)}{p(y,\theta')} P_\theta(dy) \\ &= \big\{C(\theta) - C(\theta')\big\} \int yP_\theta(dy) - \big\{B(\theta) - B(\theta')\big\} \\ &= \theta\big\{C(\theta) - C(\theta')\big\} - \big\{B(\theta) - B(\theta')\big\}; \end{aligned} \tag{2.27}$$

- *the Fisher information* $\mathbb{F}(\theta) \stackrel{\text{def}}{=} E_\theta|\ell'(Y,\theta)|^2$:

$$\mathbb{F}(\theta) = C'(\theta);$$

- *the rate function* $\mathfrak{m}(\mu,\theta,\theta') = -\log E_\theta \exp\big\{\mu\ell(Y,\theta,\theta')\big\}$:

$$\mathfrak{m}(\mu,\theta,\theta') = \mathcal{K}\big(\theta, \theta + \mu(\theta' - \theta)\big);$$

- *the variance* $\operatorname{Var}_\theta(Y)$:

$$\operatorname{Var}_\theta(Y) = 1/\mathbb{F}(\theta) = 1/C'(\theta). \tag{2.28}$$

Proof. Differentiating the equality

$$0 \equiv \int (y-\theta) P_\theta(dy) = \int (y-\theta) e^{L(y,\theta)} \mu_0(dy)$$

w.r.t. θ implies in view of Lemma 2.11.1

$$1 \equiv \mathbb{E}_\theta\bigl[(Y-\theta)\{C'(\theta)Y - B'(\theta)\}\bigr] = C'(\theta)\mathbb{E}_\theta(Y-\theta)^2.$$

This yields $\operatorname{Var}_\theta(Y) = 1/C'(\theta)$. This leads to the following representation of the Fisher information:

$$\mathbb{F}(\theta) = \operatorname{Var}_\theta\bigl[\ell'(Y,\theta)\bigr] = \operatorname{Var}_\theta[C'(\theta)Y - B'(\theta)] = \bigl[C'(\theta)\bigr]^2 \operatorname{Var}_\theta(Y) = C'(\theta).$$

Exercise 2.11.1. Check the equations for the Kullback–Leibler divergence and Fisher information from Lemma 2.11.2.

2.11.1.2 MLE and Maximum Likelihood for an EFn

Now we discuss the maximum likelihood estimation for a sample from an EFn. The log-likelihood can be represented in the form

$$\begin{aligned} L(\theta) &= \sum_{i=1}^n \log p(Y_i,\theta) = C(\theta)\sum_{i=1}^n Y_i - B(\theta)\sum_{i=1}^n 1 + \sum_{i=1}^n \log p(Y_i) \qquad (2.29) \\ &= SC(\theta) - nB(\theta) + R, \end{aligned}$$

where

$$S = \sum_{i=1}^n Y_i, \qquad R = \sum_{i=1}^n \log p(Y_i).$$

The remainder term R is unimportant because it does not depend on θ and thus it does not enter in the likelihood ratio. The MLE $\tilde{\theta}$ is defined by maximizing $L(\theta)$ w.r.t. θ, that is,

$$\tilde{\theta} = \operatorname*{argmax}_{\theta\in\Theta} L(\theta) = \operatorname*{argmax}_{\theta\in\Theta}\bigl\{SC(\theta) - nB(\theta)\bigr\}.$$

In the case of an EF with the natural parametrization, this optimization problem admits a closed form solution given by the next theorem.

Theorem 2.11.1. *Let (P_θ) be an EFn. Then the MLE $\tilde{\theta}$ fulfills*

$$\tilde{\theta} = S/n = n^{-1} \sum_{i=1}^{n} Y_i .$$

It holds

$$\mathbb{E}_\theta \tilde{\theta} = \theta, \qquad \operatorname{Var}_\theta(\tilde{\theta}) = [n\mathbb{F}(\theta)]^{-1} = [nC'(\theta)]^{-1}$$

so that $\tilde{\theta}$ is R-efficient. Moreover, the fitted log-likelihood $L(\tilde{\theta}, \theta) \stackrel{\text{def}}{=} L(\tilde{\theta}) - L(\theta)$ satisfies for any $\theta \in \Theta$:

$$L(\tilde{\theta}, \theta) = n\mathcal{K}(\tilde{\theta}, \theta). \tag{2.30}$$

Proof. Maximization of $L(\theta)$ w.r.t. θ leads to the estimating equation $nB'(\theta) - SC'(\theta) = 0$. This and the identity $B'(\theta) = \theta C'(\theta)$ yield the MLE

$$\tilde{\theta} = S/n.$$

The variance $\operatorname{Var}_\theta(\tilde{\theta})$ is computed using (2.28) from Lemma 2.11.2. The formula (2.27) for the Kullback–Leibler divergence and (2.29) yield the representation (2.30) for the fitted log-likelihood $L(\tilde{\theta}, \theta)$ for any $\theta \in \Theta$.

One can see that the estimate $\tilde{\theta}$ is the mean of the Y_i's. As for the Gaussian shift model, this estimate can be motivated by the fact that the expectation of every observation Y_i under P_θ is just θ and by the law of large numbers the empirical mean converges to its expectation as the sample size n grows.

2.11.2 *Canonical Parametrization*

Another useful representation of an EF is given by the so-called *canonical parametrization*. We say that υ is the *canonical* parameter for this EF if the density of each measure P_υ w.r.t. the dominating measure μ_0 is of the form:

$$p(y, \upsilon) \stackrel{\text{def}}{=} \frac{dP_\upsilon}{d\mu_0}(y) = p(y) \exp\{y\upsilon - d(\upsilon)\}.$$

Here $d(\upsilon)$ is a given *convex* function on Θ and $p(y)$ is a nonnegative function on $\mathcal{Y}$. The abbreviation EFc will indicate an EF with the canonical parametrization.

2.11.2.1 Some Properties of an EFc

The next relation is an obvious corollary of the definition:

Lemma 2.11.3. *An EFn (P_θ) always permits a unique canonical representation. The canonical parameter υ is related to the natural parameter θ by $\upsilon = C(\theta)$, $d(\upsilon) = B(\theta)$ and $\theta = d'(\upsilon)$.*

Proof. The first two relations follow from the definition. They imply $B'(\theta) = d'(\upsilon) \cdot d\upsilon/d\theta = d'(\upsilon) \cdot C'(\theta)$ and the last statement follows from $B'(\theta) = \theta C'(\theta)$.

The log-likelihood ratio $\ell(y, \upsilon, \upsilon_1)$ for an EFc reads as

$$\ell(Y, \upsilon, \upsilon_1) = Y(\upsilon - \upsilon_1) - d(\upsilon) + d(\upsilon_1).$$

The next lemma collects some useful facts about an EFc.

Lemma 2.11.4. *Let $\mathcal{P} = \bigl(P_\upsilon, \upsilon \in \mathcal{U}\bigr)$ be an EFc and let the function $d(\cdot)$ be two times continuously differentiable. Then it holds for any $\upsilon, \upsilon_1 \in \mathcal{U}$:*

(i). The mean $E_\upsilon Y$ and the variance $\operatorname{Var}_\upsilon(Y)$ fulfill

$$E_\upsilon Y = d'(\upsilon), \qquad \operatorname{Var}_\upsilon(Y) = E_\upsilon(Y - E_\upsilon Y)^2 = d''(\upsilon).$$

(ii). The Fisher information $\mathbb{F}(\upsilon) \stackrel{\text{def}}{=} E_\upsilon |\ell'(Y, \upsilon)|^2$ satisfies

$$\mathbb{F}(\upsilon) = d''(\upsilon).$$

(iii). The Kullback–Leibler divergence $\mathcal{K}^c(\upsilon, \upsilon_1) = E_\upsilon \ell(Y, \upsilon, \upsilon_1)$ satisfies

$$\begin{aligned}
\mathcal{K}^c(\upsilon, \upsilon_1) &= \int \log \frac{p(y, \upsilon)}{p(y, \upsilon_1)} P_\upsilon(dy) \\
&= d'(\upsilon)\bigl(\upsilon - \upsilon_1\bigr) - \bigl\{d(\upsilon) - d(\upsilon_1)\bigr\} \\
&= d''(\breve{\upsilon})\,(\upsilon_1 - \upsilon)^2/2,
\end{aligned}$$

where $\breve{\upsilon}$ is a point between υ and υ_1. Moreover, for $\upsilon \le \upsilon_1 \in \mathcal{U}$

$$\mathcal{K}^c(\upsilon, \upsilon_1) = \int_{\upsilon}^{\upsilon_1} (\upsilon_1 - u) d''(u)\, du.$$

(iv). The rate function $\mathfrak{m}(\mu, \upsilon_1, \upsilon) \stackrel{\text{def}}{=} -\log \mathbb{E}_\upsilon \exp\{\mu \ell(Y, \upsilon_1, \upsilon)\}$ fulfills

$$\mathfrak{m}(\mu, \upsilon_1, \upsilon) = \mu \mathcal{K}^c\bigl(\upsilon, \upsilon_1\bigr) - \mathcal{K}^c\bigl(\upsilon, \upsilon + \mu(\upsilon_1 - \upsilon)\bigr)$$

Table 2.1 $\upsilon(\theta)$, $d(\upsilon)$, $\mathbb{F}(\upsilon) = d''(\upsilon)$ and $\theta = \theta(\upsilon)$ for the examples from Sect. 2.9

Model	υ	$d(\upsilon)$	$I(\upsilon)$	$\theta(\upsilon)$
Gaussian regression	θ/σ^2	$\upsilon^2\sigma^2/2$	σ^2	$\sigma^2\upsilon$
Bernoulli model	$\log(\theta/(1-\theta))$	$\log(1+e^\upsilon)$	$e^\upsilon/(1+e^\upsilon)^2$	$e^\upsilon/(1+e^\upsilon)$
Poisson model	$\log\theta$	e^υ	e^υ	e^υ
Exponential model	$1/\theta$	$-\log\upsilon$	$1/\upsilon^2$	$1/\upsilon$
Volatility model	$-1/(2\theta)$	$-\frac{1}{2}\log(-2\upsilon)$	$1/(2\upsilon^2)$	$-1/(2\upsilon)$

Proof. Differentiating the equation $\int p(y,\upsilon)\mu_0(dy) = 1$ w.r.t. υ yields

$$\int \{y - d'(\upsilon)\} p(y,\upsilon)\mu_0(dy) = 0,$$

that is, $E_\upsilon Y = d'(\upsilon)$. The expression for the variance can be proved by one more differentiating of this equation. Similarly one can check (*ii*). The item (*iii*) can be checked by simple algebra and (*iv*) follows from (*i*).

Further, for any $\upsilon, \upsilon_1 \in \mathcal{U}$, it holds

$$\ell(Y,\upsilon_1,\upsilon) - E_\upsilon \ell(Y,\upsilon_1,\upsilon) = (\upsilon_1 - \upsilon)\{Y - d'(\upsilon)\}$$

and with $u = \mu(\upsilon_1 - \upsilon)$

$$\begin{aligned} &\log E_\upsilon \exp\{u(Y - d'(\upsilon))\} \\ &\quad = -ud'(\upsilon) + d(\upsilon + u) - d(\upsilon) + \log E_\upsilon \exp\{uY - d(\upsilon+u) + d(\upsilon)\} \\ &\quad = d(\upsilon+u) - d(\upsilon) - ud'(\upsilon) = \mathcal{K}^c(\upsilon, \upsilon+u), \end{aligned}$$

because

$$E_\upsilon \exp\{uY - d(\upsilon+u) + d(\upsilon)\} = E_\upsilon \frac{dP_{\upsilon+u}}{dP_\upsilon} = 1$$

and (*iv*) follows by (*iii*).

Table 2.1 presents the canonical parameter and the Fisher information for the examples of exponential families from Sect. 2.9.

Exercise 2.11.2. Check (*iii*) and (*iv*) in Lemma 2.11.4.

Exercise 2.11.3. Check the entries of Table 2.1.

Exercise 2.11.4. Check that $\mathcal{K}^c(\upsilon,\upsilon') = \mathcal{K}(\theta(\upsilon), \theta(\upsilon'))$.

Exercise 2.11.5. Plot $\mathcal{K}^c(\upsilon^*, \upsilon)$ as a function of υ for the families from Table 2.1.

2.11.2.2 Maximum Likelihood Estimation for an EFc

The structure of the log-likelihood in the case of the canonical parametrization is particularly simple:

$$\begin{aligned} L(\upsilon) &= \sum_{i=1}^{n} \log p(Y_i, \upsilon) = \upsilon \sum_{i=1}^{n} Y_i - d(\upsilon) \sum_{i=1}^{n} 1 + \sum_{i=1}^{n} \log p(Y_i) \\ &= S\upsilon - nd(\upsilon) + R \end{aligned}$$

where

$$S = \sum_{i=1}^{n} Y_i, \qquad R = \sum_{i=1}^{n} \log p(Y_i).$$

Again, as in the case of an EFn, we can ignore the remainder term R. The estimating equation $dL(\upsilon)/d\upsilon = 0$ for the maximum likelihood estimate $\tilde{\upsilon}$ reads as

$$d'(\upsilon) = S/n.$$

This and the relation $\theta = d'(\upsilon)$ lead to the following result.

Theorem 2.11.2. *The MLEs $\tilde{\theta}$ and $\tilde{\upsilon}$ for the natural and canonical parametrization are related by the equations*

$$\tilde{\theta} = d'(\tilde{\upsilon}) \qquad \tilde{\upsilon} = C(\tilde{\theta}).$$

The next result describes the structure of the fitted log-likelihood and basically repeats the result of Theorem 2.11.1.

Theorem 2.11.3. *Let (P_υ) be an EF with canonical parametrization. Then for any $\upsilon \in \mathcal{U}$ the fitted log-likelihood $L(\tilde{\upsilon}, \upsilon) \stackrel{\text{def}}{=} \max_{\upsilon'} L(\upsilon', \upsilon)$ satisfies*

$$L(\tilde{\upsilon}, \upsilon) = n\mathcal{K}^c(\tilde{\upsilon}, \upsilon).$$

Exercise 2.11.6. Check the statement of Theorem 2.11.3.

2.11.3 Deviation Probabilities for the Maximum Likelihood

Let $Y_1, \ldots, Y_n$ be i.i.d. observations from an EF $\mathcal{P}$. This section presents a probability bound for the fitted likelihood. To be more specific we assume that $\mathcal{P}$ is canonically parameterized, $\mathcal{P} = (P_\upsilon)$. However, the bound applies to the natural and any other parametrization because the value of maximum of the likelihood process

$L(\theta)$ does not depend on the choice of parametrization. The log-likelihood ratio $L(\upsilon', \upsilon)$ is given by the expression (2.29) and its maximum over υ' leads to the fitted log-likelihood $L(\tilde{\upsilon}, \upsilon) = n\mathcal{K}^c(\tilde{\upsilon}, \upsilon)$.

Our first result concerns a *deviation bound* for $L(\tilde{\upsilon}, \upsilon)$. It utilizes the representation for the fitted log-likelihood given by Theorem 2.11.1. As usual, we assume that the family $\mathcal{P}$ is regular. In addition, we require the following condition.

(*Pc*) $\mathcal{P} = (P_\upsilon, \upsilon \in \mathcal{U} \subseteq \mathbb{R})$ is a regular EF. The parameter set $\mathcal{U}$ is convex. The function $d(\upsilon)$ is two times continuously differentiable and the Fisher information $\mathbb{F}(\upsilon) = d''(\upsilon)$ satisfies $\mathbb{F}(\upsilon) > 0$ for all υ.

The condition (Pc) implies that for any compact set $\mathcal{U}_0$ there is a constant $\mathfrak{a} = \mathfrak{a}(\mathcal{U}_0) > 0$ such that

$$|\mathbb{F}(\upsilon_1)/\mathbb{F}(\upsilon_2)|^{1/2} \le \mathfrak{a}, \qquad \upsilon_1, \upsilon_2 \in \mathcal{U}_0 .$$

Theorem 2.11.4. *Let Y_i be i.i.d. from a distribution P_{υ^*} which belongs to an EFc satisfying (Pc). For any $\mathfrak{z} > 0$*

$$\mathbb{P}_{\upsilon^*}\bigl(L(\tilde{\upsilon}, \upsilon^*) > \mathfrak{z}\bigr) = \mathbb{P}_{\upsilon^*}\bigl(n\mathcal{K}^c(\tilde{\upsilon}, \upsilon^*) > \mathfrak{z}\bigr) \le 2e^{-\mathfrak{z}}.$$

Proof. The proof is based on two properties of the log-likelihood. The first one is that the expectation of the likelihood ratio is just one: $\mathbb{E}_{\upsilon^*} \exp L(\upsilon, \upsilon^*) = 1$. This and the exponential Markov inequality imply for $\mathfrak{z} \ge 0$

$$\mathbb{P}_{\upsilon^*}\bigl(L(\upsilon, \upsilon^*) \ge \mathfrak{z}\bigr) \le \mathrm{e}^{-\mathfrak{z}}. \tag{2.31}$$

The second property is specific to the considered univariate EF and is based on geometric properties of the log-likelihood function: linearity in the observations Y_i and convexity in the parameter υ. We formulate this important fact in a separate statement.

Lemma 2.11.5. *Let the EFc $\mathcal{P}$ fulfill (Pc). For given $\mathfrak{z}$ and any $\upsilon_0 \in \mathcal{U}$, there exist two values $\upsilon^+ > \upsilon_0$ and $\upsilon^- < \upsilon_0$ satisfying $\mathcal{K}^c(\upsilon^\pm, \upsilon_0) = \mathfrak{z}/n$ such that*

$$\{L(\tilde{\upsilon}, \upsilon_0) > \mathfrak{z}\} \subseteq \{L(\upsilon^+, \upsilon_0) > \mathfrak{z}\} \cup \{L(\upsilon^-, \upsilon_0) > \mathfrak{z}\}.$$

Proof. It holds

$$\begin{aligned} &\{L(\tilde{\upsilon}, \upsilon_0) > \mathfrak{z}\} = \Bigl\{\sup_{\upsilon}\bigl[S\bigl(\upsilon - \upsilon_0\bigr) - n\bigl\{d(\upsilon) - d(\upsilon_0)\bigr\}\bigr] > \mathfrak{z}\Bigr\} \\ &\subseteq \left\{ S > \inf_{\upsilon > \upsilon_0} \frac{\mathfrak{z} + n\{d(\upsilon) - d(\upsilon_0)\}}{\upsilon - \upsilon_0} \right\} \cup \left\{ -S > \inf_{\upsilon < \upsilon_0} \frac{\mathfrak{z} + n\{d(\upsilon) - d(\upsilon_0)\}}{\upsilon_0 - \upsilon} \right\}. \end{aligned}$$

Define for every $u > 0$

$$f(u) = \frac{\mathfrak{z} + n\{d(\upsilon_0 + u) - d(\upsilon_0)\}}{u}.$$

This function attains its minimum at a point u satisfying the equation

$$\mathfrak{z}/n + d(\upsilon_0 + u) - d(\upsilon_0) - d'(\upsilon_0 + u)u = 0$$

or, equivalently,

$$\mathcal{K}(\upsilon_0 + u, \upsilon_0) = \mathfrak{z}/n.$$

The condition (Pc) provides that there is only one solution $u \geq 0$ of this equation.

Exercise 2.11.7. Check that the equation $\mathcal{K}(\upsilon_0+u, \upsilon_0) = \mathfrak{z}/n$ has only one positive solution for any $\mathfrak{z} > 0$.
Hint: use that $\mathcal{K}(\upsilon_0 + u, \upsilon_0)$ is a convex function of u with minimum at $u = 0$.

Now, it holds with $\upsilon^+ = \upsilon_0 + u$

$$\left\{S > \inf_{\upsilon > \upsilon_0} \frac{\mathfrak{z} + n\big[d(\upsilon) - d(\upsilon_0)\big]}{\upsilon - \upsilon_0}\right\} = \left\{S > \frac{\mathfrak{z} + n\big[d(\upsilon^+) - d(\upsilon_0)\big]}{\upsilon^+ - \upsilon_0}\right\}$$
$$\subseteq \{L(\upsilon^+, \upsilon_0) > \mathfrak{z}\}.$$

Similarly

$$\left\{-S > \inf_{\upsilon < \upsilon_0} \frac{\mathfrak{z} + n\{d(\upsilon) - d(\upsilon_0)\}}{\upsilon_0 - \upsilon}\right\} = \left\{-S > \frac{\mathfrak{z} + n\big[d(\upsilon^-) - d(\upsilon_0)\big]}{\upsilon_0 - \upsilon^-}\right\}$$
$$\subseteq \{L(\upsilon^-, \upsilon_0) > \mathfrak{z}\}.$$

for some $\upsilon^- < \upsilon_0$.

The assertion of the theorem is now easy to obtain. Indeed,

$$\mathbb{P}_{\upsilon^*}\big(L(\tilde{\upsilon}, \upsilon^*) \geq \mathfrak{z}\big) \leq \mathbb{P}_{\upsilon^*}\big(L(\upsilon^+, \upsilon^*) \geq \mathfrak{z}\big) + \mathbb{P}_{\upsilon^*}\big(L(\upsilon^-, \upsilon^*) \geq \mathfrak{z}\big) \leq 2e^{-\mathfrak{z}}$$

yielding the result.

Exercise 2.11.8. Let (P_υ) be a Gaussian shift experiment, that is, $P_\upsilon = \mathcal{N}(\upsilon, 1)$.

- Check that $L(\tilde{\upsilon}, \upsilon) = n|\tilde{\upsilon} - \upsilon|^2/2$;
- Given $\mathfrak{z} \geq 0$, find the points υ^+ and υ^- such that

$$\{L(\tilde{\upsilon}, \upsilon^*) > \mathfrak{z}\} \subseteq \{L(\upsilon^+, \upsilon^*) > \mathfrak{z}\} \cup \{L(\upsilon^-, \upsilon^*) > \mathfrak{z}\}.$$

- Plot the mentioned sets $\{\upsilon : L(\tilde{\upsilon}, \upsilon) > \mathfrak{z}\}$, $\{\upsilon : L(\upsilon^{+}, \upsilon) > \mathfrak{z}\}$, and $\{\upsilon : L(\upsilon^{-}, \upsilon) > \mathfrak{z}\}$ as functions of υ for a fixed $S = \sum Y_i$.

Remark 2.11.1. Note that the mentioned result only utilizes the geometric structure of the univariate EFc. The most important feature of the log-likelihood ratio $L(\upsilon, \upsilon^*) = S(\upsilon - \upsilon^*) - d(\upsilon) + d(\upsilon^*)$ is its linearity w.r.t. the stochastic term S. This allows us to replace the maximum over the whole set $\mathcal{U}$ by the maximum over the set consisting of two points $\upsilon^{\pm}$. Note that the proof does not rely on the distribution of the observations Y_i. In particular, Lemma 2.11.5 continues to hold even within the quasi likelihood approach when $L(\upsilon)$ is not the true log-likelihood. However, the bound (2.31) relies on the nature of $L(\upsilon, \upsilon^*)$. Namely, it utilizes that $\mathbb{E}\exp\{L(\upsilon^{\pm}, \upsilon^*)\} = 1$, which is true under $\mathbb{P} = \mathbb{P}_{\upsilon^*}$ nut generally false in the quasi likelihood setup. Nevertheless, the exponential bound can be extended to the quasi likelihood approach under the condition of bounded exponential moments for $L(\upsilon, \upsilon^*)$: for some $\mu > 0$, it should hold $\mathbb{E}\exp\{\mu L(\upsilon, \upsilon^*)\} = C(\mu) < \infty$.

Theorem 2.11.4 yields a simple construction of a confidence interval for the parameter υ^* and the concentration property of the MLE $\tilde{\upsilon}$.

Theorem 2.11.5. *Let Y_i be i.i.d. from $P_{\upsilon^*} \in \mathcal{P}$ with $\mathcal{P}$ satisfying (Pc).*

1. *If $\mathfrak{z}_\alpha$ satisfies $e^{-\mathfrak{z}_\alpha} \le \alpha/2$, then*

$$\mathcal{E}(\mathfrak{z}_\alpha) = \bigl\{\upsilon : n\mathcal{K}^c\bigl(\tilde{\upsilon}, \upsilon\bigr) \le \mathfrak{z}_\alpha\bigr\}$$

   is an α-confidence set for the parameter υ^*.
2. *Define for any $\mathfrak{z} > 0$ the set $\mathcal{A}(\mathfrak{z}, \upsilon^*) = \{\upsilon : \mathcal{K}^c(\upsilon, \upsilon^*) \le \mathfrak{z}/n\}$. Then*

$$\mathbb{P}_{\upsilon^*}\bigl(\tilde{\upsilon} \notin \mathcal{A}(\mathfrak{z}, \upsilon^*)\bigr) \le 2e^{-\mathfrak{z}}.$$

The second assertion of the theorem claims that the estimate $\tilde{\upsilon}$ belongs with a high probability to the vicinity $\mathcal{A}(\mathfrak{z}, \upsilon^*)$ of the central point υ^* defined by the Kullback–Leibler divergence. Due to Lemma 2.11.4(iii) $\mathcal{K}^c(\upsilon, \upsilon^*) \approx \mathbb{F}(\upsilon^*)(\upsilon - \upsilon^*)^2/2$, where $\mathbb{F}(\upsilon^*)$ is the Fisher information at υ^*. This vicinity is an interval around υ^* of length of order $n^{-1/2}$. In other words, this result implies the root-n consistency of $\tilde{\upsilon}$.

The deviation bound for the fitted log-likelihood from Theorem 2.11.4 can be viewed as a bound for the normalized loss of the estimate $\tilde{\upsilon}$. Indeed, define the loss function $\wp(\upsilon', \upsilon) = \mathcal{K}^{1/2}(\upsilon', \upsilon)$. Then Theorem 2.11.4 yields that the loss is with high probability bounded by $\sqrt{\mathfrak{z}/n}$ provided that $\mathfrak{z}$ is sufficiently large. Similarly one can establish the bound for the risk.

Theorem 2.11.6. *Let Y_i be i.i.d. from the distribution P_{υ^*} which belongs to a canonically parameterized EF satisfying (Pc). The following properties hold:*

(i). For any $r > 0$ there is a constant $\mathfrak{r}_r$ such that

$$\mathbb{E}_{\upsilon^*} L^r(\tilde{\upsilon}, \upsilon^*) = n^r \mathbb{E}_{\upsilon^*} \mathcal{K}^r(\tilde{\upsilon}, \upsilon^*) \le \mathfrak{r}_r .$$

(ii). For every $\lambda < 1$

$$\mathbb{E}_{\upsilon^*} \exp\{\lambda L(\tilde{\upsilon}, \upsilon^*)\} = \mathbb{E}_{\upsilon^*} \exp\{\lambda n \mathcal{K}(\tilde{\upsilon}, \upsilon^*)\} \le (1+\lambda)/(1-\lambda).$$

Proof. By Theorem 2.11.4

$$\begin{aligned}
\mathbb{E}_{\upsilon^*} L^r(\tilde{\upsilon}, \upsilon^*) &= -\int_{\mathfrak{z}\ge 0} \mathfrak{z}^r d\mathbb{P}_{\upsilon^*}\{L(\tilde{\upsilon}, \upsilon^*) > \mathfrak{z}\} \\
&= r\int_{\mathfrak{z}\ge 0} \mathfrak{z}^{r-1} \mathbb{P}_{\upsilon^*}\{L(\tilde{\upsilon}, \upsilon^*) > \mathfrak{z}\} d\mathfrak{z} \\
&\le r\int_{\mathfrak{z}\ge 0} 2\mathfrak{z}^{r-1} e^{-\mathfrak{z}} d\mathfrak{z}
\end{aligned}$$

and the first assertion is fulfilled with $\mathfrak{r}_r = 2r\int_{\mathfrak{z}\ge 0} \mathfrak{z}^{r-1} e^{-\mathfrak{z}} d\mathfrak{z}$. The assertion *(ii)* is proved similarly.

2.11.3.1 Deviation Bound for Other Parameterizations

The results for the maximum likelihood and their corollaries have been stated for an EFc. An immediate question that arises in this respect is whether the use of the canonical parametrization is essential. The answer is "no": a similar result can be stated for any EF whatever the parametrization is used. This fact is based on the simple observation that the maximum likelihood is the value of the maximum of the likelihood process; this value does not depend on the parametrization.

Lemma 2.11.6. *Let* (P_θ) *be an EF. Then for any* θ

$$L(\tilde{\theta}, \theta) = n\mathcal{K}(P_{\tilde{\theta}}, P_\theta). \tag{2.32}$$

Exercise 2.11.9. Check the result of Lemma 2.11.6.
Hint: use that both sides of (2.32) depend only on measures $P_{\tilde{\theta}}$, P_θ and not on the parametrization.

Below we write as before $\mathcal{K}(\tilde{\theta}, \theta)$ instead of $\mathcal{K}(P_{\tilde{\theta}}, P_\theta)$. The property (2.32) and the exponential bound of Theorem 2.11.4 imply the bound for a general EF:

Theorem 2.11.7. *Let* (P_θ) *be a univariate EF. Then for any* $\mathfrak{z} > 0$

$$\mathbb{P}_{\theta^*}\bigl(L(\tilde{\theta}, \theta^*) > \mathfrak{z}\bigr) = \mathbb{P}_{\theta^*}\bigl(n\mathcal{K}(\tilde{\theta}, \theta^*) > \mathfrak{z}\bigr) \le 2e^{-\mathfrak{z}}.$$

This result allows us to build confidence sets for the parameter θ^* and concentration sets for the MLE $\tilde{\theta}$ in terms of the Kullback–Leibler divergence:

$$\mathcal{A}(\mathfrak{z},\theta^*) = \{\theta : \mathcal{K}(\theta,\theta^*) \le \mathfrak{z}/n\},$$
$$\mathcal{E}(\mathfrak{z}) = \{\theta : \mathcal{K}(\tilde{\theta},\theta) \le \mathfrak{z}/n\}.$$

Corollary 2.11.1. *Let* (P_θ) *be an EF. If* $e^{-\mathfrak{z}_\alpha} = \alpha/2$, *then*

$$\mathbb{P}_{\theta^*}\bigl(\tilde{\theta} \notin \mathcal{A}(\mathfrak{z}_\alpha,\theta^*)\bigr) \le \alpha,$$

and

$$\mathbb{P}_{\theta^*}\bigl(\mathcal{E}(\mathfrak{z}_\alpha) \not\ni \theta\bigr) \le \alpha.$$

Moreover, for any $r > 0$

$$\mathbb{E}_{\theta^*} L^r(\tilde{\theta},\theta^*) = n^r \mathbb{E}_{\theta^*} \mathcal{K}^r(\tilde{\theta},\theta^*) \le \mathfrak{r}_r\,.$$

2.11.3.2 Asymptotic Against Likelihood-Based Approach

The asymptotic approach recommends to apply symmetric confidence and concentration sets with width of order $[n\mathbb{F}(\theta^*)]^{-1/2}$:

$$\mathcal{A}_n(\mathfrak{z},\theta^*) = \{\theta : \mathbb{F}(\theta^*)\,(\theta-\theta^*)^2 \le 2\mathfrak{z}/n\},$$
$$\mathcal{E}_n(\mathfrak{z}) = \{\theta : \mathbb{F}(\theta^*)\,(\theta-\tilde{\theta})^2 \le 2\mathfrak{z}/n\},$$
$$\mathcal{E}_n'(\mathfrak{z}) = \{\theta :\ \ I(\tilde{\theta})\,(\theta-\tilde{\theta})^2 \le 2\mathfrak{z}/n\}.$$

Then asymptotically, i.e. for large n, these sets do approximately the same job as the non-asymptotic sets $\mathcal{A}(\mathfrak{z},\theta^*)$ and $\mathcal{E}(\mathfrak{z})$. However, the difference for finite samples can be quite significant. In particular, for some cases, e.g. the Bernoulli of Poisson families, the sets $\mathcal{A}_n(\mathfrak{z},\theta^*)$ and $\mathcal{E}_n'(\mathfrak{z})$ may extend beyond the parameter set Θ.

2.12 Historical Remarks and Further Reading

The main part of the chapter is inspired by the nice textbook (Borokov, 1998). The concept of *exponential families* is credited to Edwin Pitman, Georges Darmois, and Bernard Koopman in 1935–1936.

The notion of *Kullback–Leibler divergence* was originally introduced by Solomon Kullback and Richard Leibler in 1951 as the directed divergence between two distributions. Many of its useful properties are studied in monograph (Kullback, 1997).

The *Fisher information* was discussed by several early statisticians, notably Francis Edgeworth. *Maximum-likelihood estimation* was recommended, analyzed, and vastly popularized by Robert Fisher between 1912 and 1922, although it had been used earlier by Carl Gauss, Pierre-Simon Laplace, Thorvald Thiele, and Francis Edgeworth.

The *Cramér–Rao inequality* was independently obtained by Maurice Fréchet, Calyampudi Rao, and Harald Cramér around 1943–1945.

For further reading we recommend textbooks by Lehmann and Casella (1998), Borokov (1998), and Strasser (1985). The deviation bound of Theorem 2.11.4 follows Polzehl and Spokoiny (2006).

Chapter 3
Regression Estimation

This chapter discusses the estimation problem for the regression model. First a linear regression model is considered, then a generalized linear modeling is discussed. We also mention median and quantile regression.

3.1 Regression Model

The (mean) *regression model* can be written in the form $\mathbb{E}(Y|X) = f(X)$, or equivalently,

$$Y = f(X) + \varepsilon, \tag{3.1}$$

where Y is the dependent (explained) variable and X is the explanatory variable (regressor) which can be multidimensional. The target of analysis is the systematic dependence of the explained variable Y from the explanatory variable X. The *regression function* f describes the dependence of the mean of Y as a function of X. The value ε can be treated as an individual deviation (error). It is usually assumed to be random with zero mean. Below we discuss in more detail the components of the regression model (3.1).

3.1.1 Observations

In almost all practical situations, regression analysis is performed on the basis of available data (observations) given in the form of a sample of pairs (X_i, Y_i) for $i = 1, \ldots, n$, where n is the *sample size*. Here $Y_1, \ldots, Y_n$ are observed values

V. Spokoiny and T. Dickhaus, *Basics of Modern Mathematical Statistics*,
Springer Texts in Statistics, DOI 10.1007/978-3-642-39909-1_3,

of the regression variable Y and $X_1, \dots, X_n$ are the corresponding values of the explanatory variable X. For each observation Y_i, the regression model reads as:

$$Y_i = f(X_i) + \varepsilon_i$$

where ε_i is the individual ith error.

3.1.2 Design

The set $X_1, \dots, X_n$ of the regressor's values is called a *design*. The set $\mathcal{X}$ of all possible values of the regressor X is called the *design space*. If this set $\mathcal{X}$ is compact, then one speaks of a *compactly supported design*.

The nature of the design can be different for different statistical models. However, it is important to mention that the design is always observable. Two kinds of design assumptions are usually used in statistical modeling. A *deterministic* design assumes that the points $X_1, \dots, X_n$ are nonrandom and given in advance. Here are typical examples:

Example 3.1.1 (Time Series). Let $Y_{t_0}, Y_{t_0+1}, \dots, Y_T$ be a time series. The time points $t_0, t_0 + 1, \dots, T$ build a regular deterministic design. The regression function f explains the trend of the time series Y_t as a function of time.

Example 3.1.2 (Imaging). Let Y_{ij} be the observed gray value at the pixel (i, j) of an image. The coordinate X_{ij} of this pixel is the corresponding design value. The regression function $f(X_{ij})$ gives the true image value at X_{ij} which is to be recovered from the noisy observations Y_{ij}.

If the design is supported on a cube in $\mathbb{R}^d$ and the design points X_i form a grid in this cube, then the design is called *equidistant*. An important feature of such a design is that the number N_A of design points in any "massive" subset A of the unit cube is nearly the volume of this subset V_A multiplied by the sample size n: $N_A \approx nV_A$. *Design regularity* means that the value N_A is nearly proportional to nV_A, that is, $N_A \approx cnV_A$ for some positive constant c which may depend on the set A.

In some applications, it is natural to assume that the design values X_i are randomly drawn from some design distribution. Typical examples are given by sociological studies. In this case one speaks of a *random* design. The design values $X_1, \dots, X_n$ are assumed to be independent and identically distributed from a law P_X on the design space $\mathcal{X}$ which is a subset of the Euclidean space $\mathbb{R}^d$. The design variables X are also assumed to be independent of the observations Y.

One special case of random design is the *uniform* design when the design distribution is uniform on the unit cube in $\mathbb{R}^d$. The uniform design possesses a similar, important property to an equidistant design: the number of design points in a "massive" subset of the unit cube is on average close to the volume of this set multiplied by n. The random design is called *regular* on $\mathcal{X}$ if the design distribution is absolutely continuous with respect to the Lebesgue measure and the design density $\rho(x) = dP_X(x)/d\lambda$ is positive and continuous on $\mathcal{X}$. This again ensures with a probability close to one the regularity property $N_A \approx cnV_A$ with $c = \rho(x)$ for some $x \in A$.

It is worth mentioning that the case of a random design can be reduced to the case of a deterministic design by considering the conditional distribution of the data given the design variables $X_1, \ldots, X_n$.

3.1.3 Errors

The decomposition of the observed response variable Y into the systematic component $f(x)$ and the error $\boldsymbol{\varepsilon}$ in the model equation (3.1) is not formally defined and cannot be done without some assumptions on the errors ε_i. The standard approach is to assume that the mean value of every ε_i is zero. Equivalently this means that the expected value of the observation Y_i is just the regression function $f(X_i)$. This case is called *mean regression* or simply regression. It is usually assumed that the errors ε_i have finite second moments. *Homogeneous errors* case means that all the errors ε_i have the same variance $\sigma^2 = \operatorname{Var} \varepsilon_i^2$. The variance of *heterogeneous errors* ε_i may vary with i. In many applications not only the systematic component $f(X_i) = \mathbb{E} Y_i$ but also the error variance $\operatorname{Var} Y_i = \operatorname{Var} \varepsilon_i$ depend on the regressor (location) X_i. Such models are often written in the form

$$Y_i = f(X_i) + \sigma(X_i)\varepsilon_i .$$

The observation (noise) variance $\sigma^2(x)$ can be the target of analysis similarly to the mean regression function.

The assumption of zero mean noise, $\mathbb{E}\varepsilon_i = 0$, is very natural and has a clear interpretation. However, in some applications, it can cause trouble, especially if data are contaminated by outliers. In this case, the assumption of a zero mean can be replaced by a more robust assumption of a zero median. This leads to the *median regression* model which assumes $\mathbb{P}(\varepsilon_i \le 0) = 1/2$, or, equivalently

$$\mathbb{P}\bigl(Y_i - f(X_i) \le 0\bigr) = 1/2.$$

A further important assumption concerns the joint distribution of the errors ε_i. In the majority of applications the errors are assumed to be independent. However, in some situations, the dependence of the errors is quite natural. One example can be given by time series analysis. The errors ε_i are defined as the difference between the observed values Y_i and the trend function f_i at the ith time moment. These errors are often serially correlated and indicate short or long range dependence. Another example comes from imaging. The neighbor observations in an image are often correlated due to the imaging technique used for recoding the images. The correlation particularly results from the automatic movement correction.

For theoretical study one often assumes that the errors ε_i are not only independent but also identically distributed. This, of course, yields a homogeneous noise. The theoretical study can be simplified even further if the error distribution is normal. This case is called *Gaussian regression* and is denoted as $\varepsilon_i \sim \mathcal{N}(0, \sigma^2)$. This assumption is very useful and greatly simplifies the theoretical study. The main advantage of Gaussian noise is that the observations and their linear combinations are also normally distributed. This is an exclusive property of the normal law which helps to simplify the exposition and avoid technicalities.

Under the given distribution of the errors, the joint distribution of the observations Y_i is determined by the regression function $f(\cdot)$.

3.1.4 Regression Function

By Eq. (3.1), the regression variable Y can be decomposed into a systematic component and a (random) error ε. The systematic component is a deterministic function f of the explanatory variable X called the *regression* function. Classical regression theory considers the case of *linear* dependence, that is, one fits a linear relation between Y and X:

$$f(x) = a + bx$$

leading to the model equation

$$Y_i = \theta_1 + \theta_2 X_i + \varepsilon_i \, .$$

Here θ_1 and θ_2 are the parameters of the linear model. If the regressor x is multidimensional, then θ_2 is a vector from $\mathbb{R}^d$ and $\theta_2 x$ becomes the scalar product of two vectors. In many practical examples the assumption of linear dependence is too restrictive. It can be extended by several ways. One can try a more sophisticated functional dependence of Y on X, for instance polynomial. More generally, one can assume that the regression function f is known up to the finite-dimensional parameter $\boldsymbol{\theta} = (\theta_1, \ldots, \theta_p)^\top \in \mathbb{R}^p$. This situation is called *parametric regression* and denoted by $f(\cdot) = f(\cdot, \boldsymbol{\theta})$. If the function $f(\cdot, \boldsymbol{\theta})$ depends on $\boldsymbol{\theta}$ linearly, that is, $f(x, \boldsymbol{\theta}) = \theta_1 \psi_1(x) + \ldots + \theta_p \psi_p(x)$ for some given functions $\psi_1, \ldots, \psi_p$, then the model is called *linear regression*. An important special case is given by polynomial regression when $f(x)$ is a polynomial function of degree $p - 1$: $f(x) = \theta_1 + \theta_2 x + \ldots + \theta_p x^{p-1}$.

In many applications a parametric form of the regression function cannot be justified. Then one speaks of *nonparametric regression*.

3.2 Method of Substitution and M-Estimation

Observe that the parametric regression equation can be rewritten as

$$\varepsilon_i = Y_i - f(X_i, \boldsymbol{\theta}).$$

If $\tilde{\boldsymbol{\theta}}$ is an estimate of the parameter $\boldsymbol{\theta}$, then the *residuals* $\tilde{\varepsilon}_i = Y_i - f(X_i, \tilde{\boldsymbol{\theta}})$ are estimates of the individual errors ε_i. So, the idea of the method is to select the parameter estimate $\tilde{\boldsymbol{\theta}}$ in a way that the empirical distribution P_n of the residuals $\tilde{\varepsilon}_i$ mimics as well as possible certain prescribed features of the error distribution. We consider one approach called minimum contrast or M-estimation. Let $\psi(y)$ be an *influence* or *contrast* function. The main condition on the choice of this function is that

$$\mathbb{E}\psi(\varepsilon_i + z) \geq \mathbb{E}\psi(\varepsilon_i)$$

for all $i = 1, \ldots, n$ and all z. Then the true value $\boldsymbol{\theta}^*$ clearly minimizes the expectation of the sum $\sum_i \psi\bigl(Y_i - f(X_i, \boldsymbol{\theta})\bigr)$:

$$\boldsymbol{\theta}^* = \operatorname*{argmin}_{\boldsymbol{\theta}} \mathbb{E} \sum_i \psi\bigl(Y_i - f(X_i, \boldsymbol{\theta})\bigr).$$

This leads to the *M-estimate*

$$\tilde{\boldsymbol{\theta}} = \operatorname*{argmin}_{\boldsymbol{\theta} \in \Theta} \sum_i \psi\bigl(Y_i - f(X_i, \boldsymbol{\theta})\bigr).$$

This estimation method can be treated as replacing the true expectation of the errors by the empirical distribution of the residuals.

We specify this approach for regression estimation by the classical examples of least squares, least absolute deviation (LAD) and maximum likelihood estimation corresponding to $\psi(x) = x^2$, $\psi(x) = |x|$ and $\psi(x) = -\log \rho(x)$, where $\rho(x)$ is the error density. All these examples belong within framework of M-estimation and the quasi maximum likelihood approach.

3.2.1 *Mean Regression: Least Squares Estimate*

The observations Y_i are assumed to follow the model

$$Y_i = f(X_i, \boldsymbol{\theta}^*) + \varepsilon_i, \qquad \mathbb{E}\varepsilon_i = 0 \tag{3.2}$$

with an unknown target $\boldsymbol{\theta}^*$. Suppose in addition that $\sigma_i^2 = \mathbb{E}\varepsilon_i^2 < \infty$. Then for every $\boldsymbol{\theta} \in \Theta$ and every $i \le n$ due to (3.2)

$$\begin{aligned}\mathbb{E}_{\boldsymbol{\theta}^*}\{Y_i - f(X_i, \boldsymbol{\theta})\}^2 &= \mathbb{E}_{\boldsymbol{\theta}^*}\{\varepsilon_i + f(X_i, \boldsymbol{\theta}^*) - f(X_i, \boldsymbol{\theta})\}^2 \\ &= \sigma_i^2 + \bigl|f(X_i, \boldsymbol{\theta}^*) - f(X_i, \boldsymbol{\theta})\bigr|^2.\end{aligned}$$

This yields for the whole sample

$$\mathbb{E}_{\boldsymbol{\theta}^*}\sum\{Y_i - f(X_i, \boldsymbol{\theta})\}^2 = \sum\bigl\{\sigma_i^2 + \bigl|f(X_i, \boldsymbol{\theta}^*) - f(X_i, \boldsymbol{\theta})\bigr|^2\bigr\}.$$

This expression is clearly minimized at $\boldsymbol{\theta} = \boldsymbol{\theta}^*$. This leads to the idea of estimating the parameter $\boldsymbol{\theta}^*$ by maximizing its empirical counterpart. The resulting estimate is called the (ordinary) *least squares estimate* (LSE):

$$\tilde{\boldsymbol{\theta}}_{LSE} = \operatorname*{argmin}_{\boldsymbol{\theta} \in \Theta} \sum\{Y_i - f(X_i, \boldsymbol{\theta})\}^2.$$

This estimate is very natural and requires minimal information about the errors ε_i. Namely, one only needs $\mathbb{E}\varepsilon_i = 0$ and $\mathbb{E}\varepsilon_i^2 < \infty$.

3.2.2 Median Regression: LAD Estimate

Consider the same regression model as in (3.2), but the errors ε_i are not zero-mean. Instead we assume that their median is zero:

$$Y_i = f(X_i, \boldsymbol{\theta}^*) + \varepsilon_i, \qquad \operatorname{med}(\varepsilon_i) = 0.$$

As previously, the target of estimation is the parameter $\boldsymbol{\theta}^*$. Observe that $\varepsilon_i = Y_i - f(X_i, \boldsymbol{\theta}^*)$ and hence, the latter r.v. has median zero. We now use the following simple fact: if $\operatorname{med}(\varepsilon) = 0$, then for any $z \ne 0$

$$\mathbb{E}|\varepsilon + z| \ge \mathbb{E}|\varepsilon|. \tag{3.3}$$

Exercise 3.2.1. Prove (3.3).

The property (3.3) implies for every $\boldsymbol{\theta}$

$$\mathbb{E}_{\boldsymbol{\theta}^*}\sum\bigl|Y_i - f(X_i, \boldsymbol{\theta})\bigr| \ge \mathbb{E}_{\boldsymbol{\theta}^*}\sum\bigl|Y_i - f(X_i, \boldsymbol{\theta}^*)\bigr|,$$

that is, $\boldsymbol{\theta}^*$ minimizes over $\boldsymbol{\theta}$ the expectation under the true measure of the sum $\sum\bigl|Y_i - f(X_i, \boldsymbol{\theta})\bigr|$. This leads to the empirical counterpart of $\boldsymbol{\theta}^*$ given by

$$\tilde{\theta} = \underset{\theta \in \Theta}{\operatorname{argmin}} \sum \left| Y_i - f(X_i, \theta) \right|.$$

This procedure is usually referred to as *LADs* regression estimate.

3.2.3 Maximum Likelihood Regression Estimation

Let the density function $\rho(\cdot)$ of the errors ε_i be known. The regression equation (3.2) implies $\varepsilon_i = Y_i - f(X_i, \theta^*)$. Therefore, every Y_i has the density $\rho(y - f(X_i, \theta^*))$. Independence of the Y_i's implies the product structure of the density of the joint distribution:

$$\prod \rho(y_i - f(X_i, \theta)),$$

yielding the log-likelihood

$$L(\theta) = \sum \ell(Y_i - f(X_i, \theta))$$

with $\ell(t) = \log \rho(t)$. The maximum likelihood estimate (MLE) is the point of maximum of $L(\theta)$:

$$\tilde{\theta} = \underset{\theta \in \Theta}{\operatorname{argmax}} L(\theta) = \underset{\theta \in \Theta}{\operatorname{argmax}} \sum \ell(Y_i - f(X_i, \theta)).$$

A closed form solution for this equation exists only in some special cases like linear Gaussian regression. Otherwise this equation has to be solved numerically.

Consider an important special case corresponding to the i.i.d. Gaussian errors when $\rho(y)$ is the density of the normal law with mean zero and variance σ^2. Then

$$L(\theta) = -\frac{n}{2} \log(2\pi\sigma^2) - \frac{1}{2\sigma^2} \sum \left| Y_i - f(X_i, \theta) \right|^2.$$

The corresponding MLE maximizes $L(\theta)$ or, equivalently, minimizes the sum $\sum \left| Y_i - f(X_i, \theta) \right|^2$:

$$\tilde{\theta} = \underset{\theta \in \Theta}{\operatorname{argmax}} L(\theta) = \underset{\theta \in \Theta}{\operatorname{argmin}} \sum \left| Y_i - f(X_i, \theta) \right|^2. \tag{3.4}$$

This estimate has already been introduced as the *ordinary least squares* estimate (oLSE).

An extension of the previous example is given by inhomogeneous Gaussian regression, when the errors ε_i are independent Gaussian zero-mean but the variances depend on i: $\mathbb{E}\varepsilon_i^2 = \sigma_i^2$. Then the log-likelihood $L(\theta)$ is given by the sum

$$L(\boldsymbol{\theta}) = \sum \Bigl\{-\frac{|Y_i - f(X_i, \boldsymbol{\theta})|^2}{2\sigma_i^2} - \frac{1}{2}\log(2\pi\sigma_i^2)\Bigr\}.$$

Maximizing this expression w.r.t. $\boldsymbol{\theta}$ is equivalent to minimizing the weighted sum $\sum \sigma_i^{-2}|Y_i - f(X_i, \boldsymbol{\theta})|^2$:

$$\tilde{\boldsymbol{\theta}} = \operatorname*{argmin}_{\boldsymbol{\theta} \in \Theta} \sum \sigma_i^{-2}|Y_i - f(X_i, \boldsymbol{\theta})|^2.$$

Such an estimate is also called the *weighted least squares* (wLSE).

Another example corresponds to the case when the errors ε_i are i.i.d. double exponential, so that $\mathbb{P}(\pm\varepsilon_1 > t) = e^{-t/\sigma}$ for some given $\sigma > 0$. Then $\rho(y) = (2\sigma)^{-1}e^{-|y|/\sigma}$ and

$$L(\boldsymbol{\theta}) = -n\log(2\sigma) - \sigma^{-1}\sum|Y_i - f(X_i, \boldsymbol{\theta})|.$$

The MLE $\tilde{\boldsymbol{\theta}}$ maximizes $L(\boldsymbol{\theta})$ or, equivalently, minimizes the sum $\sum|Y_i - f(X_i, \boldsymbol{\theta})|$:

$$\tilde{\boldsymbol{\theta}} = \operatorname*{argmax}_{\boldsymbol{\theta} \in \Theta} L(\boldsymbol{\theta}) = \operatorname*{argmin}_{\boldsymbol{\theta} \in \Theta} \sum|Y_i - f(X_i, \boldsymbol{\theta})|.$$

So the maximum likelihood regression with Laplacian errors leads back to the *LADs* estimate.

3.2.4 Quasi Maximum Likelihood Approach

This section very briefly discusses an extension of the maximum likelihood approach. A more detailed discussion will be given in context of linear modeling in Chap. 4. To be specific, consider a regression model

$$Y_i = f(X_i) + \varepsilon_i.$$

The maximum likelihood approach requires to specify the two main ingredients of this model: a parametric class $\{f(x, \boldsymbol{\theta}), \boldsymbol{\theta} \in \Theta\}$ of regression functions and the distribution of the errors ε_i. Sometimes such information is lacking. One or even both modeling assumptions can be misspecified. In such situations one speaks of a *quasi maximum likelihood* approach, where the estimate $\tilde{\boldsymbol{\theta}}$ is defined via maximizing over $\boldsymbol{\theta}$ the random function $L(\boldsymbol{\theta})$ even though it is not necessarily the real log-likelihood. Some examples of this approach have already been given.

Below we distinguish between misspecification of the first and second kind. The first kind corresponds to the parametric assumption about the regression function: assumed is the equality $f(X_i) = f(X_i, \boldsymbol{\theta}^*)$ for some $\boldsymbol{\theta}^* \in \Theta$. In reality one can only expect a reasonable quality of approximating $f(\cdot)$ by $f(\cdot, \boldsymbol{\theta}^*)$. A typical example is given by linear (polynomial) regression. The linear structure of the regression function is useful and tractable but it can only be a rough approximation of the real relation between Y and X. The quasi maximum likelihood approach suggests to ignore this misspecification and proceed as if the parametric assumption is fulfilled. This approach raises a number of questions: what is the target of estimation and what is really estimated by such quasi ML procedure? In Chap. 4 we show in the context of linear modeling that the target of estimation can be naturally defined as the parameter $\boldsymbol{\theta}^*$ providing the best approximation of the true regression function $f(\cdot)$ by its parametric counterpart $f(\cdot, \boldsymbol{\theta})$.

The second kind of misspecification concerns the assumption about the errors ε_i. In most of the applications, the distribution of errors is unknown. Moreover, the errors can be dependent or non-identically distributed. Assumption of a specific i.i.d. structure leads to a model misspecification and thus, to the quasi maximum likelihood approach. We illustrate this situation by few examples.

Consider the regression model $Y_i = f(X_i, \boldsymbol{\theta}^*) + \varepsilon_i$ and suppose for a moment that the errors ε_i are i.i.d. normal. Then the principal term of the corresponding log-likelihood is given by the negative sum of the squared residuals: $\sum \bigl|Y_i - f(X_i, \boldsymbol{\theta})\bigr|^2$, and its maximization leads to the least squares method. So, one can say that the LSE method is the quasi MLE when the errors are assumed to be i.i.d. normal. That is, the LSE can be obtained as the MLE for the imaginary Gaussian regression model when the errors ε_i are not necessarily i.i.d. Gaussian.

If the data are contaminated or the errors have heavy tails, it could be unwise to apply the LSE method. The LAD method is known to be more robust against outliers and data contamination. At the same time, it has already been shown in Sect. 3.2.3 that the LAD estimates is the MLE when the errors are Laplacian (double exponential). In other words, LAD is the quasi MLE for the model with Laplacian errors.

Inference for the quasi ML approach is discussed in detail in Chap. 4 in the context of linear modeling.

3.3 Linear Regression

One standard way of modeling the regression relationship is based on a linear expansion of the regression function. This approach is based on the assumption that the unknown regression function $f(\cdot)$ can be represented as a linear combination of given basis functions $\psi_1(\cdot), \ldots, \psi_p(\cdot)$:

$$f(x) = \theta_1 \psi_1(x) + \ldots + \theta_p \psi_p(x).$$

A couple of popular examples are listed in this section. More examples are given below in Sect. 3.3.1 in context of projection estimation.

Example 3.3.1 (Multivariate Linear Regression). Let $\boldsymbol{x} = (x_1, \ldots, x_d)^\top$ be d-dimensional. The linear regression function $f(x)$ can be written as

$$f(x) = a + b_1 x_1 + \ldots + b_d x_d .$$

Here we have $p = d + 1$ and the basis functions are $\psi_1(x) \equiv 1$ and $\psi_m = x_{m-1}$ for $m = 2, \ldots, p$. The coefficient a is often called the *intercept* and $b_1, \ldots, b_d$ are the *slope* coefficients. The vector of coefficients $\boldsymbol{\theta} = (a, b_1, \ldots, b_d)^\top$ uniquely describes the linear relation.

Example 3.3.2 (Polynomial Regression). Let x be univariate and $f(\cdot)$ be a polynomial function of degree $p - 1$, that is,

$$f(x) = \theta_1 + \theta_2 x + \ldots + \theta_p x^{p-1} .$$

Then the basic functions are $\psi_1(x) \equiv 1$, $\psi_2(x) \equiv x$, $\psi_p(x) \equiv x^{p-1}$, while $\boldsymbol{\theta} = (\theta_1, \ldots, \theta_p)^\top$ is the corresponding vector of coefficients.

Exercise 3.3.1. Let the regressor $\boldsymbol{x}$ be d-dimensional, $\boldsymbol{x} = (x_1, \ldots, x_d)^\top$. Describe the basis system and the corresponding vector of coefficients for the case when f is a quadratic function of $\boldsymbol{x}$.

Linear regression is often described using vector–matrix notation. Let Ψ_i be the vector in $\mathbb{R}^p$ whose entries are the values $\psi_m(X_i)$ of the basis functions at the design point X_i, $m = 1, \ldots, p$. Then $f(X_i) = \Psi_i^\top \boldsymbol{\theta}^*$, and the linear regression model can be written as

$$Y_i = \Psi_i^\top \boldsymbol{\theta}^* + \varepsilon_i , \qquad i = 1, \ldots, n.$$

Denote by $\boldsymbol{Y} = (Y_1, \ldots, Y_n)^\top$ the vector of observations (responses), and $\boldsymbol{\varepsilon} = (\varepsilon_1, \ldots, \varepsilon_n)^\top$ the vector of errors. Let finally Ψ be the $p \times n$ matrix with columns $\Psi_1, \ldots, \Psi_n$, that is, $\Psi = \bigl(\psi_m(X_i)\bigr)_{m=1,\ldots,p}^{i=1,\ldots,n}$. Note that each row of Ψ is composed by the values of the corresponding basis function ψ_m at the design points X_i. Now the regression equation reads as

$$\boldsymbol{Y} = \Psi^\top \boldsymbol{\theta}^* + \boldsymbol{\varepsilon}.$$

The estimation problem for this linear model will be discussed in detail in Chap. 4.

3.3.1 Projection Estimation

Consider a (mean) regression model

$$Y_i = f(X_i) + \varepsilon_i , \qquad i = 1, \ldots, n. \tag{3.5}$$

The target of the analysis is the unknown *nonparametric* function f which has to be recovered from the noisy data $\boldsymbol{Y}$. This approach is usually considered within the *nonparametric statistical theory* because it avoids fixing any parametric specification of the model function f, and thus, of the distribution of the data $\boldsymbol{Y}$. This section discusses how this nonparametric problem can be put back into the parametric theory.

The standard way of estimating the regression function f is based on some *smoothness* assumption about this function. It enables us to expand the given function w.r.t. some given functional basis and to evaluate the accuracy of approximation by finite sums. More precisely, let $\psi_1(x), \ldots, \psi_m(x), \ldots$ be a given system of functions. Specific examples are trigonometric (Fourier, cosine), orthogonal polynomial (Chebyshev, Legendre, Jacobi), and wavelet systems among many others. The completeness of this system means that a given function f can be uniquely expanded in the form

$$f(x) = \sum_{m=1}^{\infty} \theta_m \psi_m(x). \tag{3.6}$$

A very desirable feature of the basis system is orthogonality:

$$\int \psi_m(x)\psi_{m'}(x)\mu_{\mathcal{X}}(dx) = 0, \qquad m \neq m'.$$

Here $\mu_{\mathcal{X}}$ can be some design measure on $\mathcal{X}$ or the empirical design measure $n^{-1}\sum \delta_{X_i}$. However, the expansion (3.6) is untractable because it involves infinitely many coefficients θ_m. A standard procedure is to truncate this expansion after the first p terms leading to the finite approximation

$$f(x) \approx \sum_{m=1}^{p} \theta_m \psi_m(x). \tag{3.7}$$

Accuracy of such an approximation becomes better and better as the number p of terms grows. A smoothness assumption helps to estimate the rate of convergence to zero of the remainder term $f - \theta_1\psi_1 - \ldots - \theta_p\psi_p$:

$$\left\| f - \theta_1\psi_1 - \ldots - \theta_p\psi_p \right\| \leq r_p, \tag{3.8}$$

where r_p describes the accuracy of approximation of the function f by the considered finite sums uniformly over the class of functions with the prescribed smoothness. The norm used in the definition (3.8) as well as the basis $\{\psi_m\}$ depends on the particular smoothness class. Popular examples are given by Hölder classes for the L_∞ norm, Sobolev smoothness for L_2-norm or more generally L_s-norm for some $s \geq 1$.

A choice of a proper truncation value p is one of the central problems in non-parametric function estimation. With p growing, the quality of approximation (3.7) improves in the sense that $r_p \to 0$ as $p \to \infty$. However, the growth of the parameter dimension yields the growth of model complexity, one has to estimate more and more coefficients. Section 4.7 below briefly discusses how the problem can be formalized and how one can define the optimal choice. However, a rigorous solution is postponed until the next volume. Here we suppose that the value p is fixed by some reasons and apply the quasi maximum likelihood parametric approach. Namely, the approximation (3.7) is assumed to be the exact equality: $f(x) \equiv f(x, \boldsymbol{\theta}^*) \stackrel{\text{def}}{=} \theta_1^* \psi_1(x) + \ldots + \theta_p^* \psi_p$. Model misspecification $f(\cdot) \not\equiv f(\cdot, \boldsymbol{\theta}) \stackrel{\text{def}}{=} \theta_1 \psi_1(x) + \ldots + \theta_p \psi_p$ for any vector $\boldsymbol{\theta} \in \Theta$ means the *modeling error*, or, the *modeling bias*. The parametric approach ignores this modeling error and focuses on the *error within the model* which describes the accuracy of the qMLE $\tilde{\boldsymbol{\theta}}$.

The qMLE procedure requires to specify the error distribution which appears in the log-likelihood. In the most general form, let $\mathbb{P}_0$ be the joint distribution of the error vector $\boldsymbol{\varepsilon}$, and let $\rho^{(n)}(\boldsymbol{\varepsilon})$ be its density function on $\mathbb{R}^n$. The identities $\varepsilon_i = Y_i - f(X_i, \boldsymbol{\theta})$ yield the log-likelihood

$$L(\boldsymbol{\theta}) = \log \rho^{(n)}\bigl(\boldsymbol{Y} - \boldsymbol{f}(\boldsymbol{X}, \boldsymbol{\theta})\bigr). \tag{3.9}$$

If the errors ε_i are i.i.d. with the density $\rho(y)$, then

$$L(\boldsymbol{\theta}) = \sum \log \rho(Y_i - f(X_i, \boldsymbol{\theta})). \tag{3.10}$$

The most popular least squares method (3.4) implicitly assumes Gaussian homogeneous noise: ε_i are i.i.d. $\mathcal{N}(0, \sigma^2)$. The LAD approach is based on the assumption of Laplace error distribution. Categorical data are modeled by a proper exponential family distribution; see Sect. 3.5. Below we assume that the one or another assumption about errors is fixed and the log-likelihood is described by (3.9) or (3.10). This assumption can be misspecified and the qMLE analysis has to be done under the true error distribution. Some examples of this sort for linear models are given in Sect. 4.6.

In the rest of this section we only discuss how the regression function f in (3.5) can be approximated by different series expansions. With the selected expansion and the assumption on the errors, the approximating parametric model is fixed due to (3.9). In most of the examples we only consider a univariate design with $d = 1$.

3.3.2 Polynomial Approximation

It is well known that any smooth function $f(\cdot)$ can be approximated by a polynomial. Moreover, the larger smoothness of $f(\cdot)$ is the better the accuracy of approximation. The Taylor expansion yields an approximation in the form

$$f(x) \approx \theta_0 + \theta_1 x + \theta_2 x^2 + \ldots + \theta_m x^m. \tag{3.11}$$

Such an approximation is very natural, however, it is rarely used in statistical applications. The main reason is that the different power functions $\psi_m(x) = x^m$ are highly correlated between each other. This makes difficult to identify the corresponding coefficients. Instead one can use different polynomial systems which fulfill certain orthogonality conditions.

We say that $f(x)$ is a polynomial of degree m if it can be represented in the form (3.11) with $\theta_m \neq 0$. Any sequence $1, \psi_1(x), \ldots, \psi_m(x)$ of such polynomials yields a basis in the vector space of polynomials of degree m.

Exercise 3.3.2. Let for each $j \leq m$ a polynomial of degree j be fixed. Then any polynomial $P_m(x)$ of degree m can be represented in a unique way in the form

$$P_m(x) = c_0 + c_1\psi_1(x) + \ldots + c_m\psi_m(x)$$

Hint: define $c_m = P_m^{(m)} / \psi_m^{(m)}$ and apply induction to $P_m(x) - c_m\psi_m(x)$.

3.3.3 Orthogonal Polynomials

Let μ be any measure on the real line satisfying the condition

$$\int x^m \mu(dx) < \infty, \tag{3.12}$$

for any integer m. This enables us to define the scalar product for two polynomial functions f, g by

$$\langle f, g \rangle \stackrel{\text{def}}{=} \int f(x)g(x)\mu(dx).$$

With such a Hilbert structure we aim to define an orthonormal polynomial system of polynomials ψ_m of degree m for $m = 0, 1, 2, \ldots$ such that

$$\langle \psi_j, \psi_m \rangle = \delta_{j,m} = \mathbb{I}(j = m), \qquad j, m = 0, 1, 2, \ldots.$$

Theorem 3.3.1. *Given a measure μ satisfying the condition* (3.12) *there exists unique orthonormal polynomial system $\psi_1, \psi_2, \ldots$. Any polynomial P_m of degree m can be represented as*

$$P_m(x) = a_0 + a_1\psi_1(x) + \ldots + a_m\psi_m(x)$$

with

$$a_j = \langle P_m, \psi_j \rangle. \tag{3.13}$$

Proof. We construct the function ψ_m successively. The function ψ_0 is a constant defined by

$$\psi_0^2 \int \mu(dx) = 1.$$

Suppose now that the orthonormal polynomials $\psi_1, \ldots, \psi_{m-1}$ have been already constructed. Define the coefficients

$$a_j \stackrel{\text{def}}{=} \int x^m \psi_j(x)\mu(dx), \qquad j = 0, 1, \ldots, m-1,$$

and consider the function

$$g_m(x) \stackrel{\text{def}}{=} x^m - a_0\psi_0 - a_1\psi_1(x) - \ldots - a_{m-1}\psi_{m-1}(x).$$

This is obviously a polynomial of degree m. Moreover, by orthonormality of the ψ_j's for $j < m$

$$\int g_m(x)\psi_j(x)\mu(dx) = \int x^m \psi_j(x)\mu(dx) - a_j \int \psi_j^2(x)\mu(dx) = 0.$$

So, one can define ψ_m by normalization of g_m:

$$\psi_m(x) \stackrel{\text{def}}{=} \langle g_m, g_m \rangle^{-1/2} g_m(x).$$

One can also easily see that such defined ψ_m is only polynomial of degree m which is orthogonal to ψ_j for $j < m$ and fulfills $\langle \psi_m, \psi_m \rangle = 1$, because the number of constraints is equal to the number of coefficients $\theta_0, \ldots, \theta_m$ of $\psi_m(x)$.

Let now P_m be a polynomial of degree m. Define the coefficient a_m by (3.13). Similarly to above one can show that

$$P_m(x) - \{a_0 + a_1\psi_1(x) + \ldots + a_m\psi_m(x)\} \equiv 0$$

which implies the second claim.

Exercise 3.3.3. Let $\{\psi_m\}$ be an orthonormal polynomial system. Show that for any polynomial $P_j(x)$ of degree $j < m$, it holds

$$\langle P_j, \psi_m \rangle = 0.$$

3.3.3.1 Finite Approximation and the Associated Kernel

Let f be a function satisfying

$$\int f^2(x)\mu(dx) < \infty. \tag{3.14}$$

Then the scalar product $a_j = \langle f, \psi_j \rangle$ is well defined for all $j \geq 0$ leading for each $m \geq 1$ to the following approximation:

$$f_m(x) \stackrel{\text{def}}{=} \sum_{j=0}^{m} a_j \psi_j(x) = \sum_{j=0}^{m} \int f(u)\psi_j(u)\mu(du)\psi_j(x)$$
$$= \int f(u)\Phi_m(x,u)\mu(du) \tag{3.15}$$

with

$$\Phi_m(x,u) = \sum_{j=0}^{m} \psi_j(x)\psi_j(u).$$

3.3.3.2 Completeness

The accuracy of approximation of f by f_m with m growing is one of the central questions in the approximation theory. The answer depends on the regularity of the function f and on choice of the system $\{\psi_m\}$. Let $\mathcal{F}$ be a linear space of functions f on the real line satisfying (3.14). We say that the basis system $\{\psi_m(x)\}$ is *complete* in $\mathcal{F}$ if the identities $\langle f, \psi_m \rangle = 0$ for all $m \geq 0$ imply $f \equiv 0$. As $\psi_m(x)$ is a polynomial of degree m, this definition is equivalent to the condition

$$\langle f, x^m \rangle = 0, \quad m = 0, 1, 2, \ldots \Longleftrightarrow f \equiv 0.$$

3.3.3.3 Squared Bias and Accuracy of Approximation

Let $f \in \mathcal{F}$ be a function in L_2 satisfying (3.14), and let $\{\psi_m\}$ be a complete basis. Consider the error $f(x) - f_m(x)$ of the finite approximation $f_m(x)$ from (3.15). The Parseval identity yields

$$\int f^2(x)\mu(dx) = \sum_{m=0}^{\infty} a_m^2.$$

This yields that the finite sums of $\sum_{j=0}^{m} a_j^2$ converge to the infinite sum $\sum_{m=0}^{\infty} a_m^2$ and the remainder $b_m = \sum_{j=m+1}^{\infty} a_j^2$ tends to zero with m:

$$b_m \stackrel{\text{def}}{=} \langle f - f_m \rangle = \int \left| f(x) - f_m(x) \right|^2 \mu(dx) = \sum_{j=m+1}^{\infty} a_j^2 \to 0$$

as $m \to \infty$. The value b_m is often called the *squared bias*. Below in this section we briefly overview some popular polynomial systems used in the approximation theory.

3.3.4 Chebyshev Polynomials

Chebyshev polynomials are frequently used in the approximation theory because of their very useful features. These polynomials can be defined by many ways: explicit formulas, recurrent relations, differential equations, among others.

3.3.4.1 A Trigonometric Definition

Chebyshev polynomials is usually defined in the *trigonometric* form:

$$T_m(x) = \cos\big(m \arccos(x)\big). \tag{3.16}$$

Exercise 3.3.4. Check that $T_m(x)$ from (3.16) is a polynomial of degree m.

Hint: use the formula $\cos\big((m+1)u\big) = 2\cos(u)\cos(mu) - \cos\big((m-1)u\big)$ and induction arguments.

3.3.4.2 Recurrent Formula

The trigonometric identity $\cos\big((m+1)u\big) = 2\cos(u)\cos(mu) - \cos\big((m-1)u\big)$ yields the *recurrent* relation between Chebyshev polynomials:

$$T_{m+1}(x) = 2xT_m(x) - T_{m-1}(x), \qquad m \geq 1. \tag{3.17}$$

Exercise 3.3.5. Describe the first 5 polynomials T_m.

Hint: use that $T_0(x) \equiv 1$ and $T_1(x) \equiv x$ and use the recurrent formula (3.17).

3.3.4.3 The Leading Coefficient

The recurrent relation (3.17) and the formulas $T_0(x) \equiv 1$ and $T_1(x) \equiv x$ imply that the leading coefficient of $T_m(x)$ is equal to 2^{m-1} by induction arguments. Equivalently

$$T_m^{(m)}(x) \equiv 2^{m-1} m!$$

3.3.4.4 Orthogonality and Normalization

Consider the measure $\mu(dx)$ on the open interval $(-1, 1)$ with the density $(1 - x^2)^{-1/2}$ with respect to the Lebesgue measure. By the change of variables $x = \cos(u)$ we obtain for all $j \neq m$

$$\int_{-1}^{1} T_m(x) T_j(x) \frac{dx}{\sqrt{1-x^2}} = \int_0^{\pi} \cos(mu)\cos(ju)du = 0.$$

Moreover, for $m \geq 1$

$$\int_{-1}^{1} T_m^2(x) \frac{dx}{\sqrt{1-x^2}} = \int_0^{\pi} \cos^2(mu)du = \frac{1}{2}\int_0^{\pi} \{1 + \cos(2mu)\}du = \frac{\pi}{2}.$$

Finally,

$$\int_{-1}^{1} \frac{dx}{\sqrt{1-x^2}} = \int_0^{\pi} du = \pi.$$

So, the orthonormal system can be defined by normalizing the Chebyshev polynomials $T_m(x)$:

$$\psi_0(x) \equiv \pi^{-1/2}, \qquad \psi_m(x) = \sqrt{2/\pi}\, T_m(x), \qquad m \geq 1.$$

3.3.4.5 The Moment Generating Function

The bivariate function

$$f(t, x) = \sum_{m=0}^{\infty} T_m(x) t^m \tag{3.18}$$

is called the *moment generating function*. It holds for the Chebyshev polynomials

$$f(t, x) = \frac{1 - tx}{1 - 2tx + t^2}.$$

This fact can be proven by using the recurrent formula (3.17) and the following relation:

$$\begin{aligned}
f(x,t) &= 1 + tx + t\sum_{m=1}^{\infty} T_{m+1}(x)t^m \\
&= 1 + tx + t\sum_{m=1}^{\infty} \{2xT_m(x) - T_{m-1}(x)\}t^m \\
&= 1 + tx + 2tx\{f(x,t) - 1\} - t^2 f(x,t). \qquad (3.19)
\end{aligned}$$

Exercise 3.3.6. Check (3.19) and (3.18).

3.3.4.6 Roots of T_m

The identity $\cos(\pi(k-1/2)) \equiv 0$ for all integer k yields the *roots* of the polynomial $T_m(x)$:

$$x_{k,m} = \cos\Big(\frac{\pi(k-1/2)}{m}\Big), \qquad k = 1,\dots,m. \qquad (3.20)$$

This means that $T_m(x_{k,m}) = 0$ for $k = 1,\dots,m$ and hence, $T_m(x)$ has exactly m roots on the interval $[-1, 1]$.

3.3.4.7 Discrete Orthogonality

Let $x_{1,N},\dots,x_{N,N}$ be the roots of T_N due to (3.20): $x_{k,N} = \cos\Big(\frac{\pi(2k-1)}{2N}\Big)$. Define the *discrete inner product*

$$\langle T_m, T_j\rangle_N = \sum_{k=1}^{N} T_m(x_{k,N})T_j(x_{k,N}).$$

Then it holds similarly to the continuous case

$$\langle T_m, T_j\rangle_N = \begin{cases} 0 & m \neq j, \\ N/2 & m = j \neq 0 \\ N & m = j = 0. \end{cases} \qquad (3.21)$$

Exercise 3.3.7. Prove (3.21).
Hint: use that for all $m > 0$

$$\sum_{k=1}^{N} \cos\Big(\frac{\pi m(k-1/2)}{N}\Big) = 0$$

yielding for all $m' \neq m$

$$\sum_{k=1}^{N} \cos\Big(\frac{\pi m(k-1/2)}{N}\Big)\cos\Big(\frac{\pi m'(k-1/2)}{N}\Big) = 0$$

3.3.4.8 Extremes of $T_m(x)$

Obviously $|T_m(x)| \leq 1$ because the cos-function is bounded by one in absolute value. Moreover, $\cos(k\pi) = (-1)^k$ yields the extreme points e_k with $T_m(e_k) = (-1)^k$ for

$$e_k = \cos\left(\frac{\pi k}{m}\right), \qquad k = 0, 1, \ldots, m. \tag{3.22}$$

In particular, the edge points $x = 1$ and $x = -1$ are extremes of $T_m(x)$.

Exercise 3.3.8. Check that $T_m(e_k) = (-1)^k$ for e_k from (3.22). Show that $T_m(1) = 1$ and $T_m(-1) = (-1)^m$. Show that $|T_m(x)| < 1$ for $x \neq e_k$ on $[-1, 1]$.

Hint: T_m is a polynomial of degree m, hence, it can have at most $m - 1$ extreme points inside the interval $(-1, 1)$, which are $e_1, \ldots, e_{m-1}$.

3.3.4.9 Sup-Norm

The important feature of the Chebyshev polynomials which makes them very useful for the approximation theory is that each of them minimizes the sup-norm over all polynomial of the certain degree with the fixed leading coefficient.

Theorem 3.3.2. *The scaled Chebyshev polynomial* $f_m(x) = 2^{1-m}T_m$ *minimizes the sup-norm* $\|f\|_\infty \stackrel{\text{def}}{=} \sup_{x\in[-1,1]} |f(x)|$ *over the class of all polynomials of degree m with the leading coefficient 1.*

Proof. As $|T_m(x)| \leq 1$, the sup-norm of f_m fulfills $\|f_m\|_\infty = 2^{1-m}$. Let $w(x)$ be any other polynomial with the leading coefficient one and $|w(x)| < 2^{1-m}$. Consider the difference $f_m(x) - w(x)$ at the extreme points e_k from (3.22). Then $f_m(e_k) - w(e_k) > 0$ for all even $k = 0, 2, 4, \ldots$ and $f_m(e_k) - w(e_k) < 0$ for all odd $k = 1, 3, 5, \ldots$. This means that this difference has at least m roots on $[-1, 1]$ which is impossible because it is a polynomial of degree $m - 1$.

3.3.4.10 Expansion by Chebyshev Polynomials and Discrete Cosine Transform

Let $f(x)$ be a measurable function satisfying

$$\int_{-1}^{1} f^2(x) \frac{dx}{\sqrt{1-x^2}} < \infty.$$

Then this function can be uniquely expanded by Chebyshev polynomials:

$$f(x) = \sum_{m=0}^{\infty} a_m T_m(x).$$

The coefficients a_m in this expansion can be obtained by projection

$$a_m = \langle f, T_m \rangle = \int_{-1}^{1} f(x) T_m(x) \frac{dx}{\sqrt{1-x^2}} < \infty.$$

However, this method is numerically intensive. Instead, one can use the discrete orthogonality (3.21). Let some N be fixed and $x_{k,N} = \cos\left(\frac{\pi(k-1/2)}{N}\right)$. Then for $m \geq 1$

$$a_m = \frac{1}{N} \sum_{k=1}^{N} f(x_{k,N}) \cos\left(\frac{\pi(k-1/2)}{N}\right).$$

This sum can be computed very efficiently via the discrete cosine transform.

3.3.5 Legendre Polynomials

The Legendre polynomials $P_m(x)$ are often used in physics and in harmonic analysis. It is an orthogonal polynomial system on the interval $[-1, 1]$ w.r.t. the Lebesgue measure, that is,

$$\int_{-1}^{1} P_m(x) P_{m'}(x) dx = 0, \qquad m \neq m'. \tag{3.23}$$

They also can be defined as solutions of the Legendre differential equation

$$\frac{d}{dx}\left[(1-x^2)\frac{d}{dx} P_m(x)\right] + m(m+1) P_m(x) = 0. \tag{3.24}$$

An explicit representation is given by the Rodrigues' formula

$$P_m(x) = \frac{1}{2^m m!} \frac{d^m}{dx^m}\left[(1-x^2)^m\right]. \tag{3.25}$$

Exercise 3.3.9. Check that $P_m(x)$ from (3.25) fulfills (3.24).
Hint: differentiate $m+1$ times the identity

$$(x^2-1)\frac{d}{dx}(x^2-1)^m = 2mx(x^2-1)^m$$

yielding

$$2P_m(x) + 2x\frac{d}{dx}P_m(x) + (x^2-1)\frac{d^2}{dx^2}P_m(x)$$
$$= 2mP_m(x) + 2mx\frac{d}{dx}P_m(x).$$

3.3.5.1 Orthogonality

The orthogonality property (3.23) can be checked by using the Rodrigues' formula.

Exercise 3.3.10. Check that for $m < m'$

$$\int_{-1}^{1} P_m(x)P_{m'}(x)dx = 0. \tag{3.26}$$

Hint: integrate (3.26) by part $m+1$ times with P_m from (3.25) and use that the $m+1$th derivative of P_m vanishes.

3.3.5.2 Recursive Definition

It is easy to check that $P_0(x) \equiv 1$ and $P_1(x) \equiv x$. *Bonnet's recursion formula* relates 3 subsequent Legendre polynomials: for $m \geq 1$

$$(m+1)P_{m+1}(x) = (2m+1)xP_m(x) - mP_{m-1}(x). \tag{3.27}$$

From Bonnet's recursion formula one obtains by induction the explicit representation

$$P_m(x) = \sum_{k=0}^{m}(-1)^k \binom{m}{k}^2 \left(\frac{1+x}{2}\right)^{n-k}\left(\frac{1-x}{2}\right)^k.$$

3.3.5.3 More Recursions

Further, the definition (3.25) yields the following 3-term recursion:

$$\frac{x^2-1}{m}\frac{d}{dx}P_m(x) = xP_m(x) - P_{m-1}(x) \tag{3.28}$$

Useful for the integration of Legendre polynomials is another recursion

$$(2m+1)P_m(x) = \frac{d}{dx}\big[P_{m+1}(x) - P_{m-1}(x)\big].$$

Exercise 3.3.11. Check (3.28) by using the definition (3.25).
Hint: use that

$$\begin{aligned} x\frac{d^m}{dx^m}\left[\frac{(1-x^2)^m}{m}\right] &= 2x\frac{d^{m-1}}{dx^{m-1}}\left[x(1-x^2)^{m-1}\right] \\ &= 2x^2\frac{d^{m-1}}{dx^{m-1}}(1-x^2)^{m-1} - 2x\frac{d^{m-2}}{dx^{m-2}}(1-x^2)^{m-1} \end{aligned}$$

Exercise 3.3.12. Check (3.27).
Hint: use that

$$\begin{aligned} \frac{d^m}{dx^m}\frac{d}{dx}\left[\frac{(1-x^2)^{m+1}}{m+1}\right] &= -2\frac{d^m}{dx^m}\left[x(1-x^2)^m\right] \\ &= -2x\frac{d^m}{dx^m}\left[(1-x^2)^m\right] - 2\frac{d^{m-1}}{dx^{m-1}}\left[(1-x^2)^m\right] \end{aligned}$$

3.3.5.4 Generating Function

The Legendre *generating function* is defined by

$$f(t,x) \stackrel{\text{def}}{=} \sum_{m=0}^{\infty} P_m(x)t^m.$$

It holds

$$f(t,x) = (1 - 2tx + x^2)^{-1/2}. \tag{3.29}$$

Exercise 3.3.13. Check (3.29).

3.3.6 Lagrange Polynomials

In numerical analysis, Lagrange polynomials are used for polynomial interpolation. The Lagrange polynomials are widely applied in cryptography, such as in Shamir's Secret Sharing scheme.

Given a set of $p+1$ data points $(X_0, Y_0), \ldots, (X_p, Y_p)$, where no two X_j are the same, the *interpolation polynomial in the Lagrange form* is a linear combination

$$L(x) = \sum_{m=0}^{p} \ell_m(x) Y_m \tag{3.30}$$

of Lagrange basis polynomials

$$\ell_m(x) \stackrel{\text{def}}{=} \prod_{\substack{j=0,\ldots,p,\\ j\neq m}} \frac{x - X_j}{X_m - X_j}$$

$$= \frac{x - X_0}{X_m - X_0} \cdots \frac{x - X_{m-1}}{X_m - X_{m-1}} \frac{x - X_{m+1}}{X_m - X_{m+1}} \cdots \frac{x - X_p}{X_m - X_p}.$$

This definition yields that $\ell_m(X_m) = 1$ and $\ell_m(X_j) = 0$ for $j \neq m$. Hence, $P(X_m) = Y_m$ for the polynomial $L_m(x)$ from (3.30). One can easily see that $L(x)$ is the only polynomial of degree p that fulfills $P(X_m) = Y_m$.

The main disadvantage of the Lagrange forms is that any change of the design $X_1, \ldots, X_n$ requires to change each basis function $\ell_m(x)$. Another problem is that the Lagrange basis polynomials $\ell_m(x)$ are not necessarily orthogonal. This explains why these polynomials are rarely used in statistical applications.

3.3.6.1 Barycentric Interpolation

Introduce a polynomial $\ell(x)$ of degree $p+1$ by

$$\ell(x) = (x - X_0) \ldots (x - X_p).$$

Then the Lagrange basis polynomials can be rewritten as

$$\ell_m(x) = \ell(x) \frac{w_m}{x - X_m}$$

with the *barycentric weights* w_m defined by

$$w_m^{-1} = \prod_{\substack{j=0,\ldots,p,\\ j\neq m}} (X_m - X_j)$$

which is commonly referred to as the *first form* of the *barycentric interpolation formula*. The advantage of this representation is that the interpolation polynomial may now be evaluated as

$$L(x) = \ell(x) \sum_{m=0}^{p} \frac{w_m}{x - X_m} Y_m$$

which, if the weights w_m have been pre-computed, requires only $O(p)$ operations (evaluating $\ell(x)$ and the weights $w_m/(x - X_m)$) as opposed to $O(p^2)$ for evaluating the Lagrange basis polynomials $\ell_m(x)$ individually.

The barycentric interpolation formula can also easily be updated to incorporate a new node X_{p+1} by dividing each of the w_m by $(X_m - X_{p+1})$ and constructing the new w_{p+1} as above.

We can further simplify the first form by first considering the barycentric interpolation of the constant function $g(x) \equiv 1$:

$$g(x) = \ell(x) \sum_{m=0}^{p} \frac{w_m}{x - X_m}.$$

Dividing $L(x)$ by $g(x)$ does not modify the interpolation, yet yields

$$L(x) = \frac{\sum_{m=0}^{p} w_m Y_m / (x - X_m)}{\sum_{m=0}^{p} w_m / (x - X_m)}$$

which is referred to as the *second form* or *true form* of the *barycentric interpolation formula*. This second form has the advantage that $\ell(x)$ need not be evaluated for each evaluation of $L(x)$.

3.3.7 Hermite Polynomials

The Hermite polynomials build an orthogonal system on the whole real line. The explicit representation is given by

$$H_m(x) \stackrel{\text{def}}{=} (-1)^m \mathrm{e}^{x^2} \frac{d^m}{dx^m} \mathrm{e}^{-x^2}.$$

Sometimes one uses a "probabilistic" definition

$$\breve{H}_m(x) \stackrel{\text{def}}{=} (-1)^m \mathrm{e}^{x^2/2} \frac{d^m}{dx^m} \mathrm{e}^{-x^2/2}.$$

Exercise 3.3.14. Show that each $H_m(x)$ and $\breve{H}_m(x)$ is a polynomial of degree m.

Hint: Use induction arguments to show that $\frac{d^m}{dx^m}\mathrm{e}^{-x^2}$ can be represented in the form $P_m(x)\mathrm{e}^{-x^2}$ with a polynomial $P_m(x)$ of degree m.

Exercise 3.3.15. Check that the leading coefficient of $H_m(x)$ is equal to 2^m while the leading coefficient of $\breve{H}_m(x)$ is equal to one.

3.3.7.1 Orthogonality

The Hermite polynomials are orthogonal on the whole real line with the weight function $w(x) = \mathrm{e}^{-x^2}$: for $j \neq m$

$$\int_{-\infty}^{\infty} H_m(x) H_j(x) w(x) dx = 0. \tag{3.31}$$

Note first that each $H_m(x)$ is a polynomial so the scalar product (3.31) is well defined. Suppose that $m > j$. It is obvious that it suffices to check that

$$\int_{-\infty}^{\infty} H_m(x) x^j w(x) dx = 0, \qquad j < m. \tag{3.32}$$

Define $f_m(x) \stackrel{\text{def}}{=} \frac{d^m}{dx^m} \mathrm{e}^{-x^2}$. Obviously $f'_{m-1}(x) \equiv f_m(x)$. Integration by part yields for any $j \geq 1$

$$\begin{aligned}
\int_{-\infty}^{\infty} H_m(x) x^j w(x) dx &= (-1)^m \int_{-\infty}^{\infty} x^j f_m(x) dx \\
&= (-1)^m \int_{-\infty}^{\infty} x^j f'_{m-1}(x) dx \\
&= (-1)^{m-1} j \int_{-\infty}^{\infty} x^{j-1} f_{m-1}(x) dx .
\end{aligned}$$

By the same arguments, for $m \geq 1$

$$\int_{-\infty}^{\infty} f_m(x) dx = \int_{-\infty}^{\infty} f'_{m-1}(x) dx = f'_{m-1}(\infty) - f'_{m-1}(-\infty) = 0. \tag{3.33}$$

This implies (3.32) and hence the orthogonality property (3.31).

Now we compute the scalar product of $H_m(x)$. Formula (3.33) with $j = m$ implies

$$\int_{-\infty}^{\infty} H_m(x) x^m w(x) dx = m! \int_{-\infty}^{\infty} \mathrm{e}^{-x^2} dx = \sqrt{\pi}\, m!.$$

As the leading coefficient of $H_m(x)$ is equal to 2^m, this implies

$$\int_{-\infty}^{\infty} H_m^2(x) w(x) dx = 2^m \int_{-\infty}^{\infty} H_m(x) x^m w(x) dx = \sqrt{\pi}\, 2^m m!.$$

Exercise 3.3.16. Prove the orthogonality of the probabilistic Hermite polynomials $\breve{H}_m(x)$. Compute their norm.

3.3.7.2 Recurrent Formula

The definition yields

$$H_0(x) \equiv 1, \qquad H_1(x) = 2x.$$

Further, for $m \geq 1$, we use the formula

$$\frac{d}{dx}H_m(x) = (-1)^m \frac{d}{dx}\left\{ e^{x^2} \frac{d^m}{dx^m} e^{-x^2} \right\} = 2xH_m(x) - H_{m+1}(x) \tag{3.34}$$

yielding the recurrent relation

$$H_{m+1}(x) = 2xH_m(x) - H'_m(x).$$

Moreover, integration by part, the formula (3.34), and the orthogonality property yield for $j < m - 1$

$$\int_{-\infty}^{\infty} H'_m(x)H_j(x)w(x)dx = -\int_{-\infty}^{\infty} H_m(x)H'_j(x)w(x)dx = 0$$

This means that $H'_m(x)$ is a polynomial of degree $m - 1$ and it is orthogonal to all $H_j(x)$ for $j < m - 1$. Thus, $H'_m(x)$ coincides with $H_{m-1}(x)$ up to a multiplicative factor. The leading coefficient of $H'_m(x)$ is equal to $m2^m$ while the leading coefficient of $H_{m-1}(x)$ is equal to 2^{m-1} yielding

$$H'_m(x) = 2mH_{m-1}(x).$$

This results in another recurrent equation

$$H_{m+1}(x) = 2xH_m(x) - 2mH_{m-1}(x).$$

Exercise 3.3.17. Derive the recurrent formulas for the probabilistic Hermite polynomials $\breve{H}_m(x)$.

3.3.7.3 Generating Function

The *exponential generating function* $f(x,t)$ for the Hermite polynomials is defined as

$$f(x,t) \stackrel{\text{def}}{=} \sum_{m=0}^{\infty} H_m(x)\frac{t^m}{m!}. \tag{3.35}$$

It holds

$$f(x,t) = \exp\{2xt - t^2\}.$$

It can be proved by checking the formula

$$\frac{\partial^m}{\partial t^m} f(x,t) = H_m(x-t)f(x,t) \tag{3.36}$$

Exercise 3.3.18. Check the formula (3.36) and derive (3.35).

3.3.7.4 Completeness

This property means that the system of the normalized Hermite polynomials builds an orthonormal basis in L_2 Hilbert space of functions $f(x)$ on the real line satisfying

$$\int_{-\infty}^{\infty} f^2(x)w(x)dx < \infty.$$

3.3.8 *Trigonometric Series Expansion*

The trigonometric functions are frequently used in the approximation theory, in particular due to their relation to the spectral theory. One usually applies either the Fourier basis or the cosine basis.

The Fourier basis is composed by the constant function $F_0 = 1$ and the functions $F_{2m-1}(x) = \sin(2m\pi x)$ and $F_{2m}(x) = \cos(2m\pi x)$ for $m = 1, 2, \ldots$. These functions are considered on the interval [0, 1] and are all periodic: $f(0) = f(1)$. Therefore, it can be only used for approximation of periodic functions.

The cosine basis is composed by the functions $S_0 \equiv 1$, and $S_m(x) = \cos(m\pi x)$ for $m \geq 1$. These functions are periodic for even m and antiperiodic for odd m, this allows to approximate functions which are not necessarily periodic.

3.3.8.1 Orthogonality

Trigonometric identities imply orthogonality

$$\langle F_m, F_j \rangle = \int_0^1 F_m(x)F_j(x)dx = 0, \qquad j \neq m. \tag{3.37}$$

Also

$$\int_0^1 F_m^2(x)dx = 1/2 \tag{3.38}$$

Exercise 3.3.19. Check (3.37) and (3.38).

Exercise 3.3.20. Check that

$$\int_0^1 S_j(x) S_m(x) dx = \frac{1}{2}\, \mathbb{I}(j = m).$$

Many nice features of the Chebyshev polynomials can be translated to the cosine basis by a simple change of variable: with $u = \cos(\pi x)$, it holds $S_m(x) = T_m(u)$. So, any expansion of the function $f(u)$ by the Chebyshev polynomials yields an expansion of $f\bigl(\cos(\pi x)\bigr)$ by the cosine system.

3.4 Piecewise Methods and Splines

This section discusses piecewise polynomial methods of approximation of the univariate regression functions.

3.4.1 Piecewise Constant Estimation

Any continuous function can be locally approximated by a constant. This naturally leads to the basis consisting of piecewise constant functions. Let $A_1, \ldots, A_K$ be a *non-overlapping partition* of the design space $\mathcal{X}$:

$$\mathcal{X} = \bigcup_{k=1,\ldots,K} A_k\,, \qquad A_k \cap A_{k'} = \emptyset,\ k \neq k'. \tag{3.39}$$

We approximate the function f by a finite sum

$$f(x) \approx f(x, \boldsymbol{\theta}) = \sum_{k=1}^{K} \theta_k\, \mathbb{I}(x \in A_k). \tag{3.40}$$

Here $\boldsymbol{\theta} = (\theta_1, \ldots, \theta_p)^\top$ with $p = K$. A nice feature of this approximation is that the basis indicator functions $\psi_1, \ldots, \psi_K$ are orthogonal because they have non-overlapping supports. For the case of independent errors, this makes the computation of the qMLE $\tilde{\boldsymbol{\theta}}$ very simple. In fact, every coefficient $\tilde{\theta}_k$ can be estimated independently of the others. Indeed, the general formula (3.10) yields

$$\tilde{\boldsymbol{\theta}} = \operatorname*{argmax}_{\boldsymbol{\theta}} L(\boldsymbol{\theta}) = \operatorname*{argmax}_{\boldsymbol{\theta}} \sum_{i=1}^{n} \ell(Y_i - f(X_i, \boldsymbol{\theta}))$$

$$= \operatorname*{argmax}_{\boldsymbol{\theta}=(\theta_k)} \sum_{k=1}^{M} \sum_{X_i \in A_k} \ell(Y_i - \theta_k). \tag{3.41}$$

Exercise 3.4.1. Show that $\tilde{\theta}_k$ can be obtained by the constant approximation of the data Y_i for $X_i \in A_k$:

$$\tilde{\theta}_k = \operatorname*{argmax}_{\theta_k} \sum_{X_i \in A_k} \ell(Y_i - \theta_k), \qquad k = 1, \ldots, K. \tag{3.42}$$

A similar formula can be obtained for the target $\boldsymbol{\theta}^* = (\theta_k^*) = \operatorname{argmax}_{\boldsymbol{\theta}} \mathbb{E}L(\boldsymbol{\theta})$:

$$\theta_k^* = \operatorname*{argmax}_{\theta_k} \sum_{X_i \in A_k} \mathbb{E}\ell(Y_i - \theta_k), \qquad m = 1, \ldots, K.$$

The estimator $\tilde{\boldsymbol{\theta}}$ can be computed explicitly in some special cases. In particular, if ρ corresponds a density of a normal distribution, then the resulting estimator $\tilde{\theta}_k$ is nothing but the mean of observations Y_i over the piece A_k. For the Laplacian errors, the solution is the median of the observations over A_k. First we consider the case of Gaussian likelihood.

Theorem 3.4.1. *Let $\ell(y) = -y^2/(2\sigma^2) + R$ be a log-density of a normal law. Then for every $k = 1, \ldots, K$*

$$\tilde{\theta}_k = \frac{1}{N_k} \sum_{X_i \in A_k} Y_i,$$

$$\theta_k^* = \frac{1}{N_k} \sum_{X_i \in A_k} \mathbb{E}Y_i,$$

where N_k stands for the number of design points X_i within the piece A_k:

$$N_k \stackrel{\text{def}}{=} \sum_{X_i \in A_k} 1 = \#\{i : X_i \in A_k\}.$$

Exercise 3.4.2. Check the statements of Theorem 3.4.1.

The properties of each estimator $\tilde{\theta}_k$ repeats ones of the MLE for the sample retracted to A_k; see Sect. 2.9.1.

Theorem 3.4.2. *Let $\tilde{\boldsymbol{\theta}}$ be defined by* (3.41) *for a normal density $\rho(y)$. Then with $\boldsymbol{\theta}^* = (\theta_1^*, \ldots, \theta_K)^\top = \operatorname{argmax}_{\boldsymbol{\theta}} \mathbb{E}L(\boldsymbol{\theta})$, it holds*

$$\mathbb{E}\tilde{\theta}_k = \theta_k^*, \qquad \operatorname{Var}(\tilde{\theta}_k) = \frac{1}{N_k^2} \sum_{X_i \in A_k} \operatorname{Var}(Y_i).$$

Moreover,

$$L(\tilde{\boldsymbol{\theta}}, \boldsymbol{\theta}^*) = \sum_{k=1}^{K} \frac{N_k}{2\sigma^2} (\tilde{\theta}_k - \theta_k^*)^2.$$

The statements follow by direct calculus on each interval separately.

If the errors $\varepsilon_i = Y_i - \mathbb{E}Y_i$ are normal and homogeneous, then the distribution of the maximum likelihood $L(\tilde{\boldsymbol{\theta}}, \boldsymbol{\theta}^*)$ is available.

Theorem 3.4.3. *Consider a Gaussian regression $Y_i \sim \mathcal{N}(f(X_i), \sigma^2)$ for $i = 1, \ldots, n$. Then $\tilde{\theta}_k \sim \mathcal{N}(\theta_k^*, \sigma^2/N_k)$ and*

$$L(\tilde{\boldsymbol{\theta}}, \boldsymbol{\theta}^*) = \sum_{m=1}^{K} \frac{N_k}{2\sigma^2} (\tilde{\theta}_k - \theta_k^*)^2 \sim \chi_K^2,$$

where χ_K^2 stands for the chi-squared distribution with K degrees of freedom.

This result is again a combination of the results from Sect. 2.9.1 for different pieces A_k. It is worth mentioning once again that the regression function $f(\cdot)$ is not assumed to be piecewise constant, it can be whatever function. Each $\tilde{\theta}_k$ estimates the mean θ_k^* of $f(\cdot)$ over the design points X_i withinA_k.

The results on the behavior of the maximum likelihood $L(\tilde{\boldsymbol{\theta}}, \boldsymbol{\theta}^*)$ are often used for studying the properties of the chi-squared test; see Sect. 7.1 for more details.

A choice of the partition is an important issue in the piecewise constant approximation. The presented results indicate that the accuracy of estimation of θ_k^* by $\tilde{\theta}_k$ is inversely proportional to the number of points N_k within each piece A_k. In the univariate case one usually applies the equidistant partition: the design interval is split into p equal intervals A_k leading to approximately equal values N_k. Sometimes, especially if the design is irregular, a nonuniform partition can be preferable. In general it can be recommended to split the whole design space into intervals with approximately the same number N_k of design points X_i.

A constant approximation is often not accurate enough to expand a regular regression function. One often uses a linear or polynomial approximation. The next sections explain this approach for the case of a univariate regression.

3.4.2 Piecewise Linear Univariate Estimation

The piecewise constant approximation can be naturally extended to piecewise linear and piecewise polynomial construction. The starting point is again a non-overlapping partition of $\mathcal{X}$ into intervals A_k for $k = 1, \ldots, K$. First we explain the idea for the linear approximation of the function f on each interval A_k. Any linear function on A_k can be represented in the form $a_k + c_k x$ with some coefficients a_k, c_k. This yields in total $p = 2K$ coefficients: $\boldsymbol{\theta} = (a_1, c_1, \ldots, a_K, c_K)^\top$. The corresponding function $f(\cdot, \boldsymbol{\theta})$ can be represented as

$$f(x) \approx f(x, \boldsymbol{\theta}) = \sum_{k=1}^{K} (a_k + c_k x) \, \mathbb{I}(x \in A_k).$$

The non-overlapping structure of the sets A_k yields orthogonality of basis functions for different pieces. As a corollary, one can optimize the linear approximation on every interval A_k independently of the others.

Exercise 3.4.3. Show that $\tilde{a}_k, \tilde{c}_k$ can be obtained by the linear approximation of the data Y_i for $X_i \in A_k$:

$$(\tilde{a}_k, \tilde{c}_k) = \operatorname*{argmax}_{(a_k, c_k)} \sum \ell(Y_i - a_k - c_k X_i) \, \mathbb{I}(X_i \in A_k), \qquad k = 1, \ldots, K.$$

On every piece A_k, the constant and the linear function x are not orthogonal except some very special situation. However, one can easily achieve orthogonality by a shift of the linear term.

Exercise 3.4.4. For each $k \leq K$, there exists a point x_k such that

$$\sum (X_i - x_k) \, \mathbb{I}(X_i \in A_k) = 0. \tag{3.43}$$

Introduce for each $k \leq K$ two basis functions $\phi_{j-1}(x) = \mathbb{I}(x \in A_k)$ and $\phi_j(x) = (x - x_k) \, \mathbb{I}(x \in A_k)$ with $j = 2k$.

Exercise 3.4.5. Assume (3.43) for each $k \leq K$. Check that any piecewise linear function can be uniquely represented in the form

$$f(x) = \sum_{j=1}^{p} \theta_j \phi_j(x)$$

with $p = 2K$ and the functions ϕ_j are orthogonal in the sense that for $j \neq j'$

$$\sum_{i=1}^{n} \phi_j(X_i) \phi_{j'}(X_i) = 0.$$

In addition, for each $k \leq K$

$$\|\phi_j\|^2 \stackrel{\text{def}}{=} \sum_{i=1}^{n} \phi_j^2(X_i) = \begin{cases} N_k, & j = 2k-1, \\ V_k^2 & j = 2k. \end{cases}$$

$$N_k^2 \stackrel{\text{def}}{=} \sum_{X_i \in A_k} 1, \qquad V_k^2 \stackrel{\text{def}}{=} \sum_{X_i \in A_k} (X_i - x_k)^2.$$

In the case of Gaussian regression, orthogonality of the basis helps to gain a simple closed form for the estimators $\tilde{\boldsymbol{\theta}} = (\tilde{\theta}_j)$:

$$\tilde{\theta}_j = \frac{1}{\|\psi_j\|^2} \sum_{i=1}^{n} Y_i \psi_j(X_i) = \begin{cases} \frac{1}{N_k} \sum_{X_i \in A_k} Y_i\,, & j = 2k-1, \\ \frac{1}{V_k^2} \sum_{X_i \in A_k} Y_i (X_i - x_k), & j = 2k. \end{cases}$$

see Sect. 4.2 in the next chapter for a comprehensive study.

3.4.3 Piecewise Polynomial Estimation

Local linear expansion of the function $f(x)$ on each piece A_k can be extended to a piecewise polynomial case. The basic idea is to apply a polynomial approximation of a certain degree q on each piece A_k independently. One can use for each piece a basis of the form $(x - x_k)^m \, \mathbb{I}(x \in A_k)$ for $m = 0, 1, \ldots, q$ with x_k from (3.43) yielding the approximation

$$\begin{aligned} f(x)\, \mathbb{I}(x \in A_k) &= f(x, \boldsymbol{a}_k)\, \mathbb{I}(x \in A_k) \\ &= \sum_{k=1}^{K} \{a_{0,k} + a_{1,k}(x - x_k) + \ldots + a_{q,k}(x - x_k)^q\}\, \mathbb{I}(x \in A_k) \end{aligned}$$

for $\boldsymbol{a}_k = (a_{0,k}, a_{1,k}, \ldots, a_{q,k})^\top$. This involves $q + 1$ parameter for each piece and $p = K(q+1)$ parameters in total. A nice feature of the piecewise approach is that the coefficients $\boldsymbol{a}_k$ of the piecewise polynomial approximation can be estimated independently for each piece. Namely,

$$\tilde{\boldsymbol{a}}_k = \operatorname*{argmax}_{\boldsymbol{a}} \sum_{X_i \in A_k} \{Y_i - f(X_i, \boldsymbol{a})\}^2$$

The properties of this estimator will be discussed in detail in Chap. 4.

3.4.4 Spline Estimation

The main drawback of the piecewise polynomial approximation is that the resulting function f is discontinuous at the edge points between different pieces. A natural way of improving the boundary effect is to force some conditions on the boundary behavior. One important special case is given by the spline system. Let $\mathcal{X}$ be an interval on the real line, perhaps infinite. Let also $t_0 < t_1 < \ldots < t_K$ be some ordered points in $\mathcal{X}$ such that t_0 is the left edge and t_K the right edge of $\mathcal{X}$. Such points are called *knots*. We say that a function f is a *spline* of degree q at knots (t_k) if it is polynomial on each *span* (t_{k-1}, t_k) for $k = 1, \ldots, K$ and satisfies the boundary conditions

$$f^{(m)}(t_k-) = f^{(m)}(t_k+), \qquad m = 0, \ldots, q-1, \quad k = 1, \ldots, K-1.$$

Here $f^{(m)}(t-)$ stands for the left derivative of f at t. In other words, the function f and its first $q-1$ derivatives are continuous on $\mathcal{X}$ and only the qth derivatives may have discontinuities at the knots t_k. It is obvious that the q derivative $f^{(q)}(t)$ of the spline of degree q is a piecewise constant functions on the spans $A_k = [t_{k-1}, t_k)$.

The spline is called *uniform* if the knots are equidistant, or, in other words, if all the spans A_k have equal length. Otherwise it is *nonuniform*.

Lemma 3.4.1. *The set of all splines of degree q at knots (t_k) is a linear space, that is, any linear combination of such splines is again a spline. Any function having a continuous mth derivative for $m < K$ and piecewise constant qth derivative is a q-spline.*

Splines of degree zero are just piecewise constant functions studied in Sect. 3.4.1. Linear splines are particularly transparent: this is the set of all piecewise linear continuous functions on $\mathcal{X}$. Each of them can be easily constructed from left to right or from right to left: start with a linear function $a_1 + c_1 x$ on the piece $A_1 = [t_0, t_1]$. Then $f(t_1) = a_1 + c_1 t_1$. On the piece A_2 the slope of f can be changed for c_2 leading to the function $f(x) = f(t_1) + c_2(x - t_1)$ for $x \in [t_1, t_2]$. Similarly, at t_2 the slop of fs can change for c_3 yielding $f(x) = f(t_2) + c_3(x - t_2)$ on A_3, and so on. Splines of higher order can be constructed similarly step by step: one fixes the polynomial form on the very first piece A_1 and then continues the spline function to every next piece A_k using the boundary conditions and the value of the qth derivative of f on A_k. This construction explains the next result.

Lemma 3.4.2. *Each spline f of degree q and knots (t_k) is uniquely described by the vector of coefficients $\boldsymbol{a}_1$ on the first span and the values $f^{(q)}(x)$ for each span $A_1, \ldots, A_K$.*

This result explains that the parameter dimension of the linear spline space is $q + K$. One possible basis in this space is given by polynomials x^{m-1} of degree $m = 0, 1, \ldots, q$ and the functions $\phi_k(x) \stackrel{\text{def}}{=} (x - t_k)_+^q$ for $k = 1, \ldots, K-1$.

Exercise 3.4.6. Check that $\phi_j(x)$ for $j = 1, \ldots, q + K$ form a basis in the linear spline space, and any q-spline f can be represented as

$$f(x) = \sum_{m=0}^{q} \alpha_m x^m + \sum_{k=1}^{K-1} \theta_k \phi_k(x). \tag{3.44}$$

Hint: check that the functions $\phi_j(x)$ are linearly independent and that each qth derivative $\phi_j^{(q)}(x)$ is piecewise constant.

3.4.4.1 B-Splines

Unfortunately, the basis functions $\{\phi_k(x)\}$ with $\phi_k(x) = (x - t_k)_+^q$ are only useful for theoretical study. The main problem is that the functions $\phi_j(x)$ are strongly correlated, and the recovering the coefficients θ_j in the expansion (3.44) is a hard numerical task. by this reason, one often uses another basis called B-splines. The idea is to build splines of the given degree with the minimal support. Each B-spline basis function $b_{k,q}(x)$ is only nonzero on the q neighbor spans $A_k, A_{k+1}, \ldots A_{k+q-1}$ for $k = 1, \ldots, K - q$.

Exercise 3.4.7. Let $f(x)$ be a q-spline with the support on $q' < q$ neighbor spans $A_k, A_{k+1}, \ldots A_{k+q'-1}$. Then $f(x) \equiv 0$.
Hint: consider any spline of the form $f(x) = \sum_{j=k}^{k+q'-1} c_j \phi_j(x)$. Show that the boundary conditions $f^{(m)}(t_{k+q'}) = 0$ for $m = 0, 1, \ldots, q$ yield $c_j \equiv 0$.

The basis B-spline functions can be constructed successfully. For $q = 0$, the B-splines $b_{k,0}(x)$ coincide with the functions $\phi_k(x) = \mathbb{I}(x \in A_k)$, $k = 1, \ldots, K$. Each linear B-spline $b_{k,1}(x)$ has a triangle shape on the two connected intervals A_k and A_{k+1}. It can be defined by the formula

$$b_{k,1}(x) \stackrel{\text{def}}{=} \frac{x - t_{k-1}}{t_k - t_{k-1}} b_{k,0}(x) + \frac{t_{k+1} - x}{t_{k+1} - t_k} b_{k+1,0}(x), \quad k = 1, \ldots, K - 1.$$

One can continue this way leading to the *Cox–de Boor recursion formula*

$$b_{k,m}(x) \stackrel{\text{def}}{=} \frac{x - t_{k-1}}{t_{k+m-1} - t_{k-1}} b_{k,m-1}(x) + \frac{t_{k+m} - x}{t_{k+m} - t_k} b_{k+1,m-1}(x)$$

for $k = 1, \ldots, K - m$.

Exercise 3.4.8. Check by induction for each function $b_{k,m}(x)$ the following conditions:

1. $b_{k,m}(x)$ a polynomial of degree m on each span $A_k, \ldots, A_{k+m-1}$ and zero outside;
2. $b_{k,m}(x)$ can be uniquely represented as a sum $b_{k,m}(x) = \sum_{l=0}^{m-1} c_{l,k} \phi_{k+l}(x)$;
3. $b_{k,m}(x)$ is a m-spline.

The formulas simplify for the uniform splines with equal span length $\Delta = |A_k|$:

$$b_{k,m}(x) \stackrel{\text{def}}{=} \frac{x - t_{k-1}}{m\Delta} b_{k,m-1}(x) + \frac{t_{k+m} - x}{m\Delta} b_{k+1,m-1}(x)$$

for $k = 1, \ldots, K - m$.

Exercise 3.4.9. Check that

$$b_{k,m}(x) = \sum_{l=0}^{m} \omega_{l,m} \phi_{k+l}(x)$$

with

$$\omega_{l,m} \stackrel{\text{def}}{=} \frac{(-1)^l}{\Delta^m l!(m-l)!}$$

3.4.4.2 Smoothing Splines

Such a spline system naturally arises as a solution of a penalized maximum likelihood problem. Suppose we are given the regression data (Y_i, X_i) with the univariate design $X_1 \le X_2 \le \ldots \le X_n$. Consider the mean regression model $Y_i = f(X_i) + \varepsilon_i$ with zero mean errors ε_i. The assumption of independent homogeneous Gaussian errors leads to the Gaussian log-likelihood

$$L(f) = -\sum_{i=1}^{n} |Y_i - f(X_i)|^2 / (2\sigma^2) \tag{3.45}$$

Maximization of this expression w.r.t. all possible functions f or, equivalently, all vectors $\bigl(f(X_1), \ldots, f(X_n)\bigr)^\top$ results in the trivial solution: $f(X_i) = Y_i$. This means that the full dimensional maximum likelihood perfectly reproduces the original noisy data. Some additional assumptions are needed to force any desirable feature of the reconstructed function. One popular example is given by smoothness of the function f. Degree of smoothness (or, inversely, degree of roughness) can be measured by the value

$$\mathcal{R}_q(f) \stackrel{\text{def}}{=} \int_{\mathcal{X}} |f^{(q)}(x)|^2 dx. \tag{3.46}$$

One can try to optimize the fit (3.45) subject to the constraint on the amount of roughness from (3.46). Equivalently, one can optimize the penalized log-likelihood

$$L_\lambda(f) \stackrel{\text{def}}{=} L(f) - \lambda \mathcal{R}_q(f) = -\frac{1}{2\sigma^2} \sum_{i=1}^{n} |Y_i - f(X_i)|^2 - \lambda \int_{\mathcal{X}} |f^{(q)}(x)|^2 dx,$$

where $\lambda > 0$ is a Lagrange multiplier. The corresponding maximizer is the penalized maximum likelihood estimator:

$$\tilde{f}_\lambda = \operatorname*{argmax}_f L_\lambda(f), \tag{3.47}$$

where the maximum is taken over the class of all measurable functions. It is remarkable that the solution of this optimization problem is a spline of degree q with the knots $X_1, \ldots, X_n$.

Theorem 3.4.4. *For any $\lambda > 0$ and any integer q, the problem* (3.47) *has a unique solution which is a q-spline with knots at design points (X_i).*

For the proof we refer to Green and Silverman (1994). Due to this result, one can simplify the problem and look for a spline f which minimizes the objective $L_\lambda(f)$. A solution to (3.47) is called a *smoothing spline*. If f is a q-spline, the integral $\mathcal{R}_q(f)$ can be easily computed. Indeed, $f^{(q)}(x)$ is piecewise constant, that is, $f^{(q)}(x) = c_k$ for $x \in A_k$, and

$$\mathcal{R}_q(f) = \sum_{k=1}^{K} c_k^2 \, |t_k - t_{k-1}|.$$

For the uniform design, the formula simplifies even more, and by change of the multiplier λ, one can use $\mathcal{R}_q(f) = \sum_k c_k^2$. The use of any parametric representation of a spline function f allows to represent the optimization problem (3.47) as a penalized least squares problem. Estimation and inference in such problems are studied below in Sect. 4.7.

3.5 Generalized Regression

Let the response Y_i be observed at the design point $X_i \in \mathbb{R}^d$, $i = 1, \ldots, n$. A (mean) regression model assumes that the observed values Y_i are independent and can be decomposed into the systematic component $f(X_i)$ and the individual centered stochastic error ε_i. In some cases such a decomposition is questionable. This especially concerns the case when the data Y_i are categorical, e.g. binary or discrete. Another striking example is given by nonnegative observations Y_i. In such cases one usually assumes that the distribution of Y_i belongs to some given parametric family $(P_\upsilon, \upsilon \in \mathcal{U})$ and only the parameter of this distribution depends on the design point X_i. We denote this parameter value as $f(X_i) \in \mathcal{U}$ and write the model in the form

$$Y_i \sim P_{f(X_i)} \,.$$

As previously, $f(\cdot)$ is called a *regression function* and its values at the design points X_i completely specify the joint data distribution:

$$Y \sim \prod_i P_{f(X_i)}.$$

Below we assume that (P_υ) is a univariate exponential family with the log-density $\ell(y, \upsilon)$.

The parametric modeling approach assumes that the regression function f can be specified by a finite-dimensional parameter $\boldsymbol{\theta} \in \Theta \subset \mathbb{R}^p$: $f(x) = f(x, \boldsymbol{\theta})$. As usual, by $\boldsymbol{\theta}^*$ we denote the true parameter value. The log-likelihood function for this model reads

$$L(\boldsymbol{\theta}) = \sum_i \ell\bigl(Y_i, f(X_i, \boldsymbol{\theta})\bigr).$$

The corresponding MLE $\tilde{\boldsymbol{\theta}}$ maximizes $L(\boldsymbol{\theta})$:

$$\tilde{\boldsymbol{\theta}} = \operatorname*{argmax}_{\boldsymbol{\theta}} \sum_i \ell\bigl(Y_i, f(X_i, \boldsymbol{\theta})\bigr).$$

The estimating equation $\nabla L(\boldsymbol{\theta}) = 0$ reads as

$$\sum_i \ell'\bigl(Y_i, f(X_i, \boldsymbol{\theta})\bigr) \nabla f(X_i, \boldsymbol{\theta}) = 0$$

where $\ell'(y, \upsilon) \stackrel{\text{def}}{=} \partial \ell(y, \upsilon)/\partial \upsilon$.

The approach essentially depends on the parametrization of the considered EF. Usually one applies either the natural or canonical parametrization. In the case of the natural parametrization, $\ell(y, \upsilon) = C(\upsilon)y - B(\upsilon)$, where the functions $C(\cdot)$, $B(\cdot)$ satisfy $B'(\upsilon) = \upsilon C'(\upsilon)$. This implies $\ell'(y, \upsilon) = yC'(\upsilon) - B'(\upsilon) = (y - \upsilon)C'(\upsilon)$ and the estimating equation reads as

$$\sum_i \bigl(Y_i - f(X_i, \boldsymbol{\theta})\bigr) C'\bigl(f(X_i, \boldsymbol{\theta})\bigr) \nabla f(X_i, \boldsymbol{\theta}) = 0$$

Unfortunately, a closed form solution for this equation exists only in very special cases. Even the questions of existence and uniqueness of the solution cannot be studied in whole generality. Some numerical algorithms are usually applied to solve the estimating equation.

Exercise 3.5.1. Specify the estimating equation for generalized EFn regression and find the solution for the case of the constant regression function $f(X_i, \theta) \equiv \theta$. Hint: If $f(X_i, \theta) \equiv \theta$, then the Y_i are i.i.d. from P_θ.

The equation can be slightly simplified by using the canonical parametrization. If (P_υ) is an EFc with the log-density $\ell(y,\upsilon) = y\upsilon - d(\upsilon)$, then the log-likelihood $L(\boldsymbol{\theta})$ can be represented in the form

$$L(\boldsymbol{\theta}) = \sum_i \bigl\{ Y_i f(X_i,\boldsymbol{\theta}) - d\bigl(f(X_i,\boldsymbol{\theta})\bigr) \bigr\}.$$

The corresponding estimating equation is

$$\sum_i \bigl\{ Y_i - d'\bigl(f(X_i,\boldsymbol{\theta})\bigr) \bigr\} \nabla f(X_i,\boldsymbol{\theta}) = 0.$$

Exercise 3.5.2. Specify the estimating equation for generalized EFc regression and find the solution for the case of constant regression with $f(X_i,\upsilon) \equiv \upsilon$. Relate the natural and canonical representation.

A generalized regression with a canonical link is often applied in combination with linear modeling of the regression function considered in the next section.

3.5.1 Generalized Linear Models

Consider the generalized regression model

$$Y_i \sim P_{f(X_i)} \in \mathcal{P}.$$

In addition we assume a linear (in parameters) structure of the regression function $f(X)$. Such modeling is particularly useful to combine with the canonical parametrization of the considered EF with the log-density $\ell(y,\upsilon) = y\upsilon - d(\upsilon)$. The reason is that the stochastic part in the log-likelihood of an EFc linearly depends on the parameter. So, below we assume that $\mathcal{P} = (P_\upsilon, \upsilon \in \mathcal{U})$ is an EFc.

Linear regression $f(X_i) = \Psi_i^\top \boldsymbol{\theta}$ with given feature vectors $\Psi_i \in \mathbb{R}^p$ leads to the model with the log-likelihood

$$L(\boldsymbol{\theta}) = \sum_i \bigl\{ Y_i \Psi_i^\top \boldsymbol{\theta} - d\bigl(\Psi_i^\top \boldsymbol{\theta}\bigr) \bigr\}.$$

Such a setup is called *generalized linear model* (GLM). Note that the log-likelihood can be represented as

$$L(\boldsymbol{\theta}) = S^\top \boldsymbol{\theta} - A(\boldsymbol{\theta}),$$

where

$$S = \sum_i Y_i \Psi_i, \qquad A(\boldsymbol{\theta}) = \sum_i d\bigl(\Psi_i^\top \boldsymbol{\theta}\bigr).$$

The corresponding MLE $\tilde{\boldsymbol{\theta}}$ maximizes $L(\boldsymbol{\theta})$. Again, a closed form solution only exists in special cases. However, an important advantage of the GLM approach is that the solution always exists and is unique. The reason is that the log-likelihood function $L(\boldsymbol{\theta})$ is concave in $\boldsymbol{\theta}$.

Lemma 3.5.1. *The MLE $\tilde{\boldsymbol{\theta}}$ solves the following estimating equation:*

$$\nabla L(\boldsymbol{\theta}) = S - \nabla A(\boldsymbol{\theta}) = \sum_i \Big(Y_i - d'(\Psi_i^\top \boldsymbol{\theta})\Big)\Psi_i = 0. \tag{3.48}$$

The solution exists and is unique.

Proof. Define the matrix

$$B(\boldsymbol{\theta}) = \sum_i d''(\Psi_i^\top \boldsymbol{\theta})\Psi_i \Psi_i^\top . \tag{3.49}$$

Since $d''(\upsilon)$ is strictly positive for all u, the matrix $B(\boldsymbol{\theta})$ is positively defined as well. It holds

$$\nabla^2 L(\boldsymbol{\theta}) = -\nabla^2 A(\boldsymbol{\theta}) = -\sum_i d''(\Psi_i^\top \boldsymbol{\theta})\Psi_i \Psi_i^\top = -B(\boldsymbol{\theta}).$$

Thus, the function $L(\boldsymbol{\theta})$ is strictly concave w.r.t. $\boldsymbol{\theta}$ and the estimating equation $\nabla L(\boldsymbol{\theta}) = S - \nabla A(\boldsymbol{\theta}) = 0$ has the unique solution $\tilde{\boldsymbol{\theta}}$.

The solution of (3.48) can be easily obtained numerically by the Newton–Raphson algorithm: select the initial estimate $\boldsymbol{\theta}^{(0)}$. Then for every $k \ge 1$ apply

$$\boldsymbol{\theta}^{(k+1)} = \boldsymbol{\theta}^{(k)} + \big\{B(\boldsymbol{\theta}^{(k)})\big\}^{-1}\big\{S - \nabla A(\boldsymbol{\theta}^{(k)})\big\} \tag{3.50}$$

until convergence.

Below we consider two special cases of GLMs for binary and Poissonian data.

3.5.2 *Logit Regression for Binary Data*

Suppose that the observed data Y_i are independent and binary, that is, each Y_i is either zero or one, $i = 1, \ldots, n$. Such models are often used in, e.g., sociological and medical study, two-class classification, binary imaging, among many other fields. We treat each Y_i as a Bernoulli r.v. with the corresponding parameter $f_i = f(X_i)$. This is a special case of generalized regression also called *binary response models*. The parametric modeling assumption means that the regression function $f(\cdot)$ can be represented in the form $f(X_i) = f(X_i, \boldsymbol{\theta})$ for a given class of functions $\{f(\cdot, \boldsymbol{\theta}), \boldsymbol{\theta} \in \Theta \in \mathbb{R}^p\}$. Then the log-likelihood $L(\boldsymbol{\theta})$ reads as

$$L(\boldsymbol{\theta}) = \sum_i \ell(Y_i, f(X_i, \boldsymbol{\theta})), \tag{3.51}$$

where $\ell(y, \upsilon)$ is the log-density of the Bernoulli law. For linear modeling, it is more useful to work with the canonical parametrization. Then $\ell(y, \upsilon) = y\upsilon - \log(1+e^{\upsilon})$, and the log-likelihood reads

$$L(\boldsymbol{\theta}) = \sum_i \bigl[Y_i f(X_i, \boldsymbol{\theta}) - \log\bigl(1 + \mathrm{e}^{f(X_i,\boldsymbol{\theta})}\bigr)\bigr].$$

In particular, if the regression function $f(\cdot, \boldsymbol{\theta})$ is linear, that is, $f(X_i, \boldsymbol{\theta}) = \Psi_i^\top \boldsymbol{\theta}$, then

$$L(\boldsymbol{\theta}) = \sum_i \bigl[Y_i \Psi_i^\top \boldsymbol{\theta} - \log(1 + \mathrm{e}^{\Psi_i^\top \boldsymbol{\theta}})\bigr]. \tag{3.52}$$

The corresponding estimate reads as

$$\tilde{\boldsymbol{\theta}} = \operatorname*{argmax}_{\boldsymbol{\theta}} L(\boldsymbol{\theta}) = \operatorname*{argmax}_{\boldsymbol{\theta}} \sum_i \bigl[Y_i \Psi_i^\top \boldsymbol{\theta} - \log(1 + \mathrm{e}^{\Psi_i^\top \boldsymbol{\theta}})\bigr]$$

This modeling is usually referred to as *logit regression.*

Exercise 3.5.3. Specify the estimating equation for the case of logit regression.

Exercise 3.5.4. Specify the step of the Newton–Raphson procedure for the case of logit regression.

3.5.3 Parametric Poisson Regression

Suppose that the observations Y_i are nonnegative integer numbers. The Poisson distribution is a natural candidate for modeling such data. It is supposed that the underlying Poisson parameter depends on the regressor X_i. Typical examples arise in different types of imaging including medical positron emission and magnet resonance tomography, satellite and low-luminosity imaging, queueing theory, high frequency trading, etc. The regression equation reads

$$Y_i \sim \text{Poisson}(f(X_i)).$$

The *Poisson regression* function $f(X_i)$ is usually the target of estimation. The parametric specification $f(\cdot) \in \{f(\cdot, \boldsymbol{\theta}), \boldsymbol{\theta} \in \Theta\}$ reduces this problem to estimating the parameter $\boldsymbol{\theta}$. Under the assumption of independent observations Y_i, the corresponding maximum likelihood $L(\boldsymbol{\theta})$ is given by

$$L(\boldsymbol{\theta}) = \sum \bigl[Y_i \log\{f(X_i, \boldsymbol{\theta})\} - f(X_i, \boldsymbol{\theta})\bigr] + R,$$

where the remainder R does not depend on $\boldsymbol{\theta}$ and can be omitted. Obviously, the constant function family $f(\cdot, \theta) \equiv \theta$ leads back to the case of i.i.d. modeling studied in Sect. 2.11. A further extension is given by linear Poisson regression: $f(X_i, \boldsymbol{\theta}) = \Psi_i^\top \boldsymbol{\theta}$ for some given factors Ψ_i. The regression equation reads

$$L(\boldsymbol{\theta}) = \sum_i \bigl[Y_i \log(\Psi_i^\top \boldsymbol{\theta}) - \Psi_i^\top \boldsymbol{\theta}\bigr]. \tag{3.53}$$

Exercise 3.5.5. Specify the estimating equation and the Newton–Raphson procedure for the linear Poisson regression (3.53).

An obvious problem of linear Poisson modeling is that it requires all the values $\Psi_i^\top \boldsymbol{\theta}$ to be positive. The use of canonical parametrization helps to avoid this problem. The linear structure is assumed for the canonical parameter leading to the representation $f(X_i) = \exp(\Psi_i^\top \boldsymbol{\theta})$. Then the general log-likelihood process $L(\boldsymbol{\theta})$ from (3.51) translates into

$$L(\boldsymbol{\theta}) = \sum_i \bigl[Y_i \Psi_i^\top \boldsymbol{\theta} - \exp(\Psi_i^\top \boldsymbol{\theta})\bigr]; \tag{3.54}$$

cf. with (3.52).

Exercise 3.5.6. Specify the estimating equation and the Newton–Raphson procedure for the canonical link linear Poisson regression (3.54).

If the factors Ψ_i are properly scaled, then the scalar products $\Psi_i^\top \boldsymbol{\theta}$ for all i and all $\boldsymbol{\theta} \in \Theta_0$ belong to some bounded interval. For the matrix $B(\boldsymbol{\theta})$ from (3.49), it holds

$$B(\boldsymbol{\theta}) = \sum_i \exp(\Psi_i^\top \boldsymbol{\theta}) \Psi_i \Psi_i^\top .$$

Initializing the ML optimization problem with $\boldsymbol{\theta} = 0$ leads to the oLSE

$$\tilde{\boldsymbol{\theta}}^{(0)} = \Bigl(\sum_i \Psi_i \Psi_i^\top\Bigr)^{-1} \sum_i \Psi_i Y_i .$$

The further steps of the algorithm (3.50) can be done as weighted LSE with the weights $\exp\bigl(\Psi_i^\top \tilde{\boldsymbol{\theta}}^{(k)}\bigr)$ for the estimate $\tilde{\boldsymbol{\theta}}^{(k)}$ obtained at the previous step.

3.5.4 *Piecewise Constant Methods in Generalized Regression*

Consider a generalized regression model

$$Y_i \sim P_{f(X_i)} \in (P_\upsilon)$$

for a given exponential family (P_υ). Further, let $A_1, \ldots, A_K$ be a *non-overlapping partition* of the design space $\mathcal{X}$; see (3.39). A piecewise constant approximation (3.40) of the regression function $f(\cdot)$ leads to the additive log-likelihood structure: for $\boldsymbol{\theta} = (\theta_1, \ldots, \theta_K)^\top$

$$\tilde{\boldsymbol{\theta}} = \operatorname*{argmax}_{\boldsymbol{\theta}} L(\boldsymbol{\theta}) = \operatorname*{argmax}_{\theta_1, \ldots, \theta_K} \sum_{k=1}^{K} \sum_{X_i \in A_k} \ell(Y_i, \theta_k);$$

cf. (3.41). Similarly to the mean regression case, the global optimization w.r.t. the vector $\boldsymbol{\theta}$ can be decomposed into K separated simple optimization problems:

$$\tilde{\theta}_k = \operatorname*{argmax}_{\theta} \sum_{X_i \in A_k} \ell(Y_i, \theta);$$

cf. (3.42). The same decomposition can be obtained for the target $\boldsymbol{\theta}^* = (\theta_1, \ldots, \theta_K)^\top$:

$$\boldsymbol{\theta}^* = \operatorname*{argmax}_{\boldsymbol{\theta}} \mathbb{E} L(\boldsymbol{\theta}) = \operatorname*{argmax}_{\theta_1, \ldots, \theta_K} \sum_{k=1}^{K} \sum_{X_i \in A_k} \mathbb{E}\ell(Y_i, \theta_k).$$

The properties of each estimator $\tilde{\theta}_k$ repeats ones of the qMLE for a univariate EFn; see Sect. 2.11.

Theorem 3.5.1. *Let $\ell(y, \theta) = C(\theta)y - B(\theta)$ be a density of an EFn, so that the functions $B(\theta)$ and $C(\theta)$ satisfy $B'(\theta) = \theta C'(\theta)$. Then for every $k = 1, \ldots, K$*

$$\tilde{\theta}_k = \frac{1}{N_k} \sum_{X_i \in A_k} Y_i \, ,$$

$$\theta_k^* = \frac{1}{N_k} \sum_{X_i \in A_k} \mathbb{E} Y_i \, ,$$

where N_k stands for the number of design points X_i within the piece A_k:

$$N_k \stackrel{\text{def}}{=} \sum_{X_i \in A_k} 1 = \#\{i : X_i \in A_k\}.$$

Moreover, it holds

$$\mathbb{E}\tilde{\theta}_k = \theta_k^*$$

and

$$L(\tilde{\boldsymbol{\theta}}, \boldsymbol{\theta}^*) = \sum_{k=1}^{K} N_k \mathcal{K}(\tilde{\theta}_k, \theta_k^*) \tag{3.55}$$

where $\mathcal{K}(\theta, \theta') \stackrel{\text{def}}{=} E_\theta\{\ell(Y_i, \theta) - \ell(Y_i, \theta')\}$.

These statements follow from Theorem 3.5.1 and Theorem 2.11.1 of Sect. 2.11. For the presented results, the true regression function $f(\cdot)$ can be of arbitrary structure, the true distribution of each Y_i can differ from $P_{f(X_i)}$.

Exercise 3.5.7. Check the statements of Theorem 3.5.1.

If PA is correct, that is, if f is indeed piecewise constant and the distribution of Y_i is indeed $P_{f(X_i)}$, the deviation bound for the excess $L(\tilde{\theta}_k, \theta_k^*)$ from Theorem 2.11.4 can be applied to each piece A_k yielding the following result.

Theorem 3.5.2. *Let* (P_θ) *be a EFn and let* $Y_i \sim P_{\theta_k}$ *for* $X_i \in A_k$ *and* $k = 1, \ldots, K$. *Then for any* $\mathfrak{z} > 0$

$$\mathbb{P}\bigl(L(\tilde{\boldsymbol{\theta}}, \boldsymbol{\theta}^*) > K\mathfrak{z}\bigr) \le 2K\mathrm{e}^{-\mathfrak{z}}.$$

Proof. By (3.55) and Theorem 2.11.4

$$\mathbb{P}\bigl(L(\tilde{\boldsymbol{\theta}}, \boldsymbol{\theta}^*) > K\mathfrak{z}\bigr) = \mathbb{P}\Bigl(\sum_{k=1}^{K} N_k \mathcal{K}(\tilde{\theta}_k, \theta_k^*) > K\mathfrak{z}\Bigr)$$
$$\le \sum_{k=1}^{K} \mathbb{P}\Bigl(N_k \mathcal{K}(\tilde{\theta}_k, \theta_k^*) > \mathfrak{z}\Bigr) \le 2K\mathrm{e}^{-\mathfrak{z}}$$

and the result follows.

A piecewise linear generalized regression can be treated in a similar way. The main benefit of piecewise modeling remains preserved: a global optimization over the vector $\boldsymbol{\theta}$ can be decomposed into a set of small optimization problems for each piece A_k. However, a closed form solution is available only in some special cases like Gaussian regression.

3.5.5 Smoothing Splines for Generalized Regression

Consider again the generalized regression model

$$Y_i \sim P_{f(X_i)} \in \mathcal{P}$$

for an exponential family $\mathcal{P}$ with canonical parametrization. Now we do not assume any specific parametric structure for the function f. Instead, the function f is supposed to be smooth and its smoothness is measured by the roughness $\mathcal{R}_q(f)$ from (3.46). Similarly to the regression case of Sect. 3.4.4, the function f can be estimated directly by optimizing the penalized log-likelihood $L_\lambda(f)$:

$$\begin{aligned}\tilde{f}_\lambda &= \operatorname*{argmax}_f L_\lambda(f) = \operatorname*{argmax}_f \{L(f) - \mathcal{R}_q(f)\} \\ &= \operatorname*{argmax}_f \sum_i \{Y_i f(X_i) - d\big(f(X_i)\big)\} - \int_{\mathcal{X}} \big|f^{(q)}(x)\big|^2 dx. \qquad (3.56)\end{aligned}$$

The maximum is taken over the class of all regular q-times differentiable functions. In the regression case, the function $d(\cdot)$ is quadratic and the solution is a spline functions with knots X_i. This conclusion can be extended to the case of any convex function $d(\cdot)$, thus, the problem (3.56) yields a *smoothing spline* solution. Numerically this problem is usually solved by iterations. One starts with a quadratic function $d(\upsilon) = \upsilon^2/2$ to obtain an initial approximation $\tilde{f}^{(0)}(\cdot)$ of $f(\cdot)$ by a standard smoothing spline regression. Further, at each new step $k+1$, the use of the estimate $\tilde{f}^{(k)}(\cdot)$ from the previous step k for $k \geq 0$ helps to approximate the problem (3.56) by a weighted regression. The corresponding iterations can be written in the form (3.50).

3.6 Historical Remarks and Further Reading

A nice introduction in the use of *smoothing splines* in statistics can be found in Green and Silverman (1994) and Wahba (1990). For further properties of the spline approximation and algorithmic use of splines see de Boor (2001).

Orthogonal polynomials have long stories and have been applied in many different fields of mathematics. We refer to Szegö (1939) and Chihara (2011) for the classical results and history around different polynomial systems.

Some further methods in *regression estimation* and their features are described, e.g., in Lehmann and Casella (1998), Fan and Gijbels (1996), and Wasserman (2006).

Chapter 4
Estimation in Linear Models

This chapter studies the estimation problem for a linear model. The first four sections are fairly classical and the presented results are based on the direct analysis of the linear estimation procedures. Sections 4.5 and 4.6 reproduce in a very short form the same results but now based on the likelihood analysis. The presentation is based on the celebrated chi-squared phenomenon which appears to be the fundamental fact yielding the exact likelihood-based concentration and confidence properties. The further sections are complementary and can be recommended for a more profound reading. The issues like regularization, shrinkage, smoothness, and roughness are usually studied within the nonparametric theory, here we try to fit them to the classical linear parametric setup. A special focus is on semiparametric estimation in Sect. 4.9. In particular, efficient estimation and chi-squared result are extended to the semiparametric framework.

The main tool of the study is the quasi maximum likelihood method. We especially focus on the validity of the presented results under possible model misspecification. Another important issue is the way of measuring the estimation loss and risk. We distinguish below between *response estimation* or *prediction* and the *parameter estimation*. The most advanced results like chi-squared result in Sect. 4.6 are established under the assumption of a Gaussian noise. However, a misspecification of noise structure is allowed and addressed.

4.1 Modeling Assumptions

A linear model assumes that the observations Y_i follow the equation:

$$Y_i = \Psi_i^\top \boldsymbol{\theta}^* + \varepsilon_i \tag{4.1}$$

for $i = 1, \ldots, n$, where $\boldsymbol{\theta}^* = (\theta_1^*, \ldots, \theta_p^*)^\top \in \mathbb{R}^p$ is an unknown parameter vector, Ψ_i are given vectors in $\mathbb{R}^p$, and the ε_i's are individual errors with zero mean.

V. Spokoiny and T. Dickhaus, *Basics of Modern Mathematical Statistics*, Springer Texts in Statistics, DOI 10.1007/978-3-642-39909-1_4,

A typical example is given by linear regression (see Sect. 3.3) when the vectors Ψ_i are the values of a set of functions (e.g., polynomial, trigonometric) series at the design points X_i.

A linear Gaussian model assumes in addition that the vector of errors $\boldsymbol{\varepsilon} = (\varepsilon_1, \ldots \varepsilon_n)^\top$ is normally distributed with zero mean and a covariance matrix Σ:

$$\boldsymbol{\varepsilon} \sim \mathcal{N}(0, \Sigma).$$

In this chapter we suppose that Σ is given in advance. We will distinguish between three cases:

1. the errors ε_i are i.i.d. $\mathcal{N}(0, \sigma^2)$, or equivalently, the matrix Σ is equal to $\sigma^2 \boldsymbol{I}_n$ with $\boldsymbol{I}_n$ being the unit matrix in $\mathbb{R}^n$.
2. the errors are independent but not homogeneous, that is, $\mathbb{E}\varepsilon_i^2 = \sigma_i^2$. Then the matrix Σ is diagonal: $\Sigma = \operatorname{diag}(\sigma_1^2, \ldots, \sigma_n^2)$.
3. the errors ε_i are dependent with a covariance matrix Σ.

In practical applications one mostly starts with the white Gaussian noise assumption and more general cases 2 and 3 are only considered if there are clear indications of the noise inhomogeneity or correlation. The second situation is typical, e.g., for the eigenvector decomposition in an inverse problem. The last case is the most general and includes the first two.

4.2 Quasi Maximum Likelihood Estimation

Denote by $\boldsymbol{Y} = (Y_1, \ldots, Y_n)^\top$ (resp. $\boldsymbol{\varepsilon} = (\varepsilon_1, \ldots, \varepsilon_n)^\top$) the vector of observations (resp. of errors) in $\mathbb{R}^n$ and by Ψ the $p \times n$ matrix with columns Ψ_i. Let also $\Psi^\top$ denote its transpose. Then the model equation can be rewritten as:

$$\boldsymbol{Y} = \Psi^\top \boldsymbol{\theta}^* + \boldsymbol{\varepsilon}, \qquad \boldsymbol{\varepsilon} \sim \mathcal{N}(0, \Sigma).$$

An equivalent formulation is that $\Sigma^{-1/2}(\boldsymbol{Y} - \Psi^\top \boldsymbol{\theta})$ is a standard normal vector in $\mathbb{R}^n$. The log-density of the distribution of the vector $\boldsymbol{Y} = (Y_1, \ldots, Y_n)^\top$ w.r.t. the Lebesgue measure in $\mathbb{R}^n$ is therefore of the form

$$\begin{aligned} L(\boldsymbol{\theta}) &= -\frac{n}{2}\log(2\pi) - \frac{\log(\det \Sigma)}{2} - \frac{1}{2}\|\Sigma^{-1/2}(\boldsymbol{Y} - \Psi^\top \boldsymbol{\theta})\|^2 \\ &= -\frac{n}{2}\log(2\pi) - \frac{\log(\det \Sigma)}{2} - \frac{1}{2}(\boldsymbol{Y} - \Psi^\top \boldsymbol{\theta})^\top \Sigma^{-1}(\boldsymbol{Y} - \Psi^\top \boldsymbol{\theta}). \end{aligned}$$

In case 1 this expression can be rewritten as

$$L(\boldsymbol{\theta}) = -\frac{n}{2}\log(2\pi\sigma^2) - \frac{1}{2\sigma^2}\sum_{i=1}^{n}(Y_i - \Psi_i^\top \boldsymbol{\theta})^2.$$

In case 2 the expression is similar:

$$L(\boldsymbol{\theta}) = -\sum_{i=1}^{n} \Bigl\{ \frac{1}{2} \log(2\pi\sigma_i^2) + \frac{(Y_i - \Psi_i^\top \boldsymbol{\theta})^2}{2\sigma_i^2} \Bigr\}.$$

The *maximum likelihood estimate* (MLE) $\tilde{\boldsymbol{\theta}}$ of $\boldsymbol{\theta}^*$ is defined by maximizing the log-likelihood $L(\boldsymbol{\theta})$:

$$\tilde{\boldsymbol{\theta}} = \operatorname*{argmax}_{\boldsymbol{\theta} \in \mathbb{R}^p} L(\boldsymbol{\theta}) = \operatorname*{argmin}_{\boldsymbol{\theta} \in \mathbb{R}^p} (\boldsymbol{Y} - \Psi^\top \boldsymbol{\theta})^\top \Sigma^{-1} (\boldsymbol{Y} - \Psi^\top \boldsymbol{\theta}). \tag{4.2}$$

We omit the other terms in the expression of $L(\boldsymbol{\theta})$ because they do not depend on $\boldsymbol{\theta}$. This estimate is the *least squares estimate* (LSE) because it minimizes the sum of squared distances between the observations Y_i and the linear responses $\Psi_i^\top \boldsymbol{\theta}$. Note that (4.2) is a quadratic optimization problem which has a closed form solution. Differentiating the right-hand side of (4.2) w.r.t. $\boldsymbol{\theta}$ yields the *normal equation*

$$\Psi \Sigma^{-1} \Psi^\top \tilde{\boldsymbol{\theta}} = \Psi \Sigma^{-1} \boldsymbol{Y}.$$

If the $p \times p$-matrix $\Psi \Sigma^{-1} \Psi^\top$ is non-degenerate, then the normal equation has the unique solution

$$\tilde{\boldsymbol{\theta}} = \bigl(\Psi \Sigma^{-1} \Psi^\top \bigr)^{-1} \Psi \Sigma^{-1} \boldsymbol{Y} = \mathcal{S} \boldsymbol{Y}, \tag{4.3}$$

where

$$\mathcal{S} = \bigl(\Psi \Sigma^{-1} \Psi^\top \bigr)^{-1} \Psi \Sigma^{-1}$$

is a $p \times n$ matrix. We denote by $\tilde{\theta}_m$ the entries of the vector $\tilde{\boldsymbol{\theta}}$, $m = 1, \ldots, p$.

If the matrix $\Psi \Sigma^{-1} \Psi^\top$ is degenerate, then the normal equation has infinitely many solutions. However, one can still apply the formula (4.3) where $(\Psi \Sigma^{-1} \Psi^\top)^{-1}$ is a pseudo-inverse of the matrix $\Psi \Sigma^{-1} \Psi^\top$.

The ML approach leads to the *parameter estimate* $\tilde{\boldsymbol{\theta}}$. Note that due to the model (4.1), the product $\tilde{\boldsymbol{f}} = \Psi^\top \tilde{\boldsymbol{\theta}}$ is an estimate of the mean $\boldsymbol{f}^* \stackrel{\text{def}}{=} \mathbb{E} \boldsymbol{Y}$ of the vector of observations $\boldsymbol{Y}$:

$$\tilde{\boldsymbol{f}} = \Psi^\top \tilde{\boldsymbol{\theta}} = \Psi^\top \bigl(\Psi \Sigma^{-1} \Psi^\top \bigr)^{-1} \Psi \Sigma^{-1} \boldsymbol{Y} = \Pi \boldsymbol{Y},$$

where

$$\Pi = \Psi^\top \bigl(\Psi \Sigma^{-1} \Psi^\top \bigr)^{-1} \Psi \Sigma^{-1}$$

is an $n \times n$ matrix (linear operator) in $\mathbb{R}^n$. The vector $\tilde{\boldsymbol{f}}$ is called a *prediction* or *response* regression estimate.

Below we study the properties of the estimates $\tilde{\boldsymbol{\theta}}$ and $\tilde{\boldsymbol{f}}$. In this study we try to address both types of possible model misspecification: due to a wrong assumption about the error distribution and due to a possibly wrong linear parametric structure. Namely we consider the model

$$Y_i = f_i + \varepsilon_i, \qquad \boldsymbol{\varepsilon} \sim \mathcal{N}(0, \Sigma_0). \tag{4.4}$$

The response values f_i are usually treated as the value of the regression function $f(\cdot)$ at the design points X_i. The parametric model (4.1) can be viewed as an approximation of (4.4) while Σ is an approximation of the true covariance matrix Σ_0. If $\boldsymbol{f}^*$ is indeed equal to $\Psi^\top \boldsymbol{\theta}^*$ and $\Sigma = \Sigma_0$, then $\tilde{\boldsymbol{\theta}}$ and $\tilde{\boldsymbol{f}}$ are MLEs, otherwise quasi MLEs. In our study we mostly restrict ourselves to the case 1 assumption about the noise $\boldsymbol{\varepsilon}$: $\boldsymbol{\varepsilon} \sim \mathcal{N}(0, \sigma^2 \boldsymbol{I}_n)$. The general case can be reduced to this one by a simple data transformation, namely, by multiplying the Eq. (4.4) $\boldsymbol{Y} = \boldsymbol{f}^* + \boldsymbol{\varepsilon}$ with the matrix $\Sigma^{-1/2}$, see Sect. 4.6 for more detail.

4.2.1 *Estimation Under the Homogeneous Noise Assumption*

If a homogeneous noise is assumed, that is $\Sigma = \sigma^2 \boldsymbol{I}_n$ and $\boldsymbol{\varepsilon} \sim \mathcal{N}(0, \sigma^2 \boldsymbol{I}_n)$, then the formulae for the MLEs $\tilde{\boldsymbol{\theta}}, \tilde{\boldsymbol{f}}$ slightly simplify. In particular, the variance σ^2 cancels and the resulting estimate is the *ordinary least squares* (oLSE):

$$\tilde{\boldsymbol{\theta}} = \left(\Psi \Psi^\top\right)^{-1} \Psi \boldsymbol{Y} = \mathcal{S} \boldsymbol{Y}$$

with $\mathcal{S} = \left(\Psi \Psi^\top\right)^{-1} \Psi$. Also

$$\tilde{\boldsymbol{f}} = \Psi^\top \left(\Psi \Psi^\top\right)^{-1} \Psi \boldsymbol{Y} = \Pi \boldsymbol{Y}$$

with $\Pi = \Psi^\top \left(\Psi \Psi^\top\right)^{-1} \Psi$.

Exercise 4.2.1. Derive the formulae for $\tilde{\boldsymbol{\theta}}, \tilde{\boldsymbol{f}}$ directly from the log-likelihood $L(\boldsymbol{\theta})$ for homogeneous noise.

If the assumption $\boldsymbol{\varepsilon} \sim \mathcal{N}(0, \sigma^2 \boldsymbol{I}_n)$ about the errors is not precisely fulfilled, then the oLSE can be viewed as a quasi MLE.

4.2.2 *Linear Basis Transformation*

Denote by $\boldsymbol{\psi}_1^\top, \ldots, \boldsymbol{\psi}_p^\top$ the rows of the matrix Ψ. Then the $\boldsymbol{\psi}_i$'s are vectors in $\mathbb{R}^n$ and we call them *the basis vectors*. In the linear regression case the $\boldsymbol{\psi}_i$'s are obtained as the values of the basis functions at the design points. Our linear parametric assumption simply means that the underlying vector $\boldsymbol{f}^*$ can be represented as a linear combination of the vectors $\boldsymbol{\psi}_1, \ldots, \boldsymbol{\psi}_p$:

$$\boldsymbol{f}^* = \theta_1^* \boldsymbol{\psi}_1 + \ldots + \theta_p^* \boldsymbol{\psi}_p .$$

In other words, $\boldsymbol{f}^*$ belongs to the linear subspace in $\mathbb{R}^n$ spanned by the vectors $\boldsymbol{\psi}_1, \ldots, \boldsymbol{\psi}_p$. It is clear that this assumption still holds if we select another basis in this subspace.

Let U be any linear orthogonal transformation in $\mathbb{R}^p$ with $UU^\top = \boldsymbol{I}_p$. Then the linear relation $\boldsymbol{f}^* = \Psi^\top \boldsymbol{\theta}^*$ can be rewritten as

$$\boldsymbol{f}^* = \Psi^\top U U^\top \boldsymbol{\theta}^* = \breve{\Psi}^\top \boldsymbol{u}^*$$

with $\breve{\Psi} = U^\top \Psi$ and $\boldsymbol{u}^* = U^\top \boldsymbol{\theta}^*$. Here the columns of $\breve{\Psi}$ mean the new basis vectors $\breve{\boldsymbol{\psi}}_m$ in the same subspace while $\boldsymbol{u}^*$ is the vector of coefficients describing the decomposition of the vector $\boldsymbol{f}^*$ w.r.t. this new basis:

$$\boldsymbol{f}^* = u_1^* \breve{\boldsymbol{\psi}}_1 + \ldots + u_p^* \breve{\boldsymbol{\psi}}_p .$$

The natural question is how the expression for the MLEs $\tilde{\boldsymbol{\theta}}$ and $\tilde{\boldsymbol{f}}$ changes with the change of the basis. The answer is straightforward. For notational simplicity, we only consider the case with $\Sigma = \sigma^2 \boldsymbol{I}_n$. The model can be rewritten as

$$\boldsymbol{Y} = \breve{\Psi}^\top \boldsymbol{u}^* + \boldsymbol{\varepsilon}$$

yielding the solutions

$$\tilde{\boldsymbol{u}} = \bigl(\breve{\Psi}\breve{\Psi}^\top\bigr)^{-1} \breve{\Psi} \boldsymbol{Y} = \breve{\mathcal{S}} \boldsymbol{Y}, \qquad \tilde{\boldsymbol{f}} = \breve{\Psi}^\top \bigl(\breve{\Psi}\breve{\Psi}^\top\bigr)^{-1} \breve{\Psi} \boldsymbol{Y} = \breve{\Pi} \boldsymbol{Y},$$

where $\breve{\Psi} = U^\top \Psi$ implies

$$\breve{\mathcal{S}} = \bigl(\breve{\Psi}\breve{\Psi}^\top\bigr)^{-1} \breve{\Psi} = U^\top \mathcal{S},$$
$$\breve{\Pi} = \breve{\Psi}^\top \bigl(\breve{\Psi}\breve{\Psi}^\top\bigr)^{-1} \breve{\Psi} = \Pi.$$

This yields

$$\tilde{\boldsymbol{u}} = U^\top \tilde{\boldsymbol{\theta}}$$

and moreover, the estimate $\tilde{\boldsymbol{f}}$ is not changed for any linear transformation of the basis. The first statement can be expected in view of $\boldsymbol{\theta}^* = U \boldsymbol{u}^*$, while the second one will be explained in the next section: Π is the linear projector on the subspace spanned by the basis vectors and this projector is invariant w.r.t. basis transformations.

Exercise 4.2.2. Consider univariate polynomial regression of degree $p-1$. This means that f is a polynomial function of degree $p-1$ observed at the points X_i with errors ε_i that are assumed to be i.i.d. normal. The function f can be represented as

$$f(x) = \theta_1^* + \theta_2^* x + \ldots + \theta_p^* x^{p-1}$$

using the basis functions $\psi_m(x) = x^{m-1}$ for $m = 0, \ldots, p-1$. At the same time, for any point x_0, this function can also be written as

$$f(x) = u_1^* + u_2^*(x - x_0) + \ldots + u_p^*(x - x_0)^{p-1}$$

using the basis functions $\breve{\psi}_m = (x - x_0)^{m-1}$.

- Write the matrices Ψ and $\Psi\Psi^\top$ and similarly $\breve{\Psi}$ and $\breve{\Psi}\breve{\Psi}^\top$.
- Describe the linear transformation A such that $\boldsymbol{u} = A\boldsymbol{\theta}$ for $p = 1$.
- Describe the transformation A such that $\boldsymbol{u} = A\boldsymbol{\theta}$ for $p > 1$.

Hint: use the formula

$$u_m^* = \frac{1}{(m-1)!} f^{(m-1)}(x_0), \qquad m = 1, \ldots, p$$

to identify the coefficient u_m^* via $\theta_m^*, \ldots, \theta_p^*$.

4.2.3 *Orthogonal and Orthonormal Design*

Orthogonality of the design matrix Ψ means that the basis vectors $\psi_1, \ldots, \psi_p$ are orthonormal in the sense

$$\boldsymbol{\psi}_m^\top \boldsymbol{\psi}_{m'} = \sum_{i=1}^n \psi_{m,i}\psi_{m',i} = \begin{cases} 0 & \text{if } m \neq m', \\ \lambda_m & \text{if } m = m', \end{cases}$$

for some positive values $\lambda_1, \ldots, \lambda_p$. Equivalently one can write

$$\Psi\Psi^\top = \Lambda = \operatorname{diag}(\lambda_1, \ldots, \lambda_p).$$

This feature of the design is very useful and it essentially simplifies the computation and analysis of the properties of $\tilde{\boldsymbol{\theta}}$. Indeed, $\Psi\Psi^\top = \Lambda$ implies

$$\tilde{\boldsymbol{\theta}} = \Lambda^{-1}\Psi \boldsymbol{Y}, \qquad \tilde{\boldsymbol{f}} = \Psi^\top \tilde{\boldsymbol{\theta}} = \Psi^\top \Lambda^{-1} \Psi \boldsymbol{Y}$$

with $\Lambda^{-1} = \operatorname{diag}(\lambda_1^{-1}, \ldots, \lambda_p^{-1})$. In particular, the first relation means

$$\tilde{\theta}_m = \lambda_m^{-1} \sum_{i=1}^{n} Y_i \psi_{m,i},$$

that is, $\tilde{\theta}_m$ is the scalar product of the data and the basis vector $\boldsymbol{\psi}_m$ for $m = 1, \ldots, p$. The estimate of the response $\boldsymbol{f}$ reads as

$$\tilde{\boldsymbol{f}} = \tilde{\theta}_1 \boldsymbol{\psi}_1 + \ldots + \tilde{\theta}_p \boldsymbol{\psi}_p.$$

Theorem 4.2.1. *Consider the model* $\boldsymbol{Y} = \Psi^\top \boldsymbol{\theta} + \boldsymbol{\varepsilon}$ *with homogeneous errors* $\boldsymbol{\varepsilon}$: $\mathbb{E}\boldsymbol{\varepsilon}\boldsymbol{\varepsilon}^\top = \sigma^2 \boldsymbol{I}_n$. *If the design* Ψ *is orthogonal, that is, if* $\Psi\Psi^\top = \Lambda$ *for a diagonal matrix* Λ, *then the estimated coefficients* $\tilde{\theta}_m$ *are uncorrelated:* $\operatorname{Var}(\tilde{\boldsymbol{\theta}}) = \sigma^2 \Lambda^{-1}$. *Moreover, if* $\boldsymbol{\varepsilon} \sim \mathcal{N}(0, \sigma^2 \boldsymbol{I}_n)$, *then* $\tilde{\boldsymbol{\theta}} \sim \mathcal{N}(\boldsymbol{\theta}^*, \sigma^2 \Lambda^{-1})$.

An important message of this result is that the orthogonal design allows for splitting the original multivariate problem into a collection of independent univariate problems: each coefficient θ_m^* is estimated by $\tilde{\theta}_m$ independently on the remaining coefficients.

The calculus can be further simplified in the case of an orthogonal design with $\Psi\Psi^\top = \boldsymbol{I}_p$. Then one speaks about an *orthonormal design*. This also implies that every basis function (vector) $\boldsymbol{\psi}_m$ is standardized: $\|\boldsymbol{\psi}_m\|^2 = \sum_{i=1}^{n} \psi_{m,i}^2 = 1$. In the case of an orthonormal design, the estimate $\tilde{\boldsymbol{\theta}}$ is particularly simple: $\tilde{\boldsymbol{\theta}} = \Psi \boldsymbol{Y}$. Correspondingly, the target of estimation $\boldsymbol{\theta}^*$ satisfies $\boldsymbol{\theta}^* = \Psi \boldsymbol{f}^*$. In other words, the target is the collection (θ_m^*) of the Fourier coefficients of the underlying function (vector) $\boldsymbol{f}^*$ w.r.t. the basis Ψ while the estimate $\tilde{\boldsymbol{\theta}}$ is the collection of empirical Fourier coefficients $\tilde{\theta}_m$:

$$\theta_m^* = \sum_{i=1}^{n} f_i \psi_{m,i}, \qquad \tilde{\theta}_m = \sum_{i=1}^{n} Y_i \psi_{m,i}$$

An important feature of the orthonormal design is that it preserves the noise homogeneity:

$$\operatorname{Var}(\tilde{\boldsymbol{\theta}}) = \sigma^2 I_p.$$

4.2.4 Spectral Representation

Consider a linear model

$$\boldsymbol{Y} = \Psi^\top \boldsymbol{\theta} + \boldsymbol{\varepsilon} \tag{4.5}$$

with homogeneous errors ε: $\operatorname{Var}(\varepsilon) = \sigma^2 I_n$. The rows of the matrix Ψ can be viewed as basis vectors in $\mathbb{R}^n$ and the product $\Psi^\top \theta$ is a linear combinations of these vectors with the coefficients $(\theta_1, \ldots, \theta_p)$. Effectively linear least squares estimation does a kind of projection of the data onto the subspace generated by the basis functions. This projection is of course invariant w.r.t. a basis transformation within this linear subspace. This fact can be used to reduce the model to the case of an orthogonal design considered in the previous section. Namely, one can always find a linear orthogonal transformation $U : \mathbb{R}^p \to \mathbb{R}^p$ ensuring the orthogonality of the transformed basis. This means that the rows of the matrix $\breve{\Psi} = U\Psi$ are orthogonal and the matrix $\breve{\Psi}\breve{\Psi}^\top$ is diagonal:

$$\breve{\Psi}\breve{\Psi}^\top = U\Psi\Psi^\top U^\top = \Lambda = \operatorname{diag}(\lambda_1, \ldots, \lambda_p).$$

The original model reads after this transformation in the form

$$Y = \breve{\Psi}^\top u + \varepsilon, \qquad \breve{\Psi}\breve{\Psi}^\top = \Lambda,$$

where $u = U\theta \in \mathbb{R}^p$. Within this model, the transformed parameter u can be estimated using the empirical Fourier coefficients $Z_m = \breve{\psi}_m^\top Y$, where $\breve{\psi}_m$ is the mth row of $\breve{\Psi}$, $m = 1, \ldots, p$. The original parameter vector θ can be recovered via the equation $\theta = U^\top u$. This set of equations can be written in the form

$$Z = \Lambda u + \Lambda^{1/2} \xi \tag{4.6}$$

where $Z = \breve{\Psi} Y = U\Psi Y$ is a vector in $\mathbb{R}^p$ and $\xi = \Lambda^{-1/2}\breve{\Psi}\varepsilon = \Lambda^{-1/2}U\Psi\varepsilon \in \mathbb{R}^p$. Equation (4.6) is called the *spectral representation* of the linear model (4.5). The reason is that the basic transformation U can be built by a singular value decomposition of Ψ. This representation is widely used in context of linear inverse problems; see Sect. 4.8.

Theorem 4.2.2. *Consider the model* (4.5) *with homogeneous errors* ε*, that is,* $\mathbb{E}\varepsilon\varepsilon^\top = \sigma^2 I_n$. *Then there exists an orthogonal transform* $U : \mathbb{R}^p \to \mathbb{R}^p$ *leading to the spectral representation* (4.6) *with homogeneous uncorrelated errors* ξ: $\mathbb{E}\xi\xi^\top = \sigma^2 I_p$. *If* $\varepsilon \sim \mathcal{N}(0, \sigma^2 I_n)$*, then the vector* ξ *is normal as well:* $\xi = \mathcal{N}(0, \sigma^2 I_p)$.

Exercise 4.2.3. Prove the result of Theorem 4.2.2.
Hint: select any U ensuring $U^\top \Psi\Psi^\top U = \Lambda$. Then

$$\mathbb{E}\xi\xi^\top = \Lambda^{-1/2}U\Psi\mathbb{E}\varepsilon\varepsilon^\top\Psi^\top U^\top\Lambda^{-1/2} = \sigma^2\Lambda^{-1/2}U^\top\Psi\Psi^\top U\Lambda^{-1/2} = \sigma^2 I_p.$$

A special case of the spectral representation corresponds to the orthonormal design with $\Psi\Psi^\top = I_p$. In this situation, the spectral model reads as $Z = u + \xi$, that is, we simply observe the target u corrupted with a homogeneous noise ξ. Such an equation is often called the *sequence space model* and it is intensively used in the literature for the theoretical study; cf. Sect. 4.7 below.

4.3 Properties of the Response Estimate $\tilde{f}$

This section discusses some properties of the estimate $\tilde{\boldsymbol{f}} = \Psi^\top \tilde{\boldsymbol{\theta}} = \Pi \boldsymbol{Y}$ of the response vector $\boldsymbol{f}^*$. It is worth noting that the first and essential part of the analysis does not rely on the underlying model distribution, only on our parametric assumptions that $\boldsymbol{f} = \Psi^\top \boldsymbol{\theta}^*$ and $\operatorname{Cov}(\boldsymbol{\varepsilon}) = \Sigma = \sigma^2 \boldsymbol{I}_n$. The real model only appears when studying the risk of estimation. We will comment on the cases of misspecified $\boldsymbol{f}$ and Σ.

When $\Sigma = \sigma^2 \boldsymbol{I}_n$, the operator Π in the representation $\tilde{\boldsymbol{f}} = \Pi \boldsymbol{Y}$ of the estimate $\tilde{\boldsymbol{f}}$ reads as

$$\Pi = \Psi^\top \bigl(\Psi \Psi^\top\bigr)^{-1} \Psi. \tag{4.7}$$

First we make use of the linear structure of the model (4.1) and of the estimate $\tilde{\boldsymbol{f}}$ to derive a number of its simple but important properties.

4.3.1 Decomposition into a Deterministic and a Stochastic Component

The model equation $\boldsymbol{Y} = \boldsymbol{f}^* + \boldsymbol{\varepsilon}$ yields

$$\tilde{\boldsymbol{f}} = \Pi \boldsymbol{Y} = \Pi(\boldsymbol{f}^* + \boldsymbol{\varepsilon}) = \Pi \boldsymbol{f}^* + \Pi \boldsymbol{\varepsilon}. \tag{4.8}$$

The first element of this sum, $\Pi \boldsymbol{f}^*$ is purely deterministic, but it depends on the unknown response vector $\boldsymbol{f}^*$. Moreover, it will be shown in the next lemma that $\Pi \boldsymbol{f}^* = \boldsymbol{f}^*$ if the parametric assumption holds and the vector $\boldsymbol{f}^*$ indeed can be represented as $\Psi^\top \boldsymbol{\theta}^*$. The second element is stochastic as a linear transformation of the stochastic vector $\boldsymbol{\varepsilon}$ but is independent of the model response $\boldsymbol{f}^*$. The properties of the estimate $\tilde{\boldsymbol{f}}$ heavily rely on the properties of the linear operator Π from (4.7) which we collect in the next section.

4.3.2 Properties of the Operator Π

Let $\boldsymbol{\psi}_1, \ldots, \boldsymbol{\psi}_p$ be the columns of the matrix $\Psi^\top$. These are the vectors in $\mathbb{R}^n$ also called *the basis vectors*.

Lemma 4.3.1. *Let the matrix $\Psi\Psi^\top$ be non-degenerate. Then the operator Π fulfills the following conditions:*

(i) *Π is symmetric (self-adjoint), that is, $\Pi^\top = \Pi$.*
(ii) *Π is a projector in $\mathbb{R}^n$, i.e. $\Pi^\top \Pi = \Pi^2 = \Pi$ and $\Pi(\mathbf{1}_n - \Pi) = 0$, where $\mathbf{1}_n$ means the unity operator in $\mathbb{R}^n$.*
(iii) *For an arbitrary vector v from $\mathbb{R}^n$, it holds $\|v\|^2 = \|\Pi v\|^2 + \|v - \Pi v\|^2$.*
(iv) *The trace of Π is equal to the dimension of its image, $\operatorname{tr} \Pi = p$.*
(v) *Π projects the linear space $\mathbb{R}^n$ on the linear subspace $\mathrm{L}_p = \langle \boldsymbol{\psi}_1, \ldots, \boldsymbol{\psi}_p \rangle$, which is spanned by the basis vectors $\boldsymbol{\psi}_1, \ldots \boldsymbol{\psi}_p$, that is,*

$$\|\boldsymbol{f}^* - \Pi \boldsymbol{f}^*\| = \inf_{\boldsymbol{g} \in \mathrm{L}_p} \|\boldsymbol{f}^* - \boldsymbol{g}\|.$$

(vi) *The matrix Π can be represented in the form*

$$\Pi = U^\top \Lambda_p U$$

where U is an orthonormal matrix and Λ_p is a diagonal matrix with the first p diagonal elements equal to 1 and the others equal to zero:

$$\Lambda_p = \operatorname{diag}\{\underbrace{1, \ldots, 1}_{p}, \underbrace{0, \ldots, 0}_{n-p}\}.$$

Proof. It holds

$$\left\{\Psi^\top \left(\Psi\Psi^\top\right)^{-1} \Psi\right\}^\top = \Psi^\top \left(\Psi\Psi^\top\right)^{-1} \Psi$$

and

$$\Pi^2 = \Psi^\top \left(\Psi\Psi^\top\right)^{-1} \Psi\Psi^\top \left(\Psi\Psi^\top\right)^{-1} \Psi = \Psi^\top \left(\Psi\Psi^\top\right)^{-1} \Psi = \Pi,$$

which proves the first two statements of the lemma. The third one follows directly from the first two. Next,

$$\operatorname{tr} \Pi = \operatorname{tr} \Psi^\top \left(\Psi\Psi^\top\right)^{-1} \Psi = \operatorname{tr} \Psi\Psi^\top \left(\Psi\Psi^\top\right)^{-1} = \operatorname{tr} \boldsymbol{I}_p = p.$$

The second property means that Π is a projector in $\mathbb{R}^n$ and the fourth one means that the dimension of its image space is equal to p. The basis vectors $\boldsymbol{\psi}_1, \ldots, \boldsymbol{\psi}_p$ are the rows of the matrix Ψ. It is clear that

$$\Pi \Psi^\top = \Psi^\top \left(\Psi\Psi^\top\right)^{-1} \Psi\Psi^\top = \Psi^\top.$$

Therefore, the vectors $\boldsymbol{\psi}_m$ are invariants of the operator Π and in particular, all these vectors belong to the image space of this operator. If now $\boldsymbol{g}$ is a vector in L_p, then it can be represented as $g = c_1\psi_1 + \ldots + c_p\psi_p$ and therefore, $\Pi g = g$ and $\Pi L_p = L_p$. Finally, the non-singularity of the matrix $\Psi\Psi^\top$ means that the vectors $\psi_1, \ldots, \psi_p$ forming the rows of Ψ are linearly independent. Therefore, the space L_p spanned by the vectors $\psi_1, \ldots, \psi_p$ is of dimension p, and hence it coincides with the image space of the operation Π.

The last property is the usual diagonal decomposition of a projector.

Exercise 4.3.1. Consider the case of an orthogonal design with $\Psi\Psi^\top = I_p$. Specify the projector Π of Lemma 4.3.1 for this situation, particularly its decomposition from (vi).

4.3.3 Quadratic Loss and Risk of the Response Estimation

In this section we study the quadratic risk of estimating the response f^*. The reason for studying the quadratic risk of estimating the response f^* will be made clear when we discuss the properties of the fitted likelihood in the next section.

The loss $\wp(\tilde{f}, f^*)$ of the estimate $\tilde{f}$ can be naturally defined as the squared norm of the difference $\tilde{f} - f^*$:

$$\wp(\tilde{f}, f^*) = \|\tilde{f} - f^*\|^2 = \sum_{i=1}^{n} |f_i - \tilde{f}_i|^2.$$

Correspondingly, the quadratic risk of the estimate $\tilde{f}$ is the mean of this loss

$$\mathcal{R}(\tilde{f}) = \mathbb{E}\wp(\tilde{f}, f^*) = \mathbb{E}\big[(\tilde{f} - f^*)^\top(\tilde{f} - f^*)\big]. \tag{4.9}$$

The next result describes the loss and risk decomposition for two cases: when the parametric assumption $f^* = \Psi^\top\theta^*$ is correct and in the general case.

Theorem 4.3.1. *Suppose that the errors ε_i from (4.1) are independent with $\mathbb{E}\,\varepsilon_i = 0$ and $\mathbb{E}\,\varepsilon_i^2 = \sigma^2$, i.e. $\Sigma = \sigma^2 I_n$. Then the loss $\wp(\tilde{f}, f^*) = \|\Pi Y - f^*\|^2$ and the risk $\mathcal{R}(\tilde{f})$ of the LSE $\tilde{f}$ fulfill*

$$\begin{aligned}\wp(\tilde{f}, f^*) &= \|f^* - \Pi f^*\|^2 + \|\Pi\varepsilon\|^2,\\ \mathcal{R}(\tilde{f}) &= \|f^* - \Pi f^*\|^2 + p\sigma^2.\end{aligned}$$

Moreover, if $f^* = \Psi^\top\theta^*$, then

$$\begin{aligned}\wp(\tilde{f}, f^*) &= \|\Pi\varepsilon\|^2,\\ \mathcal{R}(\tilde{f}) &= p\sigma^2.\end{aligned}$$

Proof. We apply (4.9) and the decomposition (4.8) of the estimate $\tilde{f}$. It follows

$$\begin{aligned}\wp(\tilde{f}, f^*) &= \|\tilde{f} - f^*\|^2 = \|f^* - \Pi f^* - \Pi\varepsilon\|^2\\ &= \|f^* - \Pi f^*\|^2 + 2(f^* - \Pi f^*)^\top \Pi\varepsilon + \|\Pi\varepsilon\|^2.\end{aligned}$$

This implies the decomposition for the loss of $\tilde{f}$ by Lemma 4.3.1(ii). Next we compute the mean of $\|\Pi\varepsilon\|^2$ applying again Lemma 4.3.1. Indeed

$$\begin{aligned}\mathbb{E}\|\Pi\varepsilon\|^2 &= \mathbb{E}(\Pi\varepsilon)^\top \Pi\varepsilon = \mathbb{E}\,\mathrm{tr}\bigl\{\Pi\varepsilon(\Pi\varepsilon)^\top\bigr\} = \mathbb{E}\,\mathrm{tr}\bigl(\Pi\varepsilon\varepsilon^\top \Pi^\top\bigr)\\ &= \mathrm{tr}\bigl\{\Pi\mathbb{E}(\varepsilon\varepsilon^\top)\Pi\bigr\} = \sigma^2\,\mathrm{tr}(\Pi^2) = p\sigma^2.\end{aligned}$$

Now consider the case when $f^* = \Psi^\top\theta^*$. By Lemma 4.3.1 $f^* = \Pi f^*$ and and the last two statements of the theorem clearly follow.

4.3.4 Misspecified "Colored Noise"

Here we briefly comment on the case when ε is not a white noise. So, our assumption about the errors ε_i is that they are uncorrelated and homogeneous, that is, $\Sigma = \sigma^2 I_n$ while the true covariance matrix is given by Σ_0. Many properties of the estimate $\tilde{f} = \Pi Y$ which are simply based on the linearity of the model (4.1) and of the estimate $\tilde{f}$ itself continue to apply. In particular, the loss $\wp(\tilde{f}, f^*) = \|\tilde{f} - f^*\|^2$ can again be decomposed as

$$\|\tilde{f} - f^*\|^2 = \|f^* - \Pi f^*\|^2 + \|\Pi\varepsilon\|^2.$$

Theorem 4.3.2. *Suppose that* $\mathbb{E}\varepsilon = 0$ *and* $\mathrm{Var}(\varepsilon) = \Sigma_0$. *Then the loss* $\wp(\tilde{f}, f)$ *and the risk* $\mathcal{R}(\tilde{f})$ *of the LSE* $\tilde{f}$ *fulfill*

$$\begin{aligned}\wp(\tilde{f}, f^*) &= \|f^* - \Pi f^*\|^2 + \|\Pi\varepsilon\|^2,\\ \mathcal{R}(\tilde{f}) &= \|f^* - \Pi f^*\|^2 + \mathrm{tr}\bigl(\Pi\Sigma_0\Pi\bigr).\end{aligned}$$

Moreover, if $f^* = \Psi^\top\theta^*$, *then*

$$\begin{aligned}\wp(\tilde{f}, f^*) &= \|\Pi\varepsilon\|^2,\\ \mathcal{R}(\tilde{f}) &= \mathrm{tr}\bigl(\Pi\Sigma_0\Pi\bigr).\end{aligned}$$

Proof. The decomposition of the loss from Theorem 4.3.1 only relies on the geometric properties of the projector Π and does not use the covariance structure of the noise. Hence, it only remains to check the expectation of $\|\Pi\varepsilon\|^2$. Observe that

$$\mathbb{E}\|\Pi\varepsilon\|^2 = \mathbb{E}\,\mathrm{tr}\bigl[\Pi\varepsilon(\Pi\varepsilon)^\top\bigr] = \mathrm{tr}\bigl[\Pi\mathbb{E}(\varepsilon\varepsilon^\top)\Pi\bigr] = \mathrm{tr}\bigl(\Pi\Sigma_0\Pi\bigr)$$

as required.

4.4 Properties of the MLE $\tilde{\theta}$

In this section we focus on the properties of the quasi MLE $\tilde{\theta}$ built for the idealized linear Gaussian model $Y = \Psi^\top\theta^* + \varepsilon$ with $\varepsilon \sim \mathcal{N}(0, \sigma^2 I_n)$. As in the previous section, we do not assume the parametric structure of the underlying model and consider a more general model $Y = f^* + \varepsilon$ with an unknown vector f^* and errors ε with zero mean and covariance matrix Σ_0. Due to (4.3), it holds $\tilde{\theta} = \mathcal{S}Y$ with $\mathcal{S} = \bigl(\Psi\Psi^\top\bigr)^{-1}\Psi$. An important feature of this estimate is its linear dependence on the data. The linear model equation $Y = f^* + \varepsilon$ and linear structure of the estimate $\tilde{\theta} = \mathcal{S}Y$ allow us for decomposing the vector $\tilde{\theta}$ into a deterministic and stochastic terms:

$$\tilde{\theta} = \mathcal{S}Y = \mathcal{S}\bigl(f^* + \varepsilon\bigr) = \mathcal{S}f^* + \mathcal{S}\varepsilon. \tag{4.10}$$

The first term $\mathcal{S}f^*$ is deterministic but depends on the unknown vector f^* while the second term $\mathcal{S}\varepsilon$ is stochastic but it does not involve the model response f^*. Below we study the properties of each component separately.

4.4.1 Properties of the Stochastic Component

The next result describes the distributional properties of the stochastic component $\delta = \mathcal{S}\varepsilon$ for $\mathcal{S} = \bigl(\Psi\Psi^\top\bigr)^{-1}\Psi$ and thus, of the estimate $\tilde{\theta}$.

Theorem 4.4.1. *Assume* $Y = f^* + \varepsilon$ *with* $\mathbb{E}\varepsilon = 0$ *and* $\mathrm{Var}(\varepsilon) = \Sigma_0$. *The stochastic component* $\delta = \mathcal{S}\varepsilon$ *in* (4.10) *fulfills*

$$\mathbb{E}\delta = 0, \qquad W^2 \stackrel{\text{def}}{=} \mathrm{Var}(\delta) = \mathcal{S}\Sigma_0\mathcal{S}^\top, \qquad \mathbb{E}\|\delta\|^2 = \mathrm{tr}\, W^2 = \mathrm{tr}\bigl(\mathcal{S}\Sigma_0\mathcal{S}^\top\bigr).$$

Moreover, if $\Sigma = \Sigma_0 = \sigma^2 I_n$, *then*

$$W^2 = \sigma^2\bigl(\Psi\Psi^\top\bigr)^{-1}, \qquad \mathbb{E}\|\delta\|^2 = \mathrm{tr}(W^2) = \sigma^2\,\mathrm{tr}\bigl[\bigl(\Psi\Psi^\top\bigr)^{-1}\bigr]. \tag{4.11}$$

Similarly for the estimate $\tilde{\theta}$ *it holds*

$$\mathbb{E}\tilde{\theta} = \mathcal{S}f^*, \qquad \mathrm{Var}(\tilde{\theta}) = W^2.$$

If the errors ε are Gaussian, then both $\boldsymbol{\delta}$ and $\tilde{\boldsymbol{\theta}}$ are Gaussian as well:

$$\boldsymbol{\delta} \sim \mathcal{N}(0, W^2) \qquad \tilde{\boldsymbol{\theta}} \sim \mathcal{N}(\mathcal{S}\boldsymbol{f}^*, W^2).$$

Proof. For the variance W^2 of $\boldsymbol{\delta}$ holds

$$\operatorname{Var}(\boldsymbol{\delta}) = \mathbb{E}\boldsymbol{\delta}\boldsymbol{\delta}^\top = \mathbb{E}\mathcal{S}\boldsymbol{\varepsilon}\boldsymbol{\varepsilon}^\top\mathcal{S}^\top = \mathcal{S}\Sigma_0\mathcal{S}^\top.$$

Next we use that $\mathbb{E}\|\boldsymbol{\delta}\|^2 = \mathbb{E}\boldsymbol{\delta}^\top\boldsymbol{\delta} = \mathbb{E}\operatorname{tr}(\boldsymbol{\delta}\boldsymbol{\delta}^\top) = \operatorname{tr} W^2$. If $\Sigma = \Sigma_0 = \sigma^2\boldsymbol{I}_n$, then (4.11) follows by simple algebra.

If $\boldsymbol{\varepsilon}$ is a Gaussian vector, then $\boldsymbol{\delta}$ as its linear transformation is Gaussian as well. The properties of $\tilde{\boldsymbol{\theta}}$ follow directly from the decomposition (4.10).

With $\Sigma_0 \neq \sigma^2\boldsymbol{I}_n$, the variance W^2 can be represented as

$$W^2 = \bigl(\Psi\Psi^\top\bigr)^{-1}\Psi\Sigma_0\Psi^\top\bigl(\Psi\Psi^\top\bigr)^{-1}.$$

Exercise 4.4.1. Let $\boldsymbol{\delta}$ be the stochastic component of $\tilde{\boldsymbol{\theta}}$ built for the misspecified linear model $\boldsymbol{Y} = \Psi^\top\boldsymbol{\theta}^* + \boldsymbol{\varepsilon}$ with $\operatorname{Var}(\boldsymbol{\varepsilon}) = \Sigma$. Let also the true noise variance is Σ_0. Then $\operatorname{Var}(\tilde{\boldsymbol{\theta}}) = W^2$ with

$$W^2 = \bigl(\Psi\Sigma^{-1}\Psi^\top\bigr)^{-1}\Psi\Sigma^{-1}\Sigma_0\Sigma^{-1}\Psi^\top\bigl(\Psi\Sigma^{-1}\Psi^\top\bigr)^{-1}. \tag{4.12}$$

The main finding in the presented study is that the stochastic part $\boldsymbol{\delta} = \mathcal{S}\boldsymbol{\varepsilon}$ of the estimate $\tilde{\boldsymbol{\theta}}$ is completely independent of the structure of the vector $\boldsymbol{f}^*$. In other words, the behavior of the stochastic component $\boldsymbol{\delta}$ does not change even if the linear parametric assumption is misspecified.

4.4.2 Properties of the Deterministic Component

Now we study the deterministic term starting with the parametric situation $\boldsymbol{f}^* = \Psi^\top\boldsymbol{\theta}^*$. Here we only specify the results for the case 1 with $\Sigma = \sigma^2\boldsymbol{I}_n$.

Theorem 4.4.2. *Let $\boldsymbol{f}^* = \Psi^\top\boldsymbol{\theta}^*$. Then $\tilde{\boldsymbol{\theta}} = \mathcal{S}\boldsymbol{Y}$ with $\mathcal{S} = \bigl(\Psi\Psi^\top\bigr)^{-1}\Psi$ is unbiased, that is, $\mathbb{E}\tilde{\boldsymbol{\theta}} = \mathcal{S}\boldsymbol{f}^* = \boldsymbol{\theta}^*$.*

Proof. For the proof, just observe that $\mathcal{S}\boldsymbol{f}^* = \bigl(\Psi\Psi^\top\bigr)^{-1}\Psi\Psi^\top\boldsymbol{\theta}^* = \boldsymbol{\theta}^*$.

Now we briefly discuss what happens when the linear parametric assumption is not fulfilled, that is, $\boldsymbol{f}^*$ cannot be represented as $\Psi^\top\boldsymbol{\theta}^*$. In this case it is not yet clear what $\tilde{\boldsymbol{\theta}}$ really estimates. The answer is given in the context of the general theory of minimum contrast estimation. Namely, define $\boldsymbol{\theta}^*$ as the point which maximizes the expectation of the (quasi) log-likelihood $L(\boldsymbol{\theta})$:

$$\boldsymbol{\theta}^* = \operatorname*{argmax}_{\boldsymbol{\theta}} \mathbb{E}L(\boldsymbol{\theta}). \tag{4.13}$$

Theorem 4.4.3. *The solution* $\boldsymbol{\theta}^*$ *of the optimization problem* (4.13) *is given by*

$$\boldsymbol{\theta}^* = \mathcal{S} \boldsymbol{f}^* = \left(\Psi \Psi^\top\right)^{-1} \Psi \boldsymbol{f}^*.$$

Moreover,

$$\Psi^\top \boldsymbol{\theta}^* = \Pi \boldsymbol{f}^* = \Psi^\top \left(\Psi \Psi^\top\right)^{-1} \Psi \boldsymbol{f}^*.$$

In particular, if $\boldsymbol{f}^* = \Psi^\top \boldsymbol{\theta}^*$*, then* $\boldsymbol{\theta}^*$ *follows* (4.13).

Proof. The use of the model equation $\boldsymbol{Y} = \boldsymbol{f}^* + \boldsymbol{\varepsilon}$ and of the properties of the stochastic component $\boldsymbol{\delta}$ yield by simple algebra

$$\begin{aligned}
\operatorname*{argmax}_{\boldsymbol{\theta}} \mathbb{E} L(\boldsymbol{\theta}) &= \operatorname*{argmin}_{\boldsymbol{\theta}} \mathbb{E}\left(\boldsymbol{f}^* - \Psi^\top \boldsymbol{\theta} + \boldsymbol{\varepsilon}\right)^\top \left(\boldsymbol{f}^* - \Psi^\top \boldsymbol{\theta} + \boldsymbol{\varepsilon}\right) \\
&= \operatorname*{argmin}_{\boldsymbol{\theta}} \left\{(\boldsymbol{f}^* - \Psi^\top \boldsymbol{\theta})^\top (\boldsymbol{f}^* - \Psi^\top \boldsymbol{\theta}) + \mathbb{E}(\boldsymbol{\varepsilon}^\top \boldsymbol{\varepsilon})\right\} \\
&= \operatorname*{argmin}_{\boldsymbol{\theta}} \left\{(\boldsymbol{f}^* - \Psi^\top \boldsymbol{\theta})^\top (\boldsymbol{f}^* - \Psi^\top \boldsymbol{\theta})\right\}.
\end{aligned}$$

Differentiating w.r.t. $\boldsymbol{\theta}$ leads to the equation

$$\Psi(\boldsymbol{f}^* - \Psi^\top \boldsymbol{\theta}) = 0$$

and the solution $\boldsymbol{\theta}^* = \left(\Psi \Psi^\top\right)^{-1} \Psi \boldsymbol{f}^*$ which is exactly the expected value of $\tilde{\boldsymbol{\theta}}$ by Theorem 4.4.1.

Exercise 4.4.2. State the result of Theorems 4.4.2 and 4.4.3 for the MLE $\tilde{\boldsymbol{\theta}}$ built in the model $\boldsymbol{Y} = \Psi^\top \boldsymbol{\theta}^* + \boldsymbol{\varepsilon}$ with $\operatorname{Var}(\boldsymbol{\varepsilon}) = \Sigma$.
Hint: check that the statements continue to apply with $\mathcal{S} = \left(\Psi \Sigma^{-1} \Psi^\top\right)^{-1} \Psi \Sigma^{-1}$.

The last results and the decomposition (4.10) explain the behavior of the estimate $\tilde{\boldsymbol{\theta}}$ in a very general situation. The considered model is $\boldsymbol{Y} = \boldsymbol{f}^* + \boldsymbol{\varepsilon}$. We assume a linear parametric structure and independent homogeneous noise. The estimation procedure means in fact a kind of projection of the data $\boldsymbol{Y}$ on a p-dimensional linear subspace in $\mathbb{R}^n$ spanned by the given basis vectors $\boldsymbol{\psi}_1, \ldots, \boldsymbol{\psi}_p$. This projection, as a linear operator, can be decomposed into a projection of the deterministic vector $\boldsymbol{f}^*$ and a projection of the random noise $\boldsymbol{\varepsilon}$. If the linear parametric assumption $\boldsymbol{f}^* \in \left\langle \boldsymbol{\psi}_1, \ldots, \boldsymbol{\psi}_p \right\rangle$ is correct, that is, $\boldsymbol{f}^* = \theta_1^* \boldsymbol{\psi}_1 + \ldots + \theta_p^* \boldsymbol{\psi}_p$, then this projection keeps $\boldsymbol{f}^*$ unchanged and only the random noise is reduced via this projection. If $\boldsymbol{f}^*$ cannot be exactly expanded using the basis $\boldsymbol{\psi}_1, \ldots, \boldsymbol{\psi}_p$, then the procedure recovers the projection of $\boldsymbol{f}^*$ onto this subspace. The latter projection can be written as $\Psi^\top \boldsymbol{\theta}^*$ and the vector $\boldsymbol{\theta}^*$ can be viewed as the target of estimation.

4.4.3 *Risk of Estimation: R-Efficiency*

This section briefly discusses how the obtained properties of the estimate $\tilde{\boldsymbol{\theta}}$ can be used to evaluate the risk of estimation. A particularly important question is the optimality of the MLE $\tilde{\boldsymbol{\theta}}$. The main result of the section claims that $\tilde{\boldsymbol{\theta}}$ is R-efficient if the model is correctly specified and is not if there is a misspecification.

We start with the case of a correct parametric specification $\boldsymbol{Y} = \Psi^\top \boldsymbol{\theta}^* + \boldsymbol{\varepsilon}$, that is, the linear parametric assumption $\boldsymbol{f}^* = \Psi^\top \boldsymbol{\theta}^*$ is exactly fulfilled and the noise $\boldsymbol{\varepsilon}$ is homogeneous: $\boldsymbol{\varepsilon} \sim \mathcal{N}(0, \sigma^2 \boldsymbol{I}_n)$. Later we extend the result to the case when the LPA $\boldsymbol{f}^* = \Psi^\top \boldsymbol{\theta}^*$ is not fulfilled and to the case when the noise is not homogeneous but still correctly specified. Finally we discuss the case when the noise structure is misspecified.

Under LPA $\boldsymbol{Y} = \Psi^\top \boldsymbol{\theta}^* + \boldsymbol{\varepsilon}$ with $\boldsymbol{\varepsilon} \sim \mathcal{N}(0, \sigma^2 \boldsymbol{I}_n)$, the estimate $\tilde{\boldsymbol{\theta}}$ is also normal with mean $\boldsymbol{\theta}^*$ and the variance $W^2 = \sigma^2 \mathcal{S}\mathcal{S}^\top = \sigma^2 \bigl(\Psi \Psi^\top\bigr)^{-1}$. Define a $p \times p$ symmetric matrix D by the equation

$$D^2 = \frac{1}{\sigma^2} \sum_{i=1}^{n} \Psi_i \Psi_i^\top = \frac{1}{\sigma^2} \Psi \Psi^\top .$$

Clearly $W^2 = D^{-2}$.

Now we show that $\tilde{\boldsymbol{\theta}}$ is R-efficient. Actually this fact can be derived from the Cramér–Rao Theorem because the Gaussian model is a special case of an exponential family. However, we check this statement directly by computing the Cramér–Rao efficiency bound. Recall that the Fisher information matrix $\mathbb{F}(\boldsymbol{\theta})$ for the log-likelihood $L(\boldsymbol{\theta})$ is defined as the variance of $\nabla L(\boldsymbol{\theta})$ under $\mathbb{P}_{\boldsymbol{\theta}}$.

Theorem 4.4.4 (Gauss–Markov). *Let* $\boldsymbol{Y} = \Psi^\top \boldsymbol{\theta}^* + \boldsymbol{\varepsilon}$ *with* $\boldsymbol{\varepsilon} \sim \mathcal{N}(0, \sigma^2 \boldsymbol{I}_n)$*. Then* $\tilde{\boldsymbol{\theta}}$ *is R-efficient estimate of* $\boldsymbol{\theta}^*$*:* $\mathbb{E}\tilde{\boldsymbol{\theta}} = \boldsymbol{\theta}^*$,

$$\mathbb{E}\bigl[(\tilde{\boldsymbol{\theta}} - \boldsymbol{\theta}^*)(\tilde{\boldsymbol{\theta}} - \boldsymbol{\theta}^*)^\top\bigr] = \operatorname{Var}\bigl(\tilde{\boldsymbol{\theta}}\bigr) = D^{-2},$$

and for any unbiased linear estimate $\hat{\boldsymbol{\theta}}$ *satisfying* $\mathbb{E}_{\boldsymbol{\theta}} \hat{\boldsymbol{\theta}} \equiv \boldsymbol{\theta}$*, it holds*

$$\operatorname{Var}\bigl(\hat{\boldsymbol{\theta}}\bigr) \ge \operatorname{Var}\bigl(\tilde{\boldsymbol{\theta}}\bigr) = D^{-2}.$$

Proof. Theorems 4.4.1 and 4.4.2 imply that $\tilde{\boldsymbol{\theta}} \sim \mathcal{N}(\boldsymbol{\theta}^*, W^2)$ with $W^2 = \sigma^2 (\Psi \Psi^\top)^{-1} = D^{-2}$. Next we show that for any $\boldsymbol{\theta}$

$$\operatorname{Var}\bigl[\nabla L(\boldsymbol{\theta})\bigr] = D^2,$$

that is, the Fisher information does not depend on the model function $\boldsymbol{f}^*$. The log-likelihood $L(\boldsymbol{\theta})$ for the model $\boldsymbol{Y} \sim \mathcal{N}(\Psi^\top \boldsymbol{\theta}^*, \sigma^2 \boldsymbol{I}_n)$ reads as

$$L(\boldsymbol{\theta}) = -\frac{1}{2\sigma^2}(\boldsymbol{Y} - \Psi^\top\boldsymbol{\theta})^\top(\boldsymbol{Y} - \Psi^\top\boldsymbol{\theta}) - \frac{n}{2}\log(2\pi\sigma^2).$$

This yields for its gradient $\nabla L(\boldsymbol{\theta})$:

$$\nabla L(\boldsymbol{\theta}) = \sigma^{-2}\Psi(\boldsymbol{Y} - \Psi^\top\boldsymbol{\theta})$$

and in view of $\operatorname{Var}(\boldsymbol{Y}) = \Sigma = \sigma^2 \boldsymbol{I}_n$, it holds

$$\operatorname{Var}\bigl[\nabla L(\boldsymbol{\theta})\bigr] = \sigma^{-4}\Psi\operatorname{Var}(\boldsymbol{Y})\Psi^\top = \sigma^{-2}\Psi\Psi^\top$$

as required.

The R-efficiency $\tilde{\boldsymbol{\theta}}$ follows from the Cramér–Rao efficiency bound because $\{\operatorname{Var}(\tilde{\boldsymbol{\theta}})\}^{-1} = \operatorname{Var}\{\nabla L(\boldsymbol{\theta})\}$. However, we present an independent proof of this fact. Actually we prove a sharper result that the variance of a linear unbiased estimate $\hat{\boldsymbol{\theta}}$ coincides with the variance of $\tilde{\boldsymbol{\theta}}$ only if $\hat{\boldsymbol{\theta}}$ coincides almost surely with $\tilde{\boldsymbol{\theta}}$, otherwise it is larger. The idea of the proof is quite simple. Consider the difference $\hat{\boldsymbol{\theta}} - \tilde{\boldsymbol{\theta}}$ and show that the condition $\mathbb{E}\hat{\boldsymbol{\theta}} = \mathbb{E}\tilde{\boldsymbol{\theta}} = \boldsymbol{\theta}^*$ implies orthogonality $\mathbb{E}\{\tilde{\boldsymbol{\theta}}(\hat{\boldsymbol{\theta}} - \tilde{\boldsymbol{\theta}})^\top\} = 0$. This, in turns, implies $\operatorname{Var}(\hat{\boldsymbol{\theta}}) = \operatorname{Var}(\tilde{\boldsymbol{\theta}}) + \operatorname{Var}(\hat{\boldsymbol{\theta}} - \tilde{\boldsymbol{\theta}}) \ge \operatorname{Var}(\tilde{\boldsymbol{\theta}})$. So, it remains to check the orthogonality of $\tilde{\boldsymbol{\theta}}$ and $\hat{\boldsymbol{\theta}} - \tilde{\boldsymbol{\theta}}$. Let $\hat{\boldsymbol{\theta}} = A\boldsymbol{Y}$ for a $p \times n$ matrix A and $\mathbb{E}_{\boldsymbol{\theta}}\hat{\boldsymbol{\theta}} = \boldsymbol{\theta}$ and all $\boldsymbol{\theta}$. These two equalities and $\mathbb{E}\boldsymbol{Y} = \Psi^\top\boldsymbol{\theta}^*$ imply that $A\Psi^\top\boldsymbol{\theta}^* \equiv \boldsymbol{\theta}^*$, i.e. $A\Psi^\top$ is the identity $p \times p$ matrix. The same is true for $\tilde{\boldsymbol{\theta}} = \mathcal{S}\boldsymbol{Y}$ yielding $\mathcal{S}\Psi^\top = \boldsymbol{I}_p$. Next, in view of $\mathbb{E}\hat{\boldsymbol{\theta}} = \mathbb{E}\tilde{\boldsymbol{\theta}} = \boldsymbol{\theta}^*$

$$\mathbb{E}\{(\hat{\boldsymbol{\theta}} - \tilde{\boldsymbol{\theta}})\tilde{\boldsymbol{\theta}}^\top\} = \mathbb{E}(A - \mathcal{S})\boldsymbol{\varepsilon}\boldsymbol{\varepsilon}^\top\mathcal{S}^\top = \sigma^2(A - \mathcal{S})\Psi^\top(\Psi\Psi^\top)^{-1} = 0,$$

and the assertion follows.

Exercise 4.4.3. Check the details of the proof of the theorem. Show that the statement $\operatorname{Var}(\hat{\boldsymbol{\theta}}) \ge \operatorname{Var}(\tilde{\boldsymbol{\theta}})$ only uses that $\hat{\boldsymbol{\theta}}$ is unbiased and that $\mathbb{E}\boldsymbol{Y} = \Psi^\top\boldsymbol{\theta}^*$ and $\operatorname{Var}(\boldsymbol{Y}) = \sigma^2\boldsymbol{I}_n$.

Exercise 4.4.4. Compute $\nabla^2 L(\boldsymbol{\theta})$. Check that it is non-random, does not depend on $\boldsymbol{\theta}$, and fulfills for every $\boldsymbol{\theta}$ the identity

$$\nabla^2 L(\boldsymbol{\theta}) \equiv -\operatorname{Var}\bigl[\nabla L(\boldsymbol{\theta})\bigr] = -D^2.$$

4.4.3.1 A Colored Noise

The majority of the presented results continue to apply in the case of heterogeneous and even dependent noise with $\operatorname{Var}(\boldsymbol{\varepsilon}) = \Sigma_0$. The key facts behind this extension are the decomposition (4.10) and the properties of the stochastic component $\boldsymbol{\delta}$ from Sect. 4.4.1: $\boldsymbol{\delta} \sim \mathcal{N}(0, W^2)$. In the case of a colored noise, the definition of W and D is changed for

$$D^2 \stackrel{\text{def}}{=} W^{-2} = \Psi \Sigma_0^{-1} \Psi^\top.$$

Exercise 4.4.5. State and prove the analog of Theorem 4.4.4 for the colored noise $\boldsymbol{\varepsilon} \sim \mathcal{N}(0, \Sigma_0)$.

4.4.3.2 A Misspecified LPA

An interesting feature of our results so far is that they equally apply for the correct linear specification $\boldsymbol{f}^* = \Psi^\top \boldsymbol{\theta}^*$ and for the case when the identity $\boldsymbol{f}^* = \Psi^\top \boldsymbol{\theta}$ is not precisely fulfilled whatever $\boldsymbol{\theta}$ is taken. In this situation the target of analysis is the vector $\boldsymbol{\theta}^*$ describing the best linear approximation of $\boldsymbol{f}^*$ by $\Psi^\top \boldsymbol{\theta}$. We already know from the results of Sects. 4.4.1 and 4.4.2 that the estimate $\tilde{\boldsymbol{\theta}}$ is also normal with mean $\boldsymbol{\theta}^* = \mathcal{S}\boldsymbol{f}^* = \bigl(\Psi\Psi^\top\bigr)^{-1}\Psi \boldsymbol{f}^*$ and the variance $W^2 = \sigma^2 \mathcal{S}\mathcal{S}^\top = \sigma^2 \bigl(\Psi\Psi^\top\bigr)^{-1}$.

Theorem 4.4.5. *Assume $\boldsymbol{Y} = \boldsymbol{f}^* + \boldsymbol{\varepsilon}$ with $\boldsymbol{\varepsilon} \sim \mathcal{N}(0, \sigma^2 \boldsymbol{I}_n)$. Let $\boldsymbol{\theta}^* = \mathcal{S}\boldsymbol{f}^*$. Then $\tilde{\boldsymbol{\theta}}$ is R-efficient estimate of $\boldsymbol{\theta}^*$: $\mathbb{E}\tilde{\boldsymbol{\theta}} = \boldsymbol{\theta}^*$,*

$$\mathbb{E}\bigl[(\tilde{\boldsymbol{\theta}} - \boldsymbol{\theta}^*)(\tilde{\boldsymbol{\theta}} - \boldsymbol{\theta}^*)^\top\bigr] = \operatorname{Var}(\tilde{\boldsymbol{\theta}}) = D^{-2},$$

and for any unbiased linear estimate $\hat{\boldsymbol{\theta}}$ satisfying $\mathbb{E}_{\boldsymbol{\theta}} \hat{\boldsymbol{\theta}} \equiv \boldsymbol{\theta}$, it holds

$$\operatorname{Var}(\hat{\boldsymbol{\theta}}) \ge \operatorname{Var}(\tilde{\boldsymbol{\theta}}) = D^{-2}.$$

Proof. The proofs only utilize that $\tilde{\boldsymbol{\theta}} \sim \mathcal{N}(\boldsymbol{\theta}^*, W^2)$ with $W^2 = D^{-2}$. The only small remark concerns the equality $\operatorname{Var}\bigl[\nabla L(\boldsymbol{\theta})\bigr] = D^2$ from Theorem 4.4.4.

Exercise 4.4.6. Check the identity $\operatorname{Var}\bigl[\nabla L(\boldsymbol{\theta})\bigr] = D^2$ from Theorem 4.4.4 for $\boldsymbol{\varepsilon} \sim \mathcal{N}(0, \Sigma_0)$.

4.4.4 The Case of a Misspecified Noise

Here we again consider the linear parametric assumption $\boldsymbol{Y} = \Psi^\top \boldsymbol{\theta}^* + \boldsymbol{\varepsilon}$. However, contrary to the previous section, we admit that the noise $\boldsymbol{\varepsilon}$ is not homogeneous normal: $\boldsymbol{\varepsilon} \sim \mathcal{N}(0, \Sigma_0)$ while our estimation procedure is the quasi MLE based on the assumption of noise homogeneity $\boldsymbol{\varepsilon} \sim \mathcal{N}(0, \sigma^2 \boldsymbol{I}_n)$. We already know that the estimate $\tilde{\boldsymbol{\theta}}$ is unbiased with mean $\boldsymbol{\theta}^*$ and variance $W^2 = \mathcal{S}\Sigma_0\mathcal{S}^\top$, where $\mathcal{S} = \bigl(\Psi\Psi^\top\bigr)^{-1}\Psi$. This gives

$$W^2 = \bigl(\Psi\Psi^\top\bigr)^{-1} \Psi \Sigma_0 \Psi^\top \bigl(\Psi\Psi^\top\bigr)^{-1}.$$

The question is whether the estimate $\tilde{\theta}$ based on the misspecified distributional assumption is efficient. The Cramér–Rao result delivers the lower bound for the quadratic risk in form of $\operatorname{Var}(\tilde{\theta}) \geq \left[\operatorname{Var}\bigl(\nabla L(\theta)\bigr)\right]^{-1}$. We already know that the use of the correctly specified covariance matrix of the errors leads to an R-efficient estimate $\tilde{\theta}$. The next result show that the use of a misspecified matrix Σ results in an estimate which is unbiased but not R-efficient, that is, the best estimation risk is achieved if we apply the correct model assumptions.

Theorem 4.4.6. *Let* $Y = \Psi^\top \theta^* + \varepsilon$ *with* $\varepsilon \sim \mathcal{N}(0, \Sigma_0)$. *Then*

$$\operatorname{Var}\bigl[\nabla L(\theta)\bigr] = \Psi \Sigma_0^{-1} \Psi^\top .$$

The estimate $\tilde{\theta} = \left(\Psi\Psi^\top\right)^{-1}\Psi Y$ *is unbiased, that is,* $\mathbb{E}\tilde{\theta} = \theta^*$, *but it is not R-efficient unless* $\Sigma_0 = \Sigma$.

Proof. Let $\tilde{\theta}_0$ be the MLE for the correct model specification with the noise $\varepsilon \sim \mathcal{N}(0, \Sigma_0)$. As $\tilde{\theta}$ is unbiased, the difference $\tilde{\theta} - \tilde{\theta}_0$ is orthogonal to $\tilde{\theta}_0$ and it holds for the variance of $\tilde{\theta}$

$$\operatorname{Var}(\tilde{\theta}) = \operatorname{Var}(\tilde{\theta}_0) + \operatorname{Var}(\tilde{\theta} - \tilde{\theta}_0);$$

cf. with the proof of Gauss–Markov-Theorem 4.4.4.

Exercise 4.4.7. Compare directly the variances of $\tilde{\theta}$ and of $\tilde{\theta}_0$.

4.5 Linear Models and Quadratic Log-Likelihood

Linear Gaussian modeling leads to a specific log-likelihood structure; see Sect. 4.2. Namely, the log-likelihood function $L(\theta)$ is quadratic in θ, the coefficients of the quadratic terms are deterministic and the cross term is linear both in θ and in the observations Y_i. Here we show that this geometric structure of the log-likelihood characterizes linear models. We say that $L(\theta)$ is *quadratic* if it is a quadratic function of θ and there is a deterministic symmetric matrix D^2 such that for any θ°, θ

$$L(\theta) - L(\theta^\circ) = (\theta - \theta^\circ)^\top \nabla L(\theta^\circ) - (\theta - \theta^\circ)^\top D^2 (\theta - \theta^\circ)/2. \tag{4.14}$$

Here $\nabla L(\theta) \stackrel{\text{def}}{=} \frac{dL(\theta)}{d\theta}$. As usual we define

$$\begin{aligned} \tilde{\theta} &\stackrel{\text{def}}{=} \operatorname*{argmax}_{\theta} L(\theta), \\ \theta^* &= \operatorname*{argmax}_{\theta} \mathbb{E}L(\theta). \end{aligned}$$

The next result describes some properties of the estimate $\tilde{\boldsymbol{\theta}}$ which are entirely based on the geometric (quadratic) structure of the function $L(\boldsymbol{\theta})$. All the results are stated by using the matrix D^2 and the vector $\boldsymbol{\zeta} = \nabla L(\boldsymbol{\theta}^*)$.

Theorem 4.5.1. *Let $L(\boldsymbol{\theta})$ be quadratic for a matrix $D^2 > 0$. Then for any $\boldsymbol{\theta}^\circ$*

$$\tilde{\boldsymbol{\theta}} - \boldsymbol{\theta}^\circ = D^{-2} \nabla L(\boldsymbol{\theta}^\circ). \tag{4.15}$$

In particular, with $\boldsymbol{\theta}^\circ = 0$, it holds

$$\tilde{\boldsymbol{\theta}} = D^{-2} \nabla L(0).$$

Taking $\boldsymbol{\theta}^\circ = \boldsymbol{\theta}^*$ yields

$$\tilde{\boldsymbol{\theta}} - \boldsymbol{\theta}^* = D^{-2} \boldsymbol{\zeta} \tag{4.16}$$

with $\boldsymbol{\zeta} \stackrel{\text{def}}{=} \nabla L(\boldsymbol{\theta}^*)$. Moreover, $\mathbb{E}\boldsymbol{\zeta} = 0$, and it holds with $V^2 = \operatorname{Var}(\boldsymbol{\zeta}) = \mathbb{E}\boldsymbol{\zeta}\boldsymbol{\zeta}^\top$

$$\mathbb{E}\tilde{\boldsymbol{\theta}} = \boldsymbol{\theta}^*$$
$$\operatorname{Var}\bigl(\tilde{\boldsymbol{\theta}}\bigr) = D^{-2} V^2 D^{-2}.$$

Further, for any $\boldsymbol{\theta}$,

$$L(\tilde{\boldsymbol{\theta}}) - L(\boldsymbol{\theta}) = (\tilde{\boldsymbol{\theta}} - \boldsymbol{\theta})^\top D^2 (\tilde{\boldsymbol{\theta}} - \boldsymbol{\theta})/2 = \|D(\tilde{\boldsymbol{\theta}} - \boldsymbol{\theta})\|^2/2. \tag{4.17}$$

Finally, it holds for the excess $L(\tilde{\boldsymbol{\theta}}, \boldsymbol{\theta}^*) \stackrel{\text{def}}{=} L(\tilde{\boldsymbol{\theta}}) - L(\boldsymbol{\theta}^*)$

$$2L(\tilde{\boldsymbol{\theta}}, \boldsymbol{\theta}^*) = (\tilde{\boldsymbol{\theta}} - \boldsymbol{\theta}^*)^\top D^2 (\tilde{\boldsymbol{\theta}} - \boldsymbol{\theta}^*) = \boldsymbol{\zeta}^\top D^{-2} \boldsymbol{\zeta} = \|\boldsymbol{\xi}\|^2 \tag{4.18}$$

with $\boldsymbol{\xi} = D^{-1}\boldsymbol{\zeta}$.

Proof. The extremal point equation $\nabla L(\boldsymbol{\theta}) = 0$ for the quadratic function $L(\boldsymbol{\theta})$ from (4.14) yields (4.15). Equation (4.14) with $\boldsymbol{\theta}^\circ = \boldsymbol{\theta}^*$ implies for any $\boldsymbol{\theta}$

$$\nabla L(\boldsymbol{\theta}) = \nabla L(\boldsymbol{\theta}^\circ) - D^2(\boldsymbol{\theta} - \boldsymbol{\theta}^\circ) = \boldsymbol{\zeta} - D^2(\boldsymbol{\theta} - \boldsymbol{\theta}^*). \tag{4.19}$$

Therefore, it holds for the expectation $\mathbb{E}L(\boldsymbol{\theta})$

$$\nabla \mathbb{E}L(\boldsymbol{\theta}) = \mathbb{E}\zeta - D^2(\boldsymbol{\theta} - \boldsymbol{\theta}^*),$$

and the equation $\nabla \mathbb{E}L(\boldsymbol{\theta}^*) = 0$ implies $\mathbb{E}\zeta = 0$.

To show (4.17), apply again the property (4.14) with $\boldsymbol{\theta}^\circ = \tilde{\boldsymbol{\theta}}$:

$$L(\boldsymbol{\theta}) - L(\tilde{\boldsymbol{\theta}}) = (\boldsymbol{\theta} - \tilde{\boldsymbol{\theta}})^\top \nabla L(\tilde{\boldsymbol{\theta}}) - (\boldsymbol{\theta} - \tilde{\boldsymbol{\theta}})^\top D^2 (\boldsymbol{\theta} - \tilde{\boldsymbol{\theta}})/2$$
$$= -(\tilde{\boldsymbol{\theta}} - \boldsymbol{\theta})^\top D^2 (\tilde{\boldsymbol{\theta}} - \boldsymbol{\theta})/2.$$

Here we used that $\nabla L(\tilde{\boldsymbol{\theta}}) = 0$ because $\tilde{\boldsymbol{\theta}}$ is an extreme point of $L(\boldsymbol{\theta})$. The last result (4.18) is a special case with $\boldsymbol{\theta} = \boldsymbol{\theta}^*$ in view of (4.16).

This theorem delivers an important message: the main properties of the MLE $\tilde{\boldsymbol{\theta}}$ can be explained via the geometric (quadratic) structure of the log-likelihood. An interesting question to clarify is whether a quadratic log-likelihood structure is specific for linear Gaussian model. The answer is positive: there is one-to-one correspondence between linear Gaussian models and quadratic log-likelihood functions. Indeed, the identity (4.19) with $\boldsymbol{\theta}^\circ = \boldsymbol{\theta}^*$ can be rewritten as

$$\nabla L(\boldsymbol{\theta}) + D^2\boldsymbol{\theta} \equiv \boldsymbol{\zeta} + D^2\boldsymbol{\theta}^*.$$

If we fix any $\boldsymbol{\theta}$ and define $\boldsymbol{Y} = \nabla L(\boldsymbol{\theta}) + D^2\boldsymbol{\theta}$, this yields

$$\boldsymbol{Y} = D^2\boldsymbol{\theta}^* + \boldsymbol{\zeta}.$$

Similarly, $\boldsymbol{Y} \stackrel{\text{def}}{=} D^{-1}\{\nabla L(\boldsymbol{\theta}) + D^2\boldsymbol{\theta}\}$ yields the equation

$$\boldsymbol{Y} = D\boldsymbol{\theta}^* + \boldsymbol{\xi}, \tag{4.20}$$

where $\boldsymbol{\xi} = D^{-1}\boldsymbol{\zeta}$. We can summarize as follows.

Theorem 4.5.2. *Let $L(\boldsymbol{\theta})$ be quadratic with a non-degenerated matrix D^2. Then $\boldsymbol{Y} \stackrel{\text{def}}{=} D^{-1}\{\nabla L(\boldsymbol{\theta}) + D^2\boldsymbol{\theta}\}$ does not depend on $\boldsymbol{\theta}$ and $L(\boldsymbol{\theta}) - L(\boldsymbol{\theta}^*)$ is the quasi log-likelihood ratio for the linear Gaussian model* (4.20) *with $\boldsymbol{\xi}$ standard normal. It is the true log-likelihood if and only if $\boldsymbol{\zeta} \sim \mathcal{N}(0, D^2)$.*

Proof. The model (4.20) with $\boldsymbol{\xi} \sim \mathcal{N}(0, \boldsymbol{I}_p)$ leads to the log-likelihood ratio

$$(\boldsymbol{\theta} - \boldsymbol{\theta}^*)^\top D(\boldsymbol{Y} - D\boldsymbol{\theta}^*) - \|D(\boldsymbol{\theta} - \boldsymbol{\theta}^*)\|^2/2 = (\boldsymbol{\theta} - \boldsymbol{\theta}^*)^\top \boldsymbol{\zeta} - \|D(\boldsymbol{\theta} - \boldsymbol{\theta}^*)\|^2/2$$

in view of the definition of $\boldsymbol{Y}$. The definition (4.14) implies

$$L(\boldsymbol{\theta}) - L(\boldsymbol{\theta}^*) = (\boldsymbol{\theta} - \boldsymbol{\theta}^*)^\top \nabla L(\boldsymbol{\theta}^*) - \|D(\boldsymbol{\theta} - \boldsymbol{\theta}^*)\|^2/2.$$

As these two expressions coincide, it follows that $L(\boldsymbol{\theta})$ is the true log-likelihood if and only if $\boldsymbol{\xi} = D^{-1}\boldsymbol{\zeta}$ is standard normal.

4.6 Inference Based on the Maximum Likelihood

All the results presented above for linear models were based on the explicit representation of the (quasi) MLE $\tilde{\boldsymbol{\theta}}$. Here we present the approach based on the analysis of the maximum likelihood. This approach does not require to fix any analytic expression for the point of maximum of the (quasi) likelihood process $L(\boldsymbol{\theta})$. Instead we work directly with the maximum of this process. We establish exponential inequalities for the *excess* or the *maximum likelihood* $L(\tilde{\boldsymbol{\theta}}, \boldsymbol{\theta}^*)$. We also show how these results can be used to study the accuracy of the MLE $\tilde{\boldsymbol{\theta}}$, in particular, for building confidence sets.

One more benefit of the ML-based approach is that it equally applies to a homogeneous and to a heterogeneous noise provided that the noise structure is not misspecified. The celebrated chi-squared result about the maximum likelihood $L(\tilde{\boldsymbol{\theta}}, \boldsymbol{\theta}^*)$ claims that the distribution of $2L(\tilde{\boldsymbol{\theta}}, \boldsymbol{\theta}^*)$ is chi-squared with p degrees of freedom χ_p^2 and it does not depend on the noise covariance; see Sect. 4.6.

Now we specify the setup. The starting point of the ML-approach is the linear Gaussian model assumption $\boldsymbol{Y} = \Psi^\top \boldsymbol{\theta}^* + \boldsymbol{\varepsilon}$ with $\boldsymbol{\varepsilon} \sim \mathcal{N}(0, \Sigma)$. The corresponding log-likelihood ratio $L(\boldsymbol{\theta})$ can be written as

$$L(\boldsymbol{\theta}) = -\frac{1}{2}(\boldsymbol{Y} - \Psi^\top \boldsymbol{\theta})^\top \Sigma^{-1} (\boldsymbol{Y} - \Psi^\top \boldsymbol{\theta}) + R, \tag{4.21}$$

where the remainder term R does not depend on $\boldsymbol{\theta}$. Now one can see that $L(\boldsymbol{\theta})$ is a quadratic function of $\boldsymbol{\theta}$. Moreover, $\nabla^2 L(\boldsymbol{\theta}) = \Psi \Sigma^{-1} \Psi^\top$, so that $L(\boldsymbol{\theta})$ is quadratic with $D^2 = \Psi \Sigma^{-1} \Psi^\top$. This enables us to apply the general results of Sect. 4.5 which are only based on the geometric (quadratic) structure of the log-likelihood $L(\boldsymbol{\theta})$: the true data distribution can be arbitrary.

Theorem 4.6.1. *Consider $L(\boldsymbol{\theta})$ from* (4.21)*. For any $\boldsymbol{\theta}$, it holds with $D^2 = \Psi \Sigma^{-1} \Psi^\top$*

$$L(\tilde{\boldsymbol{\theta}}, \boldsymbol{\theta}) = (\tilde{\boldsymbol{\theta}} - \boldsymbol{\theta})^\top D^2 (\tilde{\boldsymbol{\theta}} - \boldsymbol{\theta})/2. \tag{4.22}$$

In particular, if $\Sigma = \sigma^2 \boldsymbol{I}_n$ then the fitted log-likelihood is proportional to the quadratic loss $\|\tilde{\boldsymbol{f}} - \boldsymbol{f}_{\boldsymbol{\theta}}\|^2$ for $\tilde{\boldsymbol{f}} = \Psi^\top \tilde{\boldsymbol{\theta}}$ and $\boldsymbol{f}_{\boldsymbol{\theta}} = \Psi^\top \boldsymbol{\theta}$:

$$L(\tilde{\boldsymbol{\theta}}, \boldsymbol{\theta}) = \frac{1}{2\sigma^2} \left\| \Psi^\top (\tilde{\boldsymbol{\theta}} - \boldsymbol{\theta}) \right\|^2 = \frac{1}{2\sigma^2} \left\| \tilde{\boldsymbol{f}} - \boldsymbol{f}_{\boldsymbol{\theta}} \right\|^2.$$

If $\boldsymbol{\theta}^* \stackrel{\text{def}}{=} \operatorname{argmax}_{\boldsymbol{\theta}} \mathbb{E} L(\boldsymbol{\theta}) = D^{-2} \Psi \Sigma^{-1} \boldsymbol{f}^*$ for $\boldsymbol{f}^* = \mathbb{E} \boldsymbol{Y}$, then

$$2L(\tilde{\boldsymbol{\theta}}, \boldsymbol{\theta}^*) = \boldsymbol{\zeta}^\top D^{-2} \boldsymbol{\zeta} = \|\boldsymbol{\xi}\|^2 \tag{4.23}$$

with $\boldsymbol{\zeta} = \nabla L(\boldsymbol{\theta}^*)$ and $\boldsymbol{\xi} \stackrel{\text{def}}{=} D^{-1} \boldsymbol{\zeta}$.

Proof. The results (4.22) and (4.23) follow from Theorem 4.5.1; see (4.17) and (4.18).

If the model assumptions are not misspecified, one can establish the remarkable χ^2 result.

Theorem 4.6.2. *Let $L(\boldsymbol{\theta})$ from* (4.21) *be the log-likelihood for the model $\boldsymbol{Y} = \Psi^\top \boldsymbol{\theta}^* + \boldsymbol{\varepsilon}$ with $\boldsymbol{\varepsilon} \sim \mathcal{N}(0, \Sigma)$. Then $\boldsymbol{\xi} = D^{-1}\boldsymbol{\zeta} \sim \mathcal{N}(0, \boldsymbol{I}_p)$ and $2L(\tilde{\boldsymbol{\theta}}, \boldsymbol{\theta}^*) \sim \chi^2_p$ is chi-squared with p degrees of freedom.*

Proof. By direct calculus

$$\boldsymbol{\zeta} = \nabla L(\boldsymbol{\theta}^*) = \Psi\Sigma^{-1}(\boldsymbol{Y} - \Psi^\top \boldsymbol{\theta}^*) = \Psi\Sigma^{-1}\boldsymbol{\varepsilon}.$$

So, $\boldsymbol{\zeta}$ is a linear transformation of a Gaussian vector $\boldsymbol{Y}$ and thus it is Gaussian as well. By Theorem 4.5.1, $\mathbb{E}\boldsymbol{\zeta} = 0$. Moreover, $\operatorname{Var}(\boldsymbol{\varepsilon}) = \Sigma$ implies

$$\operatorname{Var}(\boldsymbol{\zeta}) = \mathbb{E}\Psi^\top\Sigma^{-1}\boldsymbol{\varepsilon}\boldsymbol{\varepsilon}^\top\Sigma^{-1}\Psi^\top = \Psi\Sigma^{-1}\Psi^\top = D^2$$

yielding that $\boldsymbol{\xi} = D^{-1}\boldsymbol{\zeta}$ is standard normal.

The last result $2L(\tilde{\boldsymbol{\theta}}, \boldsymbol{\theta}^*) \sim \chi^2_p$ is sometimes called the "chi-squared phenomenon": the distribution of the maximum likelihood only depends on the number of parameters to be estimated and is independent of the design Ψ, of the noise covariance matrix Σ, etc. This particularly explains the use of word "phenomenon" in the name of the result.

Exercise 4.6.1. Check that the linear transformation $\check{\boldsymbol{Y}} = \Sigma^{-1/2}\boldsymbol{Y}$ of the data does not change the value of the log-likelihood ratio $L(\boldsymbol{\theta}, \boldsymbol{\theta}^*)$ and hence, of the maximum likelihood $L(\tilde{\boldsymbol{\theta}}, \boldsymbol{\theta}^*)$.
Hint: use the representation

$$\begin{aligned} L(\boldsymbol{\theta}) &= \frac{1}{2}(\boldsymbol{Y} - \Psi^\top\boldsymbol{\theta})^\top\Sigma^{-1}(\boldsymbol{Y} - \Psi^\top\boldsymbol{\theta}) + R \\ &= \frac{1}{2}(\check{\boldsymbol{Y}} - \check{\Psi}^\top\boldsymbol{\theta})^\top(\check{\boldsymbol{Y}} - \check{\Psi}^\top\boldsymbol{\theta}) + R \end{aligned}$$

and check that the transformed data $\check{\boldsymbol{Y}}$ is described by the model $\check{\boldsymbol{Y}} = \check{\Psi}^\top\boldsymbol{\theta}^* + \check{\boldsymbol{\varepsilon}}$ with $\check{\Psi} = \Psi\Sigma^{-1/2}$ and $\check{\boldsymbol{\varepsilon}} = \Sigma^{-1/2}\boldsymbol{\varepsilon} \sim \mathcal{N}(0, \boldsymbol{I}_n)$ yielding the same log-likelihood ratio as in the original model.

Exercise 4.6.2. Assume homogeneous noise in (4.21) with $\Sigma = \sigma^2\boldsymbol{I}_n$. Then it holds

$$2L(\tilde{\boldsymbol{\theta}}, \boldsymbol{\theta}^*) = \sigma^{-2}\|\Pi\boldsymbol{\varepsilon}\|^2$$

where $\Pi = \Psi^\top\bigl(\Psi\Psi^\top\bigr)^{-1}\Psi$ is the projector in $\mathbb{R}^n$ on the subspace spanned by the vectors $\boldsymbol{\psi}_1, \ldots, \boldsymbol{\psi}_p$.
Hint: use that $\boldsymbol{\zeta} = \sigma^{-2}\Psi\boldsymbol{\varepsilon}$, $D^2 = \sigma^{-2}\Psi\Psi^\top$, and

$$\sigma^{-2}\|\Pi\boldsymbol{\varepsilon}\|^2 = \sigma^{-2}\boldsymbol{\varepsilon}^\top\Pi^\top\Pi\boldsymbol{\varepsilon} = \sigma^{-2}\boldsymbol{\varepsilon}^\top\Pi\boldsymbol{\varepsilon} = \boldsymbol{\zeta}^\top D^{-2}\boldsymbol{\zeta}.$$

We write the result of Theorem 4.6.1 in the form $2L(\tilde{\boldsymbol{\theta}}, \boldsymbol{\theta}^*) \sim \chi_p^2$, where χ_p^2 stands for the chi-squared distribution with p degrees of freedom. This result can be used to build likelihood-based confidence ellipsoids for the parameter $\boldsymbol{\theta}^*$. Given $\mathfrak{z} > 0$, define

$$\mathcal{E}(\mathfrak{z}) = \{\boldsymbol{\theta} : L(\tilde{\boldsymbol{\theta}}, \boldsymbol{\theta}) \le \mathfrak{z}\} = \Bigl\{\boldsymbol{\theta} : \sup_{\boldsymbol{\theta}'} L(\boldsymbol{\theta}') - L(\boldsymbol{\theta}) \le \mathfrak{z}\Bigr\}. \tag{4.24}$$

Theorem 4.6.3. *Assume $\boldsymbol{Y} = \Psi^\top \boldsymbol{\theta}^* + \boldsymbol{\varepsilon}$ with $\boldsymbol{\varepsilon} \sim \mathcal{N}(0, \Sigma)$ and consider the MLE $\tilde{\boldsymbol{\theta}}$. Define $\mathfrak{z}_\alpha$ by $P\bigl(\chi_p^2 > 2\mathfrak{z}_\alpha\bigr) = \alpha$. Then $\mathcal{E}(\mathfrak{z}_\alpha)$ from (4.24) is an α-confidence set for $\boldsymbol{\theta}^*$.*

Exercise 4.6.3. Let $D^2 = \Psi\Sigma^{-1}\Psi^\top$. Check that the likelihood-based CS $\mathcal{E}(\mathfrak{z}_\alpha)$ and estimate-based CS $E(z_\alpha) = \{\boldsymbol{\theta} : \|D(\tilde{\boldsymbol{\theta}} - \boldsymbol{\theta})\| \le z_\alpha\}$, $z_\alpha^2 = 2\mathfrak{z}_\alpha$, coincide in the case of the linear modeling:

$$\mathcal{E}(\mathfrak{z}_\alpha) = \bigl\{\boldsymbol{\theta} : \bigl\|D(\tilde{\boldsymbol{\theta}} - \boldsymbol{\theta})\bigr\|^2 \le 2\mathfrak{z}_\alpha\bigr\}.$$

Another corollary of the chi-squared result is a concentration bound for the maximum likelihood. A similar result was stated for the univariate exponential family model: the value $L(\tilde{\theta}, \theta^*)$ is stochastically bounded with exponential moments, and the bound does not depend on the particular family, parameter value, sample size, etc. Now we can extend this result to the case of a linear Gaussian model. Indeed, Theorem 4.6.1 states that the distribution of $2L(\tilde{\boldsymbol{\theta}}, \boldsymbol{\theta}^*)$ is chi-squared and only depends on the number of parameters to be estimated. The latter distribution concentrates on the ball of radius of order $p^{1/2}$ and the deviation probability is exponentially small.

Theorem 4.6.4. *Assume $\boldsymbol{Y} = \Psi^\top \boldsymbol{\theta}^* + \boldsymbol{\varepsilon}$ with $\boldsymbol{\varepsilon} \sim \mathcal{N}(0, \Sigma)$. Then for every $\mathtt{x} > 0$, it holds with $\kappa \ge 6.6$*

$$\begin{aligned} &\mathbb{P}\bigl(2L(\tilde{\boldsymbol{\theta}}, \boldsymbol{\theta}^*) > p + \sqrt{\varkappa \mathtt{x} p} \vee (\varkappa \mathtt{x})\bigr) \\ &\quad = \mathbb{P}\bigl(\bigl\|D(\tilde{\boldsymbol{\theta}} - \boldsymbol{\theta}^*)\bigr\|^2 > p + \sqrt{\varkappa \mathtt{x} p} \vee (\varkappa \mathtt{x})\bigr) \le \exp(-\mathtt{x}). \end{aligned} \tag{4.25}$$

Proof. Define $\boldsymbol{\xi} \stackrel{\text{def}}{=} D(\tilde{\boldsymbol{\theta}} - \boldsymbol{\theta}^*)$. By Theorem 4.4.4 $\boldsymbol{\xi}$ is standard normal vector in $\mathbb{R}^p$ and by Theorem 4.6.1 $2L(\tilde{\boldsymbol{\theta}}, \boldsymbol{\theta}^*) = \|\boldsymbol{\xi}\|^2$. Now the statement (4.25) follows from the general deviation bound for the Gaussian quadratic forms; see Theorem A.2.1.

The main message of this result can be explained as follows: the deviation probability that the estimate $\tilde{\boldsymbol{\theta}}$ does not belong to the elliptic set $E(z) = \{\boldsymbol{\theta} : \|D(\tilde{\boldsymbol{\theta}} - \boldsymbol{\theta})\| \le z\}$ starts to vanish when z^2 exceeds the dimensionality p of the parameter space. Similarly, the coverage probability that the true parameter $\boldsymbol{\theta}^*$ is not covered by the confidence set $\mathcal{E}(\mathfrak{z})$ starts to vanish when $2\mathfrak{z}$ exceeds p.

Corollary 4.6.1. *Assume* $Y = \Psi^\top \theta^* + \varepsilon$ *with* $\varepsilon \sim \mathcal{N}(0, \Sigma)$. *Then for every* $\mathrm{x} > 0$, *it holds with* $2\mathfrak{z} = p + \sqrt{\varkappa \mathrm{x} p} \vee (\varkappa \mathrm{x})$ *for* $\varkappa \geq 6.6$

$$\mathbb{P}\bigl(\mathcal{E}(\mathfrak{z}) \not\ni \theta^*\bigr) \leq \exp(-\mathrm{x}).$$

Exercise 4.6.4. Compute $\mathfrak{z}$ ensuring the covering of 95 % in the dimension $p = 1, 2, 10, 20$.

4.6.1 A Misspecified LPA

Now we discuss the behavior of the fitted log-likelihood for the misspecified linear parametric assumption $\mathbb{E}Y = \Psi^\top \theta^*$. Let the response function f^* not be linearly expandable as $f^* = \Psi^\top \theta^*$. Following to Theorem 4.4.3, define $\theta^* = \mathcal{S} f^*$ with $\mathcal{S} = \bigl(\Psi \Sigma^{-1} \Psi^\top\bigr)^{-1} \Psi \Sigma^{-1}$. This point provides the best approximation of the nonlinear response f^* by a linear parametric fit $\Psi^\top \theta$.

Theorem 4.6.5. *Assume* $Y = f^* + \varepsilon$ *with* $\varepsilon \sim \mathcal{N}(0, \Sigma)$. *Let* $\theta^* = \mathcal{S} f^*$. *Then* $\tilde{\theta}$ *is an R-efficient estimate of* θ^* *and*

$$2L(\tilde{\theta}, \theta^*) = \zeta^\top D^{-2} \zeta = \|\xi\|^2 \sim \chi_p^2,$$

where $D^2 = \Psi \Sigma^{-1} \Psi^\top$, $\zeta = \nabla L(\theta^*) = \Psi \Sigma^{-1} \varepsilon$, $\xi = D^{-1} \zeta$ *is standard normal vector in* $\mathbb{R}^p$ *and* χ_p^2 *is a chi-squared random variable with p degrees of freedom. In particular,* $\mathcal{E}(\mathfrak{z}_\alpha)$ *is an* α*-CS for the vector* θ^* *and the bound of Corollary 4.6.1 applies.*

Exercise 4.6.5. Prove the result of Theorem 4.6.5.

4.6.2 A Misspecified Noise Structure

This section addresses the question about the features of the maximum likelihood in the case when the likelihood is built under a wrong assumption about the noise structure. As one can expect, the chi-squared result is not valid anymore in this situation and the distribution of the maximum likelihood depends on the true noise covariance. However, the nice geometric structure of the maximum likelihood manifested by Theorems 4.6.1 and 4.6.3 does not rely on the true data distribution and it is only based on our structural assumptions on the considered model. This helps to get rigorous results about the behaviors of the maximum likelihood and particularly about its concentration properties.

Theorem 4.6.6. *Let* $\tilde{\theta}$ *be built for the model* $Y = \Psi^\top \theta^* + \varepsilon$ *with* $\varepsilon \sim \mathcal{N}(0, \Sigma)$, *while the true noise covariance is* Σ_0: $\mathbb{E}\varepsilon = 0$ *and* $\operatorname{Var}(\varepsilon) = \Sigma_0$. *Then*

$$
\begin{aligned}
\mathbb{E}\tilde{\boldsymbol{\theta}} &= \boldsymbol{\theta}^*, \\
\operatorname{Var}(\tilde{\boldsymbol{\theta}}) &= D^{-2} W^2 D^{-2},
\end{aligned}
$$

where

$$
\begin{aligned}
D^2 &= \Psi \Sigma^{-1} \Psi^\top, \\
W^2 &= \Psi \Sigma^{-1} \Sigma_0 \Sigma^{-1} \Psi^\top.
\end{aligned}
$$

Further,

$$
2L(\tilde{\boldsymbol{\theta}}, \boldsymbol{\theta}^*) = \|D(\tilde{\boldsymbol{\theta}} - \boldsymbol{\theta}^*)\|^2 = \|\boldsymbol{\xi}\|^2, \tag{4.26}
$$

where $\boldsymbol{\xi}$ is a random vector in $\mathbb{R}^p$ with $\mathbb{E}\boldsymbol{\xi} = 0$ and

$$
\operatorname{Var}(\boldsymbol{\xi}) = B \stackrel{\text{def}}{=} D^{-1} W^2 D^{-1}.
$$

Moreover, if $\boldsymbol{\varepsilon} \sim \mathcal{N}(0, \Sigma_0)$, then $\tilde{\boldsymbol{\theta}} \sim \mathcal{N}\bigl(\boldsymbol{\theta}^*, D^{-2} W^2 D^{-2}\bigr)$ and $\boldsymbol{\xi} \sim \mathcal{N}(0, B)$.

Proof. The moments of $\tilde{\boldsymbol{\theta}}$ have been computed in Theorem 4.5.1 while the equality $2L(\tilde{\boldsymbol{\theta}}, \boldsymbol{\theta}^*) = \|D(\tilde{\boldsymbol{\theta}} - \boldsymbol{\theta}^*)\|^2 = \|\boldsymbol{\xi}\|^2$ is given in Theorem 4.6.1. Next, $\boldsymbol{\zeta} = \nabla L(\boldsymbol{\theta}^*) = \Psi \Sigma^{-1} \boldsymbol{\varepsilon}$ and

$$
W^2 \stackrel{\text{def}}{=} \operatorname{Var}(\boldsymbol{\zeta}) = \Psi \Sigma^{-1} \operatorname{Var}(\boldsymbol{\varepsilon}) \Sigma^{-1} \Psi^\top = \Psi \Sigma^{-1} \Sigma_0 \Sigma^{-1} \Psi^\top.
$$

This implies that

$$
\operatorname{Var}(\boldsymbol{\xi}) = \mathbb{E}\boldsymbol{\xi}\boldsymbol{\xi}^\top = D^{-1} \operatorname{Var}(\boldsymbol{\zeta}) D^{-1} = D^{-1} W^2 D^{-1}.
$$

It remains to note that if $\boldsymbol{\varepsilon}$ is a Gaussian vector, then $\boldsymbol{\zeta} = \Psi \Sigma^{-1} \boldsymbol{\varepsilon}$, $\boldsymbol{\xi} = D^{-1} \boldsymbol{\zeta}$, and $\tilde{\boldsymbol{\theta}} - \boldsymbol{\theta}^* = D^{-2} \boldsymbol{\zeta}$ are Gaussian as well.

Exercise 4.6.6. Check that $\Sigma_0 = \Sigma$ leads back to the χ^2-result.

One can see that the chi-squared result is not valid any more if the noise structure is misspecified. An interesting question is whether the CS $\mathcal{E}(\mathfrak{z})$ can be applied in the case of a misspecified noise under some proper adjustment of the value $\mathfrak{z}$. Surprisingly, the answer is not entirely negative. The reason is that the vector $\boldsymbol{\xi}$ from (4.26) is zero mean and its norm has a similar behavior as in the case of the correct noise specification: the probability $\mathbb{P}\bigl(\|\boldsymbol{\xi}\| > z\bigr)$ starts to degenerate when z^2 exceeds $\mathbb{E}\|\boldsymbol{\xi}\|^2$. A general bound from Theorem A.2.2 in Sect. A.1 implies the following bound for the coverage probability.

Corollary 4.6.2. *Under the conditions of Theorem 4.6.6, for every $\mathrm{x} > 0$, it holds with $\mathrm{p} = \operatorname{tr}(B)$, $v^2 = 2\operatorname{tr}(B^2)$, and $a^* = \|B\|_\infty$*

$$\mathbb{P}\bigl(2L(\tilde{\boldsymbol{\theta}},\boldsymbol{\theta}^*) > \mathtt{p} + (2v\mathtt{x}^{1/2}) \vee (6a^*\mathtt{x})\bigr) \le \exp(-\mathtt{x}).$$

Exercise 4.6.7. Show that an overestimation of the noise in the sense $\Sigma \ge \Sigma_0$ preserves the coverage probability for the CS $\mathcal{E}(\mathfrak{z}_\alpha)$, that is, if $2\mathfrak{z}_\alpha$ is the $1-\alpha$ quantile of χ^2_p, then $\mathbb{P}\bigl(\mathcal{E}(\mathfrak{z}_\alpha) \not\ni \boldsymbol{\theta}^*\bigr) \le \alpha$.

4.7 Ridge Regression, Projection, and Shrinkage

This section discusses the important situation when the number of predictors $\boldsymbol{\psi}_j$ and hence the number of parameters p in the linear model $\boldsymbol{Y} = \Psi^\top\boldsymbol{\theta}^* + \boldsymbol{\varepsilon}$ is not small relative to the sample size. Then the least square or the maximum likelihood approach meets serious problems. The first one relates to the numerical issues. The definition of the LSE $\tilde{\boldsymbol{\theta}}$ involves the inversion of the $p \times p$ matrix $\Psi\Psi^\top$ and such an inversion becomes a delicate task for p large. The other problem concerns the inference for the estimated parameter $\boldsymbol{\theta}^*$. The risk bound and the width of the confidence set are proportional to the parameter dimension p and thus, with large p, the inference statements become almost uninformative. In particular, if p is of order the sample size n, even consistency is not achievable. One faces a really critical situation. We already know that the MLE is the efficient estimate in the class of all unbiased estimates. At the same time it is highly inefficient in overparametrized models. The only way out of this situation is to sacrifice the unbiasedness property in favor of reducing the model complexity: some procedures can be more efficient than MLE even if they are biased. This section discusses one way of resolving these problems by regularization or shrinkage. To be more specific, for the rest of the section we consider the following setup. The observed vector $\boldsymbol{Y}$ follows the model

$$\boldsymbol{Y} = \boldsymbol{f}^* + \boldsymbol{\varepsilon} \tag{4.27}$$

with a homogeneous error vector $\boldsymbol{\varepsilon}$: $\mathbb{E}\boldsymbol{\varepsilon} = 0$, $\operatorname{Var}(\boldsymbol{\varepsilon}) = \sigma^2 \boldsymbol{I}_n$. Noise misspecification is not considered in this section.

Furthermore, we assume a basis or a collection of basis vectors $\boldsymbol{\psi}_1, \ldots, \boldsymbol{\psi}_p$ is given with p large. This allows for approximating the response vector $\boldsymbol{f} = \mathbb{E}\boldsymbol{Y}$ in the form $\boldsymbol{f} = \Psi^\top\boldsymbol{\theta}^*$, or, equivalently,

$$\boldsymbol{f} = \theta_1^*\boldsymbol{\psi}_1 + \ldots + \theta_p^*\boldsymbol{\psi}_p .$$

In many cases we will assume that the basis is already orthogonalized: $\Psi\Psi^\top = \boldsymbol{I}_p$. The model (4.27) can be rewritten as

$$\boldsymbol{Y} = \Psi^\top\boldsymbol{\theta}^* + \boldsymbol{\varepsilon}, \qquad \operatorname{Var}(\boldsymbol{\varepsilon}) = \sigma^2 \boldsymbol{I}_n .$$

The MLE or oLSE of the parameter vector $\boldsymbol{\theta}^*$ for this model reads as

$$\tilde{\boldsymbol{\theta}} = \left(\Psi\Psi^{\top}\right)^{-1}\Psi\boldsymbol{Y}, \qquad \tilde{\boldsymbol{f}} = \Psi^{\top}\tilde{\boldsymbol{\theta}} = \Psi^{\top}\left(\Psi\Psi^{\top}\right)^{-1}\Psi\boldsymbol{Y}.$$

If the matrix $\Psi\Psi^{\top}$ is degenerate or badly posed, computing the MLE $\tilde{\boldsymbol{\theta}}$ is a hard task. Below we discuss how this problem can be treated.

4.7.1 Regularization and Ridge Regression

Let R be a positive symmetric $p \times p$ matrix. Then the sum $\Psi\Psi^{\top} + R$ is positive symmetric as well and can be inverted whatever the matrix Ψ is. This suggests to replace $\left(\Psi\Psi^{\top}\right)^{-1}$ by $\left(\Psi\Psi^{\top} + R\right)^{-1}$ leading to the regularized least squares estimate $\tilde{\boldsymbol{\theta}}_R$ of the parameter vector $\boldsymbol{\theta}$ and the corresponding response estimate $\tilde{\boldsymbol{f}}_R$:

$$\tilde{\boldsymbol{\theta}}_R \stackrel{\text{def}}{=} \left(\Psi\Psi^{\top} + R\right)^{-1}\Psi\boldsymbol{Y}, \qquad \tilde{\boldsymbol{f}}_R \stackrel{\text{def}}{=} \Psi^{\top}\left(\Psi\Psi^{\top} + R\right)^{-1}\Psi\boldsymbol{Y}. \tag{4.28}$$

Such a method is also called *ridge regression*. An example of choosing R is the multiple of the unit matrix: $R = \alpha\boldsymbol{I}_p$ where $\alpha > 0$ and $\boldsymbol{I}_p$ stands for the unit matrix. This method is also called *Tikhonov regularization* and it results in the parameter estimate $\tilde{\boldsymbol{\theta}}_{\alpha}$ and the response estimate $\tilde{\boldsymbol{f}}_{\alpha}$:

$$\tilde{\boldsymbol{\theta}}_{\alpha} \stackrel{\text{def}}{=} \left(\Psi\Psi^{\top} + \alpha\boldsymbol{I}_p\right)^{-1}\Psi\boldsymbol{Y}, \qquad \tilde{\boldsymbol{f}}_{\alpha} \stackrel{\text{def}}{=} \Psi^{\top}\left(\Psi\Psi^{\top} + \alpha\boldsymbol{I}_p\right)^{-1}\Psi\boldsymbol{Y}. \tag{4.29}$$

A proper choice of the matrix R for the ridge regression method (4.28) or the parameter α for the Tikhonov regularization (4.29) is an important issue. Below we discuss several approaches which lead to the estimate (4.28) with a specific choice of the matrix R. The properties of the estimates $\tilde{\boldsymbol{\theta}}_R$ and $\tilde{\boldsymbol{f}}_R$ will be studied in context of penalized likelihood estimation in the next section.

4.7.2 Penalized Likelihood: Bias and Variance

The estimate (4.28) can be obtained in a natural way within the (quasi) ML approach using the penalized least squares. The classical unpenalized method is based on minimizing the sum of residuals squared:

$$\tilde{\boldsymbol{\theta}} = \operatorname*{argmax}_{\boldsymbol{\theta}} L(\boldsymbol{\theta}) = \operatorname*{arginf}_{\boldsymbol{\theta}} \|\boldsymbol{Y} - \Psi^{\top}\boldsymbol{\theta}\|^2$$

with $L(\boldsymbol{\theta}) = \sigma^{-2}\|\boldsymbol{Y} - \Psi^{\top}\boldsymbol{\theta}\|^2/2$. (Here we omit the terms which do not depend on $\boldsymbol{\theta}$.) Now we introduce an additional penalty on the objective function which penalizes for the complexity of the candidate vector $\boldsymbol{\theta}$ which is expressed by the value $\|G\boldsymbol{\theta}\|^2/2$ for a given symmetric matrix G. This choice of complexity measure implicitly assumes that the vector $\boldsymbol{\theta} \equiv 0$ has the smallest complexity equal to zero and this complexity increases with the norm of $G\boldsymbol{\theta}$. Define the *penalized log-likelihood*

$$\begin{aligned} L_G(\boldsymbol{\theta}) &\stackrel{\text{def}}{=} L(\boldsymbol{\theta}) - \|G\boldsymbol{\theta}\|^2/2 \\ &= -(2\sigma^2)^{-1}\|\boldsymbol{Y} - \Psi^\top\boldsymbol{\theta}\|^2 - \|G\boldsymbol{\theta}\|^2/2 - (n/2)\log(2\pi\sigma^2). \end{aligned} \tag{4.30}$$

The penalized MLE reads as

$$\tilde{\boldsymbol{\theta}}_G = \operatorname*{argmax}_{\boldsymbol{\theta}} L_G(\boldsymbol{\theta}) = \operatorname*{argmin}_{\boldsymbol{\theta}}\bigl\{(2\sigma^2)^{-1}\|\boldsymbol{Y} - \Psi^\top\boldsymbol{\theta}\|^2 + \|G\boldsymbol{\theta}\|^2/2\bigr\}.$$

A straightforward calculus leads to the expression (4.28) for $\tilde{\boldsymbol{\theta}}_G$ with $R = \sigma^2 G^2$:

$$\tilde{\boldsymbol{\theta}}_G \stackrel{\text{def}}{=} \bigl(\Psi\Psi^\top + \sigma^2 G^2\bigr)^{-1}\Psi\boldsymbol{Y}. \tag{4.31}$$

We see that $\tilde{\boldsymbol{\theta}}_G$ is again a linear estimate: $\tilde{\boldsymbol{\theta}}_G = \mathcal{S}_G\boldsymbol{Y}$ with $\mathcal{S}_G = \bigl(\Psi\Psi^\top + \sigma^2 G^2\bigr)^{-1}\Psi$. The results of Sect. 4.4 explain that $\tilde{\boldsymbol{\theta}}_G$ in fact estimates the value $\boldsymbol{\theta}_G$ defined by

$$\begin{aligned} \boldsymbol{\theta}_G &= \operatorname*{argmax}_{\boldsymbol{\theta}} \mathbb{E}L_G(\boldsymbol{\theta}) \\ &= \operatorname*{arginf}_{\boldsymbol{\theta}} \mathbb{E}\bigl\{\|\boldsymbol{Y} - \Psi^\top\boldsymbol{\theta}\|^2 + \sigma^2\|G\boldsymbol{\theta}\|^2\bigr\} \\ &= \bigl(\Psi\Psi^\top + \sigma^2 G^2\bigr)^{-1}\Psi\boldsymbol{f}^* = \mathcal{S}_G\boldsymbol{f}^*. \end{aligned} \tag{4.32}$$

In particular, if $\boldsymbol{f}^* = \Psi^\top\boldsymbol{\theta}^*$, then

$$\boldsymbol{\theta}_G = \bigl(\Psi\Psi^\top + \sigma^2 G^2\bigr)^{-1}\Psi\Psi^\top\boldsymbol{\theta}^* \tag{4.33}$$

and $\boldsymbol{\theta}_G \neq \boldsymbol{\theta}^*$ unless $G = 0$. In other words, the penalized MLE $\tilde{\boldsymbol{\theta}}_G$ is biased.

Exercise 4.7.1. Check that $\mathbb{E}\tilde{\boldsymbol{\theta}}_\alpha = \boldsymbol{\theta}_\alpha$ for $\boldsymbol{\theta}_\alpha = \bigl(\Psi\Psi^\top + \alpha \boldsymbol{I}_p\bigr)^{-1}\Psi\Psi^\top\boldsymbol{\theta}^*$, the bias $\|\boldsymbol{\theta}_\alpha - \boldsymbol{\theta}^*\|$ grows with the regularization parameter α.

The penalized MLE $\tilde{\boldsymbol{\theta}}_G$ leads to the response estimate $\tilde{\boldsymbol{f}}_G = \Psi^\top\tilde{\boldsymbol{\theta}}_G$.

Exercise 4.7.2. Check that the penalized ML approach leads to the response estimate

$$\tilde{\boldsymbol{f}}_G = \Psi^\top\tilde{\boldsymbol{\theta}}_G = \Psi^\top\bigl(\Psi\Psi^\top + \sigma^2 G^2\bigr)^{-1}\Psi\boldsymbol{Y} = \Pi_G\boldsymbol{Y}$$

with $\Pi_G = \Psi^\top\bigl(\Psi\Psi^\top + \sigma^2 G^2\bigr)^{-1}\Psi$. Show that Π_G is a sub-projector in the sense that $\|\Pi_G\boldsymbol{u}\| \le \|\boldsymbol{u}\|$ for any $\boldsymbol{u} \in \mathbb{R}^n$.

Exercise 4.7.3. Let Ψ be orthonormal: $\Psi\Psi^{\top} = \boldsymbol{I}_p$. Then the penalized MLE $\tilde{\boldsymbol{\theta}}_G$ can be represented as

$$\tilde{\boldsymbol{\theta}}_G = (\boldsymbol{I}_p + \sigma^2 G^2)^{-1} \boldsymbol{Z},$$

where $\boldsymbol{Z} = \Psi \boldsymbol{Y}$ is the vector of empirical Fourier coefficients. Specify the result for the case of a diagonal matrix $G = \operatorname{diag}(g_1, \ldots, g_p)$ and describe the corresponding response estimate $\tilde{\boldsymbol{f}}_G$.

The previous results indicate that introducing the penalization leads to some bias of estimation. One can ask about a benefit of using a penalized procedure. The next result shows that penalization decreases the variance of estimation and thus makes the procedure more stable.

Theorem 4.7.1. *Let* $\tilde{\boldsymbol{\theta}}_G$ *be a penalized MLE from* (4.31)*. Then* $\mathbb{E}\tilde{\boldsymbol{\theta}}_G = \boldsymbol{\theta}_G$*, see* (4.33)*, and under noise homogeneity* $\operatorname{Var}(\boldsymbol{\varepsilon}) = \sigma^2 \boldsymbol{I}_n$*, it holds*

$$\begin{aligned}\operatorname{Var}(\tilde{\boldsymbol{\theta}}_G) &= \bigl(\sigma^{-2}\Psi\Psi^{\top} + G^2\bigr)^{-1} \sigma^{-2}\Psi\Psi^{\top} \bigl(\sigma^{-2}\Psi\Psi^{\top} + G^2\bigr)^{-1} \\ &= D_G^{-2} D^2 D_G^{-2}\end{aligned}$$

with $D_G^2 = \sigma^{-2}\Psi\Psi^{\top} + G^2$*. In particular,* $\operatorname{Var}(\tilde{\boldsymbol{\theta}}_G) \le D_G^{-2}$*. If* $\boldsymbol{\varepsilon} \sim \mathcal{N}(0, \sigma^2 \boldsymbol{I}_n)$*, then* $\tilde{\boldsymbol{\theta}}_G$ *is also normal:* $\tilde{\boldsymbol{\theta}}_G \sim \mathcal{N}(\boldsymbol{\theta}_G, D_G^{-2} D^2 D_G^{-2})$*.*

Moreover, the bias $\|\boldsymbol{\theta}_G - \boldsymbol{\theta}^*\|$ *monotonously increases in* G^2 *while the variance monotonously decreases with the penalization* G*.*

Proof. The first two moments of $\tilde{\boldsymbol{\theta}}_G$ are computed from $\tilde{\boldsymbol{\theta}}_G = \mathcal{S}_G \boldsymbol{Y}$. Monotonicity of the bias and variance of $\tilde{\boldsymbol{\theta}}_G$ is proved below in Exercise 4.7.6.

Exercise 4.7.4. Let Ψ be orthonormal: $\Psi\Psi^{\top} = \boldsymbol{I}_p$. Describe $\operatorname{Var}(\tilde{\boldsymbol{\theta}}_G)$. Show that the variance decreases with the penalization G in the sense that $G_1 \ge G$ implies $\operatorname{Var}(\tilde{\boldsymbol{\theta}}_{G_1}) \le \operatorname{Var}(\tilde{\boldsymbol{\theta}}_G)$.

Exercise 4.7.5. Let $\Psi\Psi^{\top} = \boldsymbol{I}_p$ and let $G = \operatorname{diag}(g_1, \ldots, g_p)$ be a diagonal matrix. Compute the squared bias $\|\boldsymbol{\theta}_G - \boldsymbol{\theta}^*\|^2$ and show that it monotonously increases in each g_j for $j = 1, \ldots, p$.

Exercise 4.7.6. Let G be a symmetric matrix and $\tilde{\boldsymbol{\theta}}_G$ the corresponding penalized MLE. Show that the variance $\operatorname{Var}(\tilde{\boldsymbol{\theta}}_G)$ decreases while the bias $\|\boldsymbol{\theta}_G - \boldsymbol{\theta}^*\|$ increases in G^2.
Hint: with $D^2 = \sigma^{-2}\Psi\Psi^{\top}$, show that for any vector $\boldsymbol{w} \in \mathbb{R}^p$ and $\boldsymbol{u} = D^{-1}\boldsymbol{w}$, it holds

$$\boldsymbol{w}^{\top} \operatorname{Var}(\tilde{\boldsymbol{\theta}}_G) \boldsymbol{w} = \boldsymbol{u}^{\top} (\boldsymbol{I}_p + D^{-1} G^2 D^{-1})^{-2} \boldsymbol{u}$$

and this value decreases with G^2 because $\boldsymbol{I}_p + D^{-1} G^2 D^{-1}$ increases. Show in a similar way that

$$\|\boldsymbol{\theta}_G - \boldsymbol{\theta}^*\|^2 = \|(D^2 + G^2)^{-1} G^2 \boldsymbol{\theta}^*\|^2 = \boldsymbol{\theta}^{*\top} \Gamma^{-1} \boldsymbol{\theta}^*$$

with $\Gamma = \bigl(\boldsymbol{I}_p + G^{-2} D^2\bigr)\bigl(\boldsymbol{I}_p + D^2 G^{-2}\bigr)$. Show that the matrix Γ monotonously increases and thus Γ^{-1} monotonously decreases as a function of the symmetric matrix $B = G^{-2}$.

Putting together the results about the bias and the variance of $\tilde{\boldsymbol{\theta}}_G$ yields the statement about the quadratic risk.

Theorem 4.7.2. *Assume the model* $\boldsymbol{Y} = \Psi^\top \boldsymbol{\theta}^* + \boldsymbol{\varepsilon}$ *with* $\operatorname{Var}(\boldsymbol{\varepsilon}) = \sigma^2 \boldsymbol{I}_n$. *Then the estimate* $\tilde{\boldsymbol{\theta}}_G$ *fulfills*

$$\mathbb{E}\|\tilde{\boldsymbol{\theta}}_G - \boldsymbol{\theta}^*\|^2 = \|\boldsymbol{\theta}_G - \boldsymbol{\theta}^*\|^2 + \operatorname{tr}\bigl(D_G^{-2} D^2 D_G^{-2}\bigr).$$

This result is called the *bias-variance decomposition*. The choice of a proper regularization is usually based on this decomposition: one selects a regularization from a given class to provide the minimal possible risk. This approach is referred to as *bias-variance trade-off*.

4.7.3 Inference for the Penalized MLE

Here we discuss some properties of the penalized MLE $\tilde{\boldsymbol{\theta}}_G$. In particular, we focus on the construction of confidence and concentration sets based on the penalized log-likelihood. We know that the regularized estimate $\tilde{\boldsymbol{\theta}}_G$ is the empirical counterpart of the value $\boldsymbol{\theta}_G$ which solves the regularized deterministic problem (4.32). We also know that the key results are expressed via the value of the supremum $\sup_{\boldsymbol{\theta}} L_G(\boldsymbol{\theta}) - L_G(\boldsymbol{\theta}_G)$. The next result extends Theorem 4.6.1 to the penalized likelihood.

Theorem 4.7.3. *Let* $L_G(\boldsymbol{\theta})$ *be the penalized log-likelihood from* (4.30). *Then*

$$2L_G(\tilde{\boldsymbol{\theta}}_G, \boldsymbol{\theta}_G) = \bigl(\tilde{\boldsymbol{\theta}}_G - \boldsymbol{\theta}_G\bigr)^\top D_G^2 \bigl(\tilde{\boldsymbol{\theta}}_G - \boldsymbol{\theta}_G\bigr) \tag{4.34}$$

$$= \sigma^{-2} \boldsymbol{\varepsilon}^\top \Pi_G \, \boldsymbol{\varepsilon} \tag{4.35}$$

with $\Pi_G = \Psi^\top \bigl(\Psi\Psi^\top + \sigma^2 G^2\bigr)^{-1} \Psi$.

In general the matrix Π_G is not a projector and hence, $\sigma^{-2}\boldsymbol{\varepsilon}^\top \Pi_G \boldsymbol{\varepsilon}$ is not χ^2-distributed, the chi-squared result does not apply.

Exercise 4.7.7. Prove (4.34).
Hint: apply the Taylor expansion to $L_G(\boldsymbol{\theta})$ at $\tilde{\boldsymbol{\theta}}_G$. Use that $\nabla L_G(\boldsymbol{\theta}_G) = 0$ and $-\nabla^2 L_G(\boldsymbol{\theta}) \equiv \sigma^{-2}\Psi\Psi^\top + G^2$.

Exercise 4.7.8. Prove (4.35).
Hint: show that $\tilde{\boldsymbol{\theta}}_G - \boldsymbol{\theta}_G = \mathcal{S}_G \boldsymbol{\varepsilon}$ with $\mathcal{S}_G = \bigl(\Psi\Psi^\top + \sigma^2 G^2\bigr)^{-1} \Psi$.

The straightforward corollaries of Theorem 4.7.3 are the concentration and confidence probabilities. Define the confidence set $\mathcal{E}_G(\mathfrak{z})$ for $\boldsymbol{\theta}_G$ as

$$\mathcal{E}_G(\mathfrak{z}) \stackrel{\text{def}}{=} \{\boldsymbol{\theta} : L_G(\tilde{\boldsymbol{\theta}}_G, \boldsymbol{\theta}) \le \mathfrak{z}\}.$$

The definition implies the following result for the coverage probability:

$$\mathbb{P}\bigl(\mathcal{E}_G(\mathfrak{z}) \not\ni \boldsymbol{\theta}_G\bigr) \le \mathbb{P}\bigl(L_G(\tilde{\boldsymbol{\theta}}_G, \boldsymbol{\theta}_G) > \mathfrak{z}\bigr).$$

Now the representation (4.35) for $L_G(\tilde{\boldsymbol{\theta}}_G, \boldsymbol{\theta}_G)$ reduces the problem to a deviation bound for a quadratic form. We apply the general result of Sect. A.1.

Theorem 4.7.4. *Let $L_G(\boldsymbol{\theta})$ be the penalized log-likelihood from* (4.30) *and let $\boldsymbol{\varepsilon} \sim \mathcal{N}(0, \sigma^2 \boldsymbol{I}_n)$. Then it holds with $\mathtt{p}_G = \operatorname{tr}(\Pi_G)$ and $v_G^2 = 2\operatorname{tr}(\Pi_G^2)$ that*

$$\mathbb{P}\bigl(2L_G(\tilde{\boldsymbol{\theta}}_G, \boldsymbol{\theta}_G) > \mathtt{p}_G + (2v_G\mathtt{x}^{1/2}) \vee (6\mathtt{x})\bigr) \le \exp(-\mathtt{x}).$$

Similarly one can state the concentration result. With $D_G^2 = \sigma^{-2}\Psi\Psi^\top + G^2$

$$2L_G(\tilde{\boldsymbol{\theta}}_G, \boldsymbol{\theta}_G) = \bigl\| D_G(\tilde{\boldsymbol{\theta}}_G - \boldsymbol{\theta}_G)\bigr\|^2$$

and the result of Theorem 4.7.4 can be restated as the concentration bound:

$$\mathbb{P}\bigl(\|D_G(\tilde{\boldsymbol{\theta}}_G - \boldsymbol{\theta}_G)\|^2 > \mathtt{p}_G + (2v_G\mathtt{x}^{1/2}) \vee (6\mathtt{x})\bigr) \le \exp(-\mathtt{x}).$$

In other words, $\tilde{\boldsymbol{\theta}}_G$ concentrates on the set $\mathcal{A}(\mathfrak{z}, \boldsymbol{\theta}_G) = \bigl\{\boldsymbol{\theta} : \|\boldsymbol{\theta} - \boldsymbol{\theta}_G\|^2 \le 2\mathfrak{z}\bigr\}$ for $2\mathfrak{z} > \mathtt{p}_G$.

4.7.4 Projection and Shrinkage Estimates

Consider a linear model $\boldsymbol{Y} = \Psi^\top\boldsymbol{\theta}^* + \boldsymbol{\varepsilon}$ in which the matrix Ψ is orthonormal in the sense $\Psi\Psi^\top = \boldsymbol{I}_p$. Then the multiplication with Ψ maps this model in the sequence space model $\boldsymbol{Z} = \boldsymbol{\theta}^* + \boldsymbol{\xi}$, where $\boldsymbol{Z} = \Psi\boldsymbol{Y} = (z_1, \ldots, z_p)^\top$ is the vector of empirical Fourier coefficients $z_j = \boldsymbol{\psi}_j^\top\boldsymbol{Y}$. The noise $\boldsymbol{\xi} = \Psi\boldsymbol{\varepsilon}$ borrows the feature of the original noise $\boldsymbol{\varepsilon}$: if $\boldsymbol{\varepsilon}$ is zero mean and homogeneous, the same applies to $\boldsymbol{\xi}$. The number of coefficients p can be large or even infinite. To get a sensible estimate, one has to apply some regularization method. The simplest one is called *projection*: one just considers the first m empirical coefficients $z_1, \ldots, z_m$ and drop the others. The corresponding parameter estimate $\tilde{\boldsymbol{\theta}}_m$ reads as

$$\tilde{\theta}_{m,j} = \begin{cases} z_j & \text{if } j \le m, \\ 0 & \text{otherwise.} \end{cases}$$

The response vector $\boldsymbol{f}^* = \mathbb{E}\boldsymbol{Y}$ is estimated by $\Psi^\top \tilde{\boldsymbol{\theta}}_m$ leading to the representation

$$\tilde{\boldsymbol{f}}_m = z_1 \boldsymbol{\psi}_1 + \ldots + z_m \boldsymbol{\psi}_m$$

with $z_j = \boldsymbol{\psi}_j^\top \boldsymbol{Y}$. In other words, $\tilde{\boldsymbol{f}}_m$ is just a projection of the observed vector $\boldsymbol{Y}$ onto the subspace L_m spanned by the first m basis vectors $\boldsymbol{\psi}_1, \ldots, \boldsymbol{\psi}_m$: $\mathrm{L}_m = \langle \boldsymbol{\psi}_1, \ldots, \boldsymbol{\psi}_m \rangle$. This explains the name of the method. Clearly one can study the properties of $\tilde{\boldsymbol{\theta}}_m$ or $\tilde{\boldsymbol{f}}_m$ using the methods of previous sections. However, one more question for this approach is still open: a proper choice of m. The standard way of accessing this issue is based on the analysis of the quadratic risk.

Consider first the prediction risk defined as $\mathcal{R}(\tilde{\boldsymbol{f}}_m) = \mathbb{E}\|\tilde{\boldsymbol{f}}_m - \boldsymbol{f}^*\|^2$. Below we focus on the case of a homogeneous noise with $\mathrm{Var}(\boldsymbol{\varepsilon}) = \sigma^2 \boldsymbol{I}_p$. An extension to the colored noise is possible. Recall that $\tilde{\boldsymbol{f}}_m$ effectively estimates the vector $\boldsymbol{f}_m = \Pi_m \boldsymbol{f}^*$, where Π_m is the projector on L_m; see Sect. 4.3.3. Moreover, the quadratic risk $\mathcal{R}(\tilde{\boldsymbol{f}}_m)$ can be decomposed as

$$\mathcal{R}(\tilde{\boldsymbol{f}}_m) = \|\boldsymbol{f}^* - \Pi_m \boldsymbol{f}^*\|^2 + \sigma^2 m = \sigma^2 m + \sum_{j=m+1}^{p} \theta_j^{*2}.$$

Obviously the squared bias $\|\boldsymbol{f}^* - \Pi_m \boldsymbol{f}^*\|^2$ decreases with m while the variance $\sigma^2 m$ linearly grows with m. Risk minimization leads to the so-called *bias-variance trade-off*: one selects m which minimizes the risk $\mathcal{R}(\tilde{\boldsymbol{f}}_m)$ over all possible m:

$$m^* \stackrel{\mathrm{def}}{=} \operatorname*{argmin}_m \mathcal{R}(\tilde{\boldsymbol{f}}_m) = \operatorname*{argmin}_m \bigl\{ \|\boldsymbol{f}^* - \Pi_m \boldsymbol{f}^*\|^2 + \sigma^2 m \bigr\}.$$

Unfortunately this choice requires some information about the bias $\|\boldsymbol{f}^* - \Pi_m \boldsymbol{f}^*\|$ which depends on the unknown vector $\boldsymbol{f}$. As this information is not available in typical situation, the value m^* is also called an *oracle* choice. A data-driven choice of m is one of the central issues in the nonparametric statistics.

The situation is not changed if we consider the estimation risk $\mathbb{E}\|\tilde{\boldsymbol{\theta}}_m - \boldsymbol{\theta}^*\|^2$. Indeed, the basis orthogonality $\Psi\Psi^\top = \boldsymbol{I}_p$ implies for $\boldsymbol{f}^* = \Psi^\top \boldsymbol{\theta}^*$

$$\|\tilde{\boldsymbol{f}}_m - \boldsymbol{f}^*\|^2 = \|\Psi^\top \tilde{\boldsymbol{\theta}}_m - \Psi^\top \boldsymbol{\theta}^*\|^2 = \|\tilde{\boldsymbol{\theta}}_m - \boldsymbol{\theta}^*\|^2$$

and minimization of the estimation risk coincides with minimization of the prediction risk.

A disadvantage of the projection method is that it either keeps each empirical coefficient z_m or completely discards it. An extension of the projection method is called *shrinkage*: one multiplies every empirical coefficient z_j with a factor $\alpha_j \in (0, 1)$. This leads to the *shrinkage* estimate $\tilde{\boldsymbol{\theta}}_{\boldsymbol{\alpha}}$ with

$$\tilde{\theta}_{\boldsymbol{\alpha},j} = \alpha_j z_j .$$

Here $\boldsymbol{\alpha}$ stands for the vector of coefficients α_j for $j = 1, \ldots, p$. A projection method is a special case of this shrinkage with α_j equal to one or zero. Another popular choice of the coefficients α_j is given by

$$\alpha_j = (1 - j/m)^{\beta} \mathbf{1}(j \leq m) \tag{4.36}$$

for some $\beta > 0$ and $m \leq p$. This choice ensures that the coefficients α_j smoothly approach zero as j approaches the value m, and α_j vanish for $j > m$. In this case, the vector $\boldsymbol{\alpha}$ is completely specified by two parameters m and β. The projection method corresponds to $\beta = 0$. The design orthogonality $\Psi\Psi^{\top} = \boldsymbol{I}_p$ yields again that the estimation risk $\mathbb{E}\|\tilde{\boldsymbol{\theta}}_{\boldsymbol{\alpha}} - \boldsymbol{\theta}^*\|^2$ coincides with the prediction risk $\mathbb{E}\|\tilde{\boldsymbol{f}}_{\boldsymbol{\alpha}} - \boldsymbol{f}^*\|^2$.

Exercise 4.7.9. Let $\operatorname{Var}(\boldsymbol{\varepsilon}) = \sigma^2 \boldsymbol{I}_p$. The risk $\mathcal{R}(\tilde{\boldsymbol{f}}_{\boldsymbol{\alpha}})$ of the shrinkage estimate $\tilde{\boldsymbol{f}}_{\boldsymbol{\alpha}}$ fulfills

$$\mathcal{R}(\tilde{\boldsymbol{f}}_{\boldsymbol{\alpha}}) \stackrel{\text{def}}{=} \mathbb{E}\|\tilde{\boldsymbol{f}}_{\boldsymbol{\alpha}} - \boldsymbol{f}^*\|^2 = \sum_{j=1}^{p} \theta_j^{*2}(1 - \alpha_j)^2 + \sum_{j=1}^{p} \alpha_j^2 \sigma^2.$$

Specify the cases of $\boldsymbol{\alpha} = \boldsymbol{\alpha}(m, \beta)$ from (4.36). Evaluate the variance term $\sum_j \alpha_j^2 \sigma^2$. Hint: approximate the sum over j by the integral $\int (1 - x/m)_+^{2\beta} dx$.

The oracle choice is again defined by risk minimization:

$$\boldsymbol{\alpha}^* \stackrel{\text{def}}{=} \operatorname*{argmin}_{\boldsymbol{\alpha}} \mathcal{R}(\tilde{\boldsymbol{f}}_{\boldsymbol{\alpha}}),$$

where minimization is taken over the class of all considered coefficient vectors $\boldsymbol{\alpha}$.

One way of obtaining a shrinkage estimate in the sequence space model $\boldsymbol{Z} = \boldsymbol{\theta}^* + \boldsymbol{\xi}$ is by using a roughness penalization. Let G be a symmetric matrix. Consider the regularized estimate $\tilde{\boldsymbol{\theta}}_G$ from (4.28). The next result claims that if G is a diagonal matrix, then $\tilde{\boldsymbol{\theta}}_G$ is a shrinkage estimate. Moreover, a general penalized MLE can be represented as shrinkage by an orthogonal basis transformation.

Theorem 4.7.5. *Let G be a diagonal matrix, $G = \operatorname{diag}(g_1, \ldots, g_p)$. The penalized MLE $\tilde{\boldsymbol{\theta}}_G$ in the sequence space model $\boldsymbol{Z} = \boldsymbol{\theta}^* + \boldsymbol{\xi}$ with $\boldsymbol{\xi} \sim \mathcal{N}(0, \sigma^2 \boldsymbol{I}_p)$ coincides with the shrinkage estimate $\tilde{\boldsymbol{\theta}}_{\boldsymbol{\alpha}}$ for $\alpha_j = (1 + \sigma^2 g_j^2)^{-1} \leq 1$. Moreover, a penalized MLE $\tilde{\boldsymbol{\theta}}_G$ for a general matrix G can be reduced to a shrinkage estimate by a basis transformation in the sequence space model.*

Proof. The first statement for a diagonal matrix G follows from the representation $\tilde{\boldsymbol{\theta}}_G = (\boldsymbol{I}_p + \sigma^2 G^2)^{-1} \boldsymbol{Z}$. Next, let U be an orthogonal transform leading to the diagonal representation $G^2 = U^{\top} D^2 U$ with $D^2 = \operatorname{diag}(g_1, \ldots, g_p)$. Then

$$U\tilde{\boldsymbol{\theta}}_G = (\boldsymbol{I}_p + \sigma^2 D^2)^{-1} U\boldsymbol{Z}$$

that is, $U\tilde{\boldsymbol{\theta}}_G$ is a shrinkage estimate in the transformed model $U\boldsymbol{Z} = U\boldsymbol{\theta}^* + U\boldsymbol{\xi}$.

In other words, roughness penalization results in some kind of shrinkage. Interestingly, the inverse statement holds as well.

Exercise 4.7.10. Let $\tilde{\boldsymbol{\theta}}_{\boldsymbol{\alpha}}$ is a shrinkage estimate for a vector $\boldsymbol{\alpha} = (\alpha_j)$. Then there is a diagonal penalty matrix G such that $\tilde{\boldsymbol{\theta}}_{\boldsymbol{\alpha}} = \tilde{\boldsymbol{\theta}}_G$.
Hint: define the jth diagonal entry g_j by the equation $\alpha_j = (1 + \sigma^2 g_j^2)^{-1}$.

4.7.5 Smoothness Constraints and Roughness Penalty Approach

Another way of reducing the complexity of the estimation procedure is based on smoothness constraints. The notion of smoothness originates from regression estimation. A nonlinear regression function f is expanded using a Fourier or some other functional basis and $\boldsymbol{\theta}^*$ is the corresponding vector of coefficients. Smoothness properties of the regression function imply certain rate of decay of the corresponding Fourier coefficients: the larger frequency is, the fewer amount of information about the regression function is contained in the related coefficient. This leads to the natural idea to replace the original optimization problem over the whole parameter space with the constrained optimization over a subset of "smooth" parameter vectors. Here we consider one popular example of Sobolev smoothness constraints which effectively means that the sth derivative of the function f^* has a bounded L_2-norm. A general Sobolev ball can be defined using a diagonal matrix G:

$$\mathcal{B}_G(R) \stackrel{\text{def}}{=} \|G\boldsymbol{\theta}\| \le R.$$

Now we consider a constrained ML problem:

$$\tilde{\boldsymbol{\theta}}_{G,R} = \operatorname*{argmax}_{\boldsymbol{\theta}\in\mathcal{B}_G(R)} L(\boldsymbol{\theta}) = \operatorname*{argmin}_{\boldsymbol{\theta}\in\Theta:\,\|G\boldsymbol{\theta}\|\le R} \|\boldsymbol{Y} - \Psi^\top\boldsymbol{\theta}\|^2. \tag{4.37}$$

The Lagrange multiplier method leads to an unconstrained problem

$$\tilde{\boldsymbol{\theta}}_{G,\lambda} = \operatorname*{argmin}_{\boldsymbol{\theta}} \bigl\{\|\boldsymbol{Y} - \Psi^\top\boldsymbol{\theta}\|^2 + \lambda\|G\boldsymbol{\theta}\|^2\bigr\}.$$

A proper choice of λ ensures that the solution $\tilde{\boldsymbol{\theta}}_{G,\lambda}$ belongs to $\mathcal{B}_G(R)$ and solves also the problem (4.37). So, the approach based on a Sobolev smoothness assumption leads back to regularization and shrinkage.

4.8 Shrinkage in a Linear Inverse Problem

This section extends the previous approaches to the situation with indirect observations. More precisely, we focus on the model

$$Y = Af^* + \varepsilon, \tag{4.38}$$

where A is a given linear operator (matrix) and f^* is the target of analysis. With the obvious change of notation this problem can be put back in the general linear setup $Y = \Psi^\top \theta + \varepsilon$. The special focus is due to the facts that the target can be high dimensional or even functional and that the product $A^\top A$ is usually badly posed and its inversion is a hard task. Below we consider separately the cases when the spectral representation for this problem is available and the general case.

4.8.1 *Spectral Cut-Off and Spectral Penalization: Diagonal Estimates*

Suppose that the eigenvectors of the matrix $A^\top A$ are available. This allows for reducing the model to the spectral representation by an orthogonal change of the coordinate system: $Z = \Lambda u + \Lambda^{1/2}\xi$ with a diagonal matrix $\Lambda = \operatorname{diag}\{\lambda_1, \ldots, \lambda_p\}$ and a homogeneous noise $\operatorname{Var}(\xi) = \sigma^2 I_p$; see Sect. 4.2.4. Below we assume without loss of generality that the eigenvalues λ_j are ordered and decrease with j. This spectral representation means that one observes empirical Fourier coefficients z_m described by the equation $z_j = \lambda_j u_j + \lambda_j^{1/2}\xi_j$ for $j = 1, \ldots, p$. The LSE or qMLE estimate of the spectral parameter u is given by

$$\tilde{u} = \Lambda^{-1} Z = (\lambda_1^{-1} z_1, \ldots, \lambda_p^{-1} z_p)^\top.$$

Exercise 4.8.1. Consider the spectral representation $Z = \Lambda u + \Lambda^{1/2}\xi$. The LSE $\tilde{u}$ reads as $\tilde{u} = \Lambda^{-1} Z$.

If the dimension p of the model is high or, specifically, if the spectral values λ_j rapidly go to zero, it might be useful to only track few coefficients $u_1, \ldots, u_m$ and to set all the remaining ones to zero. The corresponding estimate $\tilde{u}_m = (\tilde{u}_{m,1}, \ldots, \tilde{u}_{m,p})^\top$ reads as

$$\tilde{u}_{m,j} \stackrel{\text{def}}{=} \begin{cases} \lambda_j^{-1} z_j & \text{if } j \le m, \\ 0 & \text{otherwise.} \end{cases}$$

It is usually referred to as a *spectral cut-off* estimate.

Exercise 4.8.2. Consider the linear model $Y = Af^* + \varepsilon$. Let U be an orthogonal transform in $\mathbb{R}^p$ providing $UA^\top AU^\top = \Lambda$ with a diagonal matrix Λ leading to the spectral representation for $\boldsymbol{Z} = UA\boldsymbol{Y}$. Write the corresponding spectral cut-off estimate $\tilde{\boldsymbol{f}}_m$ for the original vector $\boldsymbol{f}^*$. Show that computing this estimate only requires to know the first m eigenvalues and eigenvectors of the matrix $A^\top A$.

Similarly to the direct case, a spectral cut-off can be extended to *spectral shrinkage*: one multiplies every empirical coefficient z_j with a factor $\alpha_j \in (0,1)$. This leads to the *spectral shrinkage* estimate $\tilde{\boldsymbol{u}}_{\boldsymbol{\alpha}}$ with $\tilde{u}_{\alpha,j} = \alpha_j \lambda_j^{-1} z_j$. Here $\boldsymbol{\alpha}$ stands for the vector of coefficients α_j for $j = 1, \ldots, p$. A spectral cut-off method is a special case of this shrinkage with α_j equal to one or zero.

Exercise 4.8.3. Specify the spectral shrinkage $\tilde{\boldsymbol{u}}_{\boldsymbol{\alpha}}$ with a given vector $\boldsymbol{\alpha}$ for the situation of Exercise 4.8.2.

The spectral cut-off method can be described as follows. Let $\boldsymbol{\psi}_1, \boldsymbol{\psi}_2, \ldots$ be the intrinsic orthonormal basis of the problem composed of the standardized eigenvectors of $A^\top A$ and leading to the spectral representation $\boldsymbol{Z} = \Lambda \boldsymbol{u} + \Lambda^{1/2}\boldsymbol{\xi}$ with the target vector $\boldsymbol{u}$. In terms of the original target $\boldsymbol{f}^*$, one is looking for a solution or an estimate in the form $\boldsymbol{f} = \sum_j u_j \boldsymbol{\psi}_j$. The design orthogonality allows to estimate every coefficient u_j independently of the others using the empirical Fourier coefficient $\boldsymbol{\psi}_j^\top \boldsymbol{Y}$. Namely, $\tilde{u}_j = \lambda_j^{-1}\boldsymbol{\psi}_j^\top \boldsymbol{Y} = \lambda_j^{-1} z_j$. The LSE procedure tries to recover $\boldsymbol{f}$ as the full sum $\tilde{\boldsymbol{f}} = \sum_j \tilde{u}_j \boldsymbol{\psi}_j$. The projection method suggests to cut this sum at the index m: $\tilde{\boldsymbol{f}}_m = \sum_{j \le m} \tilde{u}_j \boldsymbol{\psi}_j$, while the shrinkage procedure is based on downweighting the empirical coefficients $\tilde{u}_j$: $\tilde{\boldsymbol{f}}_{\boldsymbol{\alpha}} = \sum_j \alpha_j \tilde{u}_j \boldsymbol{\psi}_j$.
Next we study the risk of the shrinkage method. Orthonormality of the basis $\boldsymbol{\psi}_j$ allows to represent the loss as $\|\tilde{\boldsymbol{u}}_{\boldsymbol{\alpha}} - \boldsymbol{u}^*\|^2 = \|\tilde{\boldsymbol{f}}_{\boldsymbol{\alpha}} - \boldsymbol{f}^*\|^2$. Under the noise homogeneity one obtains the following result.

Theorem 4.8.1. *Let* $\boldsymbol{Z} = \Lambda \boldsymbol{u}^* + \Lambda^{1/2}\boldsymbol{\xi}$ *with* $\operatorname{Var}(\boldsymbol{\xi}) = \sigma^2 \boldsymbol{I}_p$. *It holds for the shrinkage estimate* $\tilde{\boldsymbol{u}}_{\boldsymbol{\alpha}}$

$$\mathcal{R}(\tilde{\boldsymbol{u}}_{\boldsymbol{\alpha}}) \stackrel{\text{def}}{=} \mathbb{E}\|\tilde{\boldsymbol{u}}_{\boldsymbol{\alpha}} - \boldsymbol{u}^*\|^2 = \sum_{j=1}^{p} |\alpha_j - 1|^2 u_j^{*2} + \sum_{j=1}^{p} \alpha_j^2 \sigma^2 \lambda_j^{-1}.$$

Proof. The empirical Fourier coefficients z_j are uncorrelated and $\mathbb{E} z_j = \lambda_j u_j^*$, $\operatorname{Var} z_j = \sigma^2 \lambda_j$. This implies

$$\mathbb{E}\|\tilde{\boldsymbol{u}}_{\boldsymbol{\alpha}} - \boldsymbol{u}^*\|^2 = \sum_{j=1}^{p} \mathbb{E}|\alpha_j \lambda_j^{-1} z_j - u_j^*|^2 = \sum_{j=1}^{p} \bigl\{ |\alpha_j - 1|^2 u_j^{*2} + \alpha_j^2 \sigma^2 \lambda_j^{-1} \bigr\}$$

as required.

Risk minimization leads to the oracle choice of the vector $\boldsymbol{\alpha}$ or

$$\boldsymbol{\alpha}^* = \operatorname*{argmin}_{\boldsymbol{\alpha}} \mathcal{R}(\tilde{\boldsymbol{u}}_{\boldsymbol{\alpha}})$$

where the minimum is taken over the set of all admissible vectors $\boldsymbol{\alpha}$.

Similar analysis can be done for the spectral cut-off method.

Exercise 4.8.4. The risk of the spectral cut-off estimate $\tilde{\boldsymbol{u}}_m$ fulfills

$$\mathcal{R}(\tilde{\boldsymbol{u}}_m) = \sum_{j=1}^{m} \lambda_j^{-1}\sigma^2 + \sum_{j=m+1}^{p} u_j^{*2}.$$

Specify the choice of the oracle cut-off index m^*.

4.8.2 Galerkin Method

A general problem with the spectral shrinkage approach is that it requires to precisely know the intrinsic basis $\boldsymbol{\psi}_1, \boldsymbol{\psi}_2, \ldots$ or equivalently the eigenvalue decomposition of A leading to the spectral representation. After this basis is fixed, one can apply the projection or shrinkage method using the corresponding Fourier coefficients. In some situations this basis is hardly available or difficult to compute. A possible way out of this problem is to take some other orthogonal basis $\boldsymbol{\phi}_1, \boldsymbol{\phi}_2, \ldots$ which is tractable and convenient but does not lead to the spectral representation of the model. The Galerkin method is based on projecting the original high dimensional problem to a lower dimensional problem in terms of the new basic $\{\boldsymbol{\phi}_j\}$. Namely, without loss of generality suppose that the target function $\boldsymbol{f}^*$ can be decomposed as

$$\boldsymbol{f}^* = \sum_j u_j \boldsymbol{\phi}_j .$$

This can be achieved, e.g., if $\boldsymbol{f}^*$ belongs to some Hilbert space and $\{\boldsymbol{\phi}_j\}$ is an orthonormal basis in this space. Now we cut this sum and replace this exact decomposition by a finite approximation

$$\boldsymbol{f}^* \approx \boldsymbol{f}_m = \sum_{j \le m} u_j \boldsymbol{\phi}_j = \Phi_m^\top \boldsymbol{u}_m ,$$

where $\boldsymbol{u}_m = (u_1, \ldots, u_m)^\top$ and the matrix Φ_m is built of the vectors $\boldsymbol{\phi}_1, \ldots, \boldsymbol{\phi}_m$: $\Phi_m = (\boldsymbol{\phi}_1, \ldots, \boldsymbol{\phi}_m)$. Now we plug this decomposition in the original equation $\boldsymbol{Y} = A\boldsymbol{f}^* + \boldsymbol{\varepsilon}$. This leads to the linear model $\boldsymbol{Y} = A\Phi_m^\top \boldsymbol{u}_m + \boldsymbol{\varepsilon} = \Psi_m^\top \boldsymbol{u}_m + \boldsymbol{\varepsilon}$ with $\Psi_m = \Phi_m A^\top$. The corresponding (quasi) MLE reads

$$\tilde{\boldsymbol{u}}_m = \bigl(\Psi_m \Psi_m^\top\bigr)^{-1} \Psi_m \boldsymbol{Y}.$$

Note that for computing this estimate one only needs to evaluate the action of the operator A on the basis functions $\boldsymbol{\phi}_1, \ldots, \boldsymbol{\phi}_m$ and on the data $\boldsymbol{Y}$. With this estimate $\tilde{\boldsymbol{u}}_m$ of the vector $\boldsymbol{u}^*$, one obtains the response estimate $\tilde{\boldsymbol{f}}_m$ of the form

$$\tilde{f}_m = \Phi_m^\top \tilde{u}_m = \tilde{u}_1 \phi_1 + \ldots + \tilde{u}_m \phi_m .$$

The properties of this estimate can be studied in the same way as for a general qMLE in a linear model: the true data distribution follows (4.38) while we use the approximating model $Y = A f_m^* + \varepsilon$ with $\varepsilon \sim \mathcal{N}(0, \sigma^2 I)$ for building the quasi likelihood.

A further extension of the qMLE approach concerns the case when the operator A is not precisely known. Instead, an approximation or an estimate $\tilde{A}$ is available. The pragmatic way of tackling this problem is to use the model $Y = \tilde{A} f_m^* + \varepsilon$ for building the quasi likelihood. The use of the Galerkin method is quite natural in this situation because the spectral representation for $\tilde{A}$ will not necessarily result in a similar representation for the true operator A.

4.9 Semiparametric Estimation

This section discusses the situation when the target of estimation does not coincide with the parameter vector. This problem is usually referred to as *semiparametric estimation*. One typical example is the problem of estimating a part of the parameter vector. More generally one can try to estimate a given function/functional of the unknown parameter. We focus here on linear modeling, that is, the considered model and the considered mapping of the parameter space to the target space are linear. For the ease of presentation we assume everywhere the homogeneous noise with $\mathrm{Var}(\varepsilon) = \sigma^2 I_n$.

4.9.1 (θ, η)- and υ-Setup

This section presents two equivalent descriptions of the semiparametric problem. The first one assumes that the total parameter vector can be decomposed into the *target* parameter θ and the *nuisance* parameter η. The second one operates with the total parameter υ and the target θ is a linear mapping of υ.

We start with the (θ, η)-setup. Let the response Y be modeled in dependence of two sets of factors: $\{\psi_j, j = 1, \ldots, p\}$ and $\{\phi_m, m = 1, \ldots, p_1\}$. We are mostly interested in understanding the impact of the first set $\{\psi_j\}$ but we cannot ignore the influence of the $\{\phi_m\}$'s. Otherwise the model would be incomplete. This situation can be described by the linear model

$$Y = \Psi^\top \theta^* + \Phi^\top \eta^* + \varepsilon, \tag{4.39}$$

where Ψ is the $p \times n$ matrix with the columns ψ_j, while Φ is the $p_1 \times n$-matrix with the columns ϕ_m. We primarily aim at recovering the vector θ^*, while the coefficients η^* are of secondary importance. The corresponding (quasi) log-likelihood reads as

$$\mathcal{L}(\boldsymbol{\theta}, \boldsymbol{\eta}) = -(2\sigma^2)^{-1} \| \boldsymbol{Y} - \Psi^\top \boldsymbol{\theta} - \Phi^\top \boldsymbol{\eta} \|^2 + R,$$

where R denotes the remainder term which does not depend on the parameters $\boldsymbol{\theta}, \boldsymbol{\eta}$. The more general $\boldsymbol{\upsilon}$-setup considers a general linear model

$$\boldsymbol{Y} = \Upsilon^\top \boldsymbol{\upsilon}^* + \boldsymbol{\varepsilon}, \tag{4.40}$$

where Υ is $p^* \times n$ matrix of p^* factors, and the target of estimation is a linear mapping $\boldsymbol{\theta}^* = P\boldsymbol{\upsilon}^*$ for a given operator P from $\mathbb{R}^{p^*}$ to $\mathbb{R}^p$. Obviously the $(\boldsymbol{\theta}, \boldsymbol{\eta})$-setup is a special case of the $\boldsymbol{\upsilon}$-setup. However, a general $\boldsymbol{\upsilon}$-setup can be reduced back to the $(\boldsymbol{\theta}, \boldsymbol{\eta})$-setup by a change of variable.

Exercise 4.9.1. Consider the sequence space model $\boldsymbol{Y} = \boldsymbol{\upsilon}^* + \boldsymbol{\xi}$ in $\mathbb{R}^p$ and let the target of estimation be the sum of the coefficients $\upsilon_1^* + \ldots + \upsilon_p^*$. Describe the υ-setup for the problem. Reduce to $(\boldsymbol{\theta}, \boldsymbol{\eta})$-setup by an orthogonal change of the basis.

In the $\boldsymbol{\upsilon}$-setup, the (quasi) log-likelihood reads as

$$\mathcal{L}(\boldsymbol{\upsilon}) = -(2\sigma^2)^{-1} \| \boldsymbol{Y} - \Upsilon^\top \boldsymbol{\upsilon} \|^2 + R,$$

where R is the remainder which does not depend on $\boldsymbol{\upsilon}$. It implies quadraticity of the log-likelihood $\mathcal{L}(\boldsymbol{\upsilon})$: given by

$$\mathcal{D}^2 = -\nabla^2 \mathcal{L}(\boldsymbol{\upsilon}) = \operatorname{Var}\{\nabla \mathcal{L}(\boldsymbol{\upsilon})\} = \sigma^{-2} \Upsilon \Upsilon^\top. \tag{4.41}$$

Exercise 4.9.2. Check the statements in (4.41).

Exercise 4.9.3. Show that for the model (4.39) holds with $\Upsilon = \binom{\Psi}{\Phi}$

$$\mathcal{D}^2 = \sigma^{-2} \Upsilon \Upsilon^\top = \sigma^{-2} \begin{pmatrix} \Psi\Psi^\top & \Psi\Phi^\top \\ \Phi\Psi^\top & \Phi\Phi^\top \end{pmatrix}, \tag{4.42}$$

$$\nabla \mathcal{L}(\boldsymbol{\upsilon}^*) = \sigma^{-2} \Upsilon \boldsymbol{\varepsilon} = \sigma^{-2} \begin{pmatrix} \Psi \boldsymbol{\varepsilon} \\ \Phi \boldsymbol{\varepsilon} \end{pmatrix}.$$

4.9.2 *Orthogonality and Product Structure*

Consider the model (4.39) under the *orthogonality* condition $\Psi\Phi^\top = 0$. This condition effectively means that the factors of interest $\{\boldsymbol{\psi}_j\}$ are orthogonal to the nuisance factors $\{\boldsymbol{\phi}_m\}$. An important feature of this orthogonal case is that the model has the product structure leading to the additive form of the log-likelihood. Consider the partial $\boldsymbol{\theta}$-model $\boldsymbol{Y} = \Psi^\top \boldsymbol{\theta} + \boldsymbol{\varepsilon}$ with the (quasi) log-likelihood

$$L(\boldsymbol{\theta}) = -(2\sigma^2)^{-1}\|\boldsymbol{Y} - \Psi^\top \boldsymbol{\theta}\|^2 + R$$

Similarly $L_1(\boldsymbol{\eta}) = -(2\sigma^2)^{-1}\|\boldsymbol{Y} - \Phi^\top \boldsymbol{\eta}\|^2 + R_1$ denotes the log-likelihood in the partial $\boldsymbol{\eta}$-model $\boldsymbol{Y} = \Phi^\top \boldsymbol{\theta} + \boldsymbol{\varepsilon}$.

Theorem 4.9.1. *Assume the condition* $\Psi\Phi^\top = 0$. *Then*

$$\mathcal{L}(\boldsymbol{\theta}, \boldsymbol{\eta}) = L(\boldsymbol{\theta}) + L_1(\boldsymbol{\eta}) + R(\boldsymbol{Y}), \tag{4.43}$$

where $R(\boldsymbol{Y})$ *is independent of* $\boldsymbol{\theta}$ *and* $\boldsymbol{\eta}$. *This implies the block diagonal structure of the matrix* $\mathcal{D}^2 = \sigma^{-2}\Upsilon\Upsilon^\top$:

$$\mathcal{D}^2 = \sigma^{-2}\begin{pmatrix} \Psi\Psi^\top & 0 \\ 0 & \Phi\Phi^\top \end{pmatrix} = \begin{pmatrix} D^2 & 0 \\ 0 & H^2 \end{pmatrix}, \tag{4.44}$$

with $D^2 = \sigma^{-2}\Psi\Psi^\top$, $H^2 = \sigma^{-2}\Phi\Phi^\top$.

Proof. The formula (4.44) follows from (4.42) and the orthogonality condition. The statement (4.43) follows if we show that the difference

$$\mathcal{L}(\boldsymbol{\theta}, \boldsymbol{\eta}) - L(\boldsymbol{\theta}) - L_1(\boldsymbol{\eta})$$

does not depend on $\boldsymbol{\theta}$ and $\boldsymbol{\eta}$. This is a quadratic expression in $\boldsymbol{\theta}, \boldsymbol{\eta}$, so it suffices to check its first and the second derivative w.r.t. the parameters. For the first derivative, it holds by the orthogonality condition

$$\nabla_{\boldsymbol{\theta}}\mathcal{L}(\boldsymbol{\theta}, \boldsymbol{\eta}) \stackrel{\text{def}}{=} \partial\mathcal{L}(\boldsymbol{\theta}, \boldsymbol{\eta})/\partial\boldsymbol{\theta} = \sigma^{-2}\Psi(\boldsymbol{Y} - \Psi^\top\boldsymbol{\theta} - \Phi^\top\boldsymbol{\eta}) = \sigma^{-2}\Psi(\boldsymbol{Y} - \Psi^\top\boldsymbol{\theta})$$

that coincides with $\nabla L(\boldsymbol{\theta})$. Similarly $\nabla_{\boldsymbol{\theta}}\mathcal{L}(\boldsymbol{\theta}, \boldsymbol{\eta}) = \nabla L_1(\boldsymbol{\eta})$ yielding

$$\nabla\bigl\{\mathcal{L}(\boldsymbol{\theta}, \boldsymbol{\eta}) - L(\boldsymbol{\theta}) - L_1(\boldsymbol{\eta})\bigr\} = 0.$$

The identities (4.41) and (4.42) imply that $\nabla^2\bigl\{\mathcal{L}(\boldsymbol{\theta}, \boldsymbol{\eta}) - L(\boldsymbol{\theta}) - L_1(\boldsymbol{\eta})\bigr\} = 0$. This implies the desired assertion.

Exercise 4.9.4. Check the statement (4.43) by direct computations. Describe the term $R(\boldsymbol{Y})$.

Now we demonstrate how the general case can be reduced to the orthogonal one by a linear transformation of the nuisance parameter. Let C be a $p \times p_1$ matrix. Define $\breve{\boldsymbol{\eta}} = \boldsymbol{\eta} + C^\top\boldsymbol{\theta}$. Then the model equation $\boldsymbol{Y} = \Psi^\top\boldsymbol{\theta} + \Phi^\top\boldsymbol{\eta} + \boldsymbol{\varepsilon}$ can be rewritten as

$$\boldsymbol{Y} = \Psi^\top\boldsymbol{\theta} + \Phi^\top(\breve{\boldsymbol{\eta}} - C^\top\boldsymbol{\theta}) + \boldsymbol{\varepsilon} = (\Psi - C\Phi)^\top\boldsymbol{\theta} + \Phi^\top\breve{\boldsymbol{\eta}} + \boldsymbol{\varepsilon}.$$

Now we select C to ensure the orthogonality. This leads to the equation

$$(\Psi - C\Phi)\Phi^\top = 0$$

or $C = \Psi\Phi^\top(\Phi\Phi^\top)^{-1}$. So, the original model can be rewritten as

$$\begin{aligned} \boldsymbol{Y} &= \breve{\Psi}^\top\boldsymbol{\theta} + \Phi^\top\breve{\boldsymbol{\eta}} + \boldsymbol{\varepsilon}, \\ \breve{\Psi} &= \Psi - C\Phi = \Psi(I - \Pi_\eta), \end{aligned} \tag{4.45}$$

where $\Pi_\eta = \Phi^\top(\Phi\Phi^\top)^{-1}\Phi$ being the projector on the linear subspace spanned by the nuisance factors $\{\boldsymbol{\phi}_m\}$. This construction has a natural interpretation: correction the $\boldsymbol{\theta}$-factors $\boldsymbol{\psi}_1, \ldots, \boldsymbol{\psi}_p$ by removing their interaction with the nuisance factors $\boldsymbol{\phi}_1, \ldots, \boldsymbol{\phi}_{p_1}$ reduces the general case to the orthogonal one. We summarize:

Theorem 4.9.2. *The linear model* (4.39) *can be represented in the orthogonal form*

$$\boldsymbol{Y} = \breve{\Psi}^\top\boldsymbol{\theta} + \Phi^\top\breve{\boldsymbol{\eta}} + \boldsymbol{\varepsilon},$$

where $\breve{\Psi}$ *from* (4.45) *satisfies* $\breve{\Psi}\Phi^\top = 0$ *and* $\breve{\boldsymbol{\eta}} = \boldsymbol{\eta} + C^\top\boldsymbol{\theta}$ *for* $C = \Psi\Phi^\top(\Phi\Phi^\top)^{-1}$. *Moreover, it holds for* $\boldsymbol{\upsilon} = (\boldsymbol{\theta}, \boldsymbol{\eta})$

$$\mathcal{L}(\boldsymbol{\upsilon}) = \breve{L}(\boldsymbol{\theta}) + L_1(\breve{\boldsymbol{\eta}}) + R(\boldsymbol{Y}) \tag{4.46}$$

with

$$\begin{aligned} \breve{L}(\boldsymbol{\theta}) &= -(2\sigma^2)^{-1}\|\boldsymbol{Y} - \breve{\Psi}^\top\boldsymbol{\theta}\|^2 + R, \\ L_1(\boldsymbol{\eta}) &= -(2\sigma^2)^{-1}\|\boldsymbol{Y} - \Phi^\top\boldsymbol{\eta}\|^2 + R_1. \end{aligned}$$

Exercise 4.9.5. Show that for $C = \Psi\Phi^\top(\Phi\Phi^\top)^{-1}$

$$\nabla\breve{L}(\boldsymbol{\theta}) = \nabla_\theta\mathcal{L}(\boldsymbol{\upsilon}) - C\nabla_\eta\mathcal{L}(\boldsymbol{\upsilon}).$$

Exercise 4.9.6. Show that the remainder term $R(\boldsymbol{Y})$ in the Eq. (4.46) is the same as in the orthogonal case (4.43).

Exercise 4.9.7. Show that $\breve{\Psi}\breve{\Psi}^\top < \Psi\Psi^\top$ if $\Psi\Phi^\top \neq 0$.
Hint: for any vector $\boldsymbol{\gamma} \in \mathbb{R}^p$, it holds with $\boldsymbol{h} = \Psi^\top\boldsymbol{\gamma}$

$$\boldsymbol{\gamma}^\top\breve{\Psi}\breve{\Psi}^\top\boldsymbol{\gamma} = \|(\Psi - \Psi\Pi_\eta)^\top\boldsymbol{\gamma}\|^2 = \|\boldsymbol{h} - \Pi_\eta\boldsymbol{h}\|^2 \le \|\boldsymbol{h}\|^2.$$

Moreover, the equality here for any $\boldsymbol{\gamma}$ is only possible if

$$\Pi_\eta\boldsymbol{h} = \Phi^\top(\Phi\Phi^\top)^{-1}\Phi\Psi^\top\boldsymbol{\gamma} \equiv 0,$$

that is, if $\Phi\Psi^\top = 0$.

4.9.3 Partial Estimation

This section explains the important notion of partial estimation which is quite natural and transparent in the $(\boldsymbol{\theta}, \boldsymbol{\eta})$-setup. Let some value $\boldsymbol{\eta}^\circ$ of the nuisance parameter be fixed. A particular case of this sort is just ignoring the factors $\{\boldsymbol{\phi}_m\}$ corresponding to the nuisance component, that is, one uses $\boldsymbol{\eta}^\circ \equiv 0$. This approach is reasonable in certain situation, e.g. in context of projection method or spectral cut-off.

Define the estimate $\tilde{\boldsymbol{\theta}}(\boldsymbol{\eta}^\circ)$ by partial optimization of the joint log-likelihood $\mathcal{L}(\boldsymbol{\theta}, \boldsymbol{\eta}^\circ)$ w.r.t. the first parameter $\boldsymbol{\theta}$:

$$\tilde{\boldsymbol{\theta}}(\boldsymbol{\eta}^\circ) = \operatorname*{argmax}_{\boldsymbol{\theta}} \mathcal{L}(\boldsymbol{\theta}, \boldsymbol{\eta}^\circ).$$

Obviously $\tilde{\boldsymbol{\theta}}(\boldsymbol{\eta}^\circ)$ is the MLE in the residual model $\boldsymbol{Y} - \Phi^\top \boldsymbol{\eta}^\circ = \Psi^\top \boldsymbol{\theta}^* + \boldsymbol{\varepsilon}$:

$$\tilde{\boldsymbol{\theta}}(\boldsymbol{\eta}^\circ) = \bigl(\Psi\Psi^\top\bigr)^{-1}\Psi(\boldsymbol{Y} - \Phi^\top \boldsymbol{\eta}^\circ).$$

This allows for describing the properties of the partial estimate $\tilde{\boldsymbol{\theta}}(\boldsymbol{\eta}^\circ)$ similarly to the usual parametric situation.

Theorem 4.9.3. *Consider the model* (4.39). *Then the partial estimate $\tilde{\boldsymbol{\theta}}(\boldsymbol{\eta}^\circ)$ fulfills*

$$\mathbb{E}\tilde{\boldsymbol{\theta}}(\boldsymbol{\eta}^\circ) = \boldsymbol{\theta}^* + \bigl(\Psi\Psi^\top\bigr)^{-1}\Psi\Phi^\top(\boldsymbol{\eta}^* - \boldsymbol{\eta}^\circ), \qquad \operatorname{Var}\bigl\{\tilde{\boldsymbol{\theta}}(\boldsymbol{\eta}^\circ)\bigr\} = \sigma^2\bigl(\Psi\Psi^\top\bigr)^{-1}.$$

In other words, $\tilde{\boldsymbol{\theta}}(\boldsymbol{\eta})$ has the same variance as the MLE in the partial model $\boldsymbol{Y} = \Psi^\top \boldsymbol{\theta}^* + \boldsymbol{\varepsilon}$ but it is biased if $\Psi\Phi^\top(\boldsymbol{\eta}^* - \boldsymbol{\eta}^\circ) \neq 0$. The ideal situation corresponds to the case when $\boldsymbol{\eta}^\circ = \boldsymbol{\eta}^*$. Then $\tilde{\boldsymbol{\theta}}(\boldsymbol{\eta}^*)$ is the MLE in the correctly specified $\boldsymbol{\theta}$-model: with $\boldsymbol{Y}(\boldsymbol{\eta}^*) \stackrel{\text{def}}{=} \boldsymbol{Y} - \Phi^\top \boldsymbol{\eta}^*$,

$$\boldsymbol{Y}(\boldsymbol{\eta}^*) = \Psi^\top \boldsymbol{\theta}^* + \boldsymbol{\varepsilon}.$$

An interesting and natural question is a legitimation of the partial estimation method: under which conditions it is justified and does not produce any estimation bias. The answer is given by Theorem 4.9.1: the orthogonality condition $\Psi\Phi^\top = 0$ would ensure the desired feature because of the decomposition (4.43).

Theorem 4.9.4. *Assume orthogonality $\Psi\Phi^\top = 0$. Then the partial estimate $\tilde{\boldsymbol{\theta}}(\boldsymbol{\eta}^\circ)$ does not depend on the nuisance parameter $\boldsymbol{\eta}^\circ$ used:*

$$\tilde{\boldsymbol{\theta}} = \tilde{\boldsymbol{\theta}}(\boldsymbol{\eta}^\circ) = \tilde{\boldsymbol{\theta}}(\boldsymbol{\eta}^*) = \bigl(\Psi\Psi^\top\bigr)^{-1}\Psi\boldsymbol{Y}.$$

In particular, one can ignore the nuisance parameter and estimate $\boldsymbol{\theta}^*$ from the partial incomplete model $\boldsymbol{Y} = \Psi^\top \boldsymbol{\theta}^* + \boldsymbol{\varepsilon}$.

Exercise 4.9.8. Check that the partial derivative $\frac{\partial}{\partial\boldsymbol{\theta}}\mathcal{L}(\boldsymbol{\theta},\boldsymbol{\eta})$ does not depend on $\boldsymbol{\eta}$ under the orthogonality condition.

The partial estimation can be considered in context of estimating the nuisance parameter $\boldsymbol{\eta}$ by inverting the role of $\boldsymbol{\theta}$ and $\boldsymbol{\eta}$. Namely, given a fixed value $\boldsymbol{\theta}^{\circ}$, one can optimize the joint log-likelihood $\mathcal{L}(\boldsymbol{\theta},\boldsymbol{\eta})$ w.r.t. the second argument $\boldsymbol{\eta}$ leading to the estimate

$$\tilde{\boldsymbol{\eta}}(\boldsymbol{\theta}^{\circ}) \stackrel{\text{def}}{=} \operatorname*{argmax}_{\boldsymbol{\eta}} \mathcal{L}(\boldsymbol{\theta}^{\circ},\boldsymbol{\eta})$$

In the orthogonal situation the initial point $\boldsymbol{\theta}^{\circ}$ is not important and one can use the partial incomplete model $\boldsymbol{Y} = \Phi^{\top}\boldsymbol{\eta}^{*} + \boldsymbol{\varepsilon}$.

4.9.4 *Profile Estimation*

This section discusses one general *profile likelihood* method of estimating the target parameter $\boldsymbol{\theta}$ in the semiparametric situation. Later we show its optimality and R-efficiency. The method suggests to first estimate the entire parameter vector $\boldsymbol{\upsilon}$ by using the (quasi) ML method. Then the operator P is applied to the obtained estimate $\tilde{\boldsymbol{\upsilon}}$ to produce the estimate $\tilde{\boldsymbol{\theta}}$. One can describe this method as

$$\tilde{\boldsymbol{\upsilon}} = \operatorname*{argmax}_{\boldsymbol{\upsilon}} \mathcal{L}(\boldsymbol{\upsilon}), \qquad \tilde{\boldsymbol{\theta}} = P\tilde{\boldsymbol{\upsilon}}. \tag{4.47}$$

The first step here is the usual LS estimation of $\boldsymbol{\upsilon}^{*}$ in the linear model (4.40):

$$\tilde{\boldsymbol{\upsilon}} = \operatorname*{arginf}_{\boldsymbol{\upsilon}} \|\boldsymbol{Y} - \Upsilon^{\top}\boldsymbol{\upsilon}\|^{2} = \bigl(\Upsilon\Upsilon^{\top}\bigr)^{-1}\Upsilon\boldsymbol{Y}.$$

The estimate $\tilde{\boldsymbol{\theta}}$ is obtained by applying P to $\tilde{\boldsymbol{\upsilon}}$:

$$\tilde{\boldsymbol{\theta}} = P\tilde{\boldsymbol{\upsilon}} = P\bigl(\Upsilon\Upsilon^{\top}\bigr)^{-1}\Upsilon\boldsymbol{Y} = \mathcal{S}\boldsymbol{Y} \tag{4.48}$$

with $\mathcal{S} = P\bigl(\Upsilon\Upsilon^{\top}\bigr)^{-1}\Upsilon$. The properties of this estimate can be studied using the decomposition $\boldsymbol{Y} = \boldsymbol{f}^{*} + \boldsymbol{\varepsilon}$ with $\boldsymbol{f}^{*} = \mathbb{E}\boldsymbol{Y}$; cf. Sect. 4.4. In particular, it holds

$$\mathbb{E}\tilde{\boldsymbol{\theta}} = \mathcal{S}\boldsymbol{f}^{*}, \qquad \operatorname{Var}(\tilde{\boldsymbol{\theta}}) = \mathcal{S}\operatorname{Var}(\boldsymbol{\varepsilon})\mathcal{S}^{\top}. \tag{4.49}$$

If the noise $\boldsymbol{\varepsilon}$ is homogeneous with $\operatorname{Var}(\boldsymbol{\varepsilon}) = \sigma^{2}\boldsymbol{I}_{n}$, then

$$\operatorname{Var}(\tilde{\boldsymbol{\theta}}) = \sigma^{2}\mathcal{S}\mathcal{S}^{\top} = \sigma^{2}P\bigl(\Upsilon\Upsilon^{\top}\bigr)^{-1}P^{\top}. \tag{4.50}$$

The next theorem summarizes our findings.

Theorem 4.9.5. *Consider the model* (4.40) *with homogeneous error* $\mathrm{Var}(\boldsymbol{\varepsilon}) = \sigma^2 \boldsymbol{I}_n$. *The profile MLE* $\tilde{\boldsymbol{\theta}}$ *follows* (4.48). *Its means and variance are given by* (4.49) *and* (4.50).

The profile MLE is usually written in the $(\boldsymbol{\theta}, \boldsymbol{\eta})$-setup. Let $\boldsymbol{v} = (\boldsymbol{\theta}, \boldsymbol{\eta})$. Then the target $\boldsymbol{\theta}$ is obtained by projecting the MLE $(\tilde{\boldsymbol{\theta}}, \tilde{\boldsymbol{\eta}})$ on the $\boldsymbol{\theta}$-coordinates. This procedure can be formalized as

$$\tilde{\boldsymbol{\theta}} = \operatorname*{argmax}_{\boldsymbol{\theta}} \max_{\boldsymbol{\eta}} \mathcal{L}(\boldsymbol{\theta}, \boldsymbol{\eta}).$$

Another way of describing the profile MLE is based on the partial optimization considered in the previous section. Define for each $\boldsymbol{\theta}$ the value $\breve{L}(\boldsymbol{\theta})$ by optimizing the log-likelihood $\mathcal{L}(\boldsymbol{v})$ under the condition $P\boldsymbol{v} = \boldsymbol{\theta}$:

$$\breve{L}(\boldsymbol{\theta}) \stackrel{\text{def}}{=} \sup_{\boldsymbol{v}:\, P\boldsymbol{v} = \boldsymbol{\theta}} \mathcal{L}(\boldsymbol{v}) = \sup_{\boldsymbol{\eta}} \mathcal{L}(\boldsymbol{\theta}, \boldsymbol{\eta}). \tag{4.51}$$

Then $\tilde{\boldsymbol{\theta}}$ is defined by maximizing the partial fit $\breve{L}(\boldsymbol{\theta})$:

$$\tilde{\boldsymbol{\theta}} \stackrel{\text{def}}{=} \operatorname*{argmax}_{\boldsymbol{\theta}} \breve{L}(\boldsymbol{\theta}). \tag{4.52}$$

Exercise 4.9.9. Check that (4.47) and (4.52) lead to the same estimate $\tilde{\boldsymbol{\theta}}$.

We use for the function $\breve{L}(\boldsymbol{\theta})$ obtained by partial optimization (4.51) the same notation as for the function obtained by the orthogonal decomposition (4.43) in Sect. 4.9.2. Later we show that these two functions indeed coincide. This helps in understanding the structure of the profile estimate $\tilde{\boldsymbol{\theta}}$.

Consider first the orthogonal case $\Psi \Phi^{\top} = 0$. This assumption gradually simplifies the study. In particular, the result of Theorem 4.9.4 for partial estimation can be obviously extended to the profile method in view of product structure (4.43): when estimating the parameter $\boldsymbol{\theta}$, one can ignore the nuisance parameter $\boldsymbol{\eta}$ and proceed as if the partial model $\boldsymbol{Y} = \Psi^{\top}\boldsymbol{\theta}^* + \boldsymbol{\varepsilon}$ were correct. Theorem 4.9.1 implies:

Theorem 4.9.6. *Assume that* $\Psi \Phi^{\top} = 0$ *in the model* (4.39). *Then the profile MLE* $\tilde{\boldsymbol{\theta}}$ *from* (4.52) *coincides with the MLE from the partial model* $\boldsymbol{Y} = \Psi^{\top}\boldsymbol{\theta}^* + \boldsymbol{\varepsilon}$:

$$\tilde{\boldsymbol{\theta}} = \operatorname*{argmax}_{\boldsymbol{\theta}} L(\boldsymbol{\theta}) = \operatorname*{argmin}_{\boldsymbol{\theta}} \|\boldsymbol{Y} - \Psi^{\top}\boldsymbol{\theta}\|^2 = \bigl(\Psi\Psi^{\top}\bigr)^{-1}\Psi\boldsymbol{Y}.$$

It holds $\mathbb{E}\tilde{\boldsymbol{\theta}} = \boldsymbol{\theta}^*$ *and*

$$\tilde{\boldsymbol{\theta}} - \boldsymbol{\theta}^* = D^{-2}\boldsymbol{\zeta} = D^{-1}\boldsymbol{\xi}$$

with $D^2 = \sigma^{-2}\Psi\Psi^{\top}$, $\boldsymbol{\zeta} = \sigma^{-2}\Psi\boldsymbol{\varepsilon}$, *and* $\boldsymbol{\xi} = D^{-1}\boldsymbol{\zeta}$. *Finally,* $\breve{L}(\boldsymbol{\theta})$ *from* (4.51) *fulfills*

$$2\{\breve{L}(\tilde{\theta}) - \breve{L}(\theta^*)\} = \|D(\tilde{\theta} - \theta^*)\|^2 = \zeta^\top D^{-2}\zeta = \|\xi\|^2.$$

Moreover, if $\mathrm{Var}(\varepsilon) = \sigma^2 I_n$*, then* $\mathrm{Var}(\xi) = \boldsymbol{I}_p$*. If* $\varepsilon \sim \mathcal{N}(0, \sigma^2 \boldsymbol{I}_n)$*, then* ξ *is standard normal in* $\mathbb{R}^p$*.*

The general case can be reduced to the orthogonal one by the construction from Theorem 4.9.2. Let

$$\breve{\Psi} = \Psi - \Psi\Pi_\eta = \Psi - \Psi\Phi^\top\big(\Phi\Phi^\top\big)^{-1}\Phi$$

be the corrected Ψ-factors after removing their interactions with the Φ-factors.

Theorem 4.9.7. *Consider the model* (4.39)*, and let the matrix* $\breve{D}^2 = \sigma^{-2}\breve{\Psi}\breve{\Psi}^\top$ *is non-degenerated. Then the profile MLE* $\tilde{\theta}$ *reads as*

$$\tilde{\theta} = \operatorname*{argmin}_{\theta} \|Y - \breve{\Psi}^\top\theta\|^2 = \big(\breve{\Psi}\breve{\Psi}^\top\big)^{-1}\breve{\Psi}Y. \tag{4.53}$$

It holds $\mathbb{E}\tilde{\theta} = \theta^*$ *and*

$$\tilde{\theta} - \theta^* = \big(\breve{\Psi}\breve{\Psi}^\top\big)^{-1}\breve{\Psi}\varepsilon = \breve{D}^{-2}\breve{\zeta} = \breve{D}^{-1}\breve{\xi} \tag{4.54}$$

with $\breve{D}^2 = \sigma^{-2}\breve{\Psi}\breve{\Psi}^\top$, $\breve{\zeta} = \sigma^{-2}\breve{\Psi}\varepsilon$*, and* $\breve{\xi} = \breve{D}^{-1}\breve{\zeta}$*. Furthermore,* $\breve{L}(\theta)$ *from* (4.51) *fulfills*

$$2\{\breve{L}(\tilde{\theta}) - \breve{L}(\theta^*)\} = \|\breve{D}(\tilde{\theta} - \theta^*)\|^2 = \breve{\zeta}^\top\breve{D}^{-2}\breve{\zeta} = \|\breve{\xi}\|^2. \tag{4.55}$$

Moreover, if $\mathrm{Var}(\varepsilon) = \sigma^2 I_n$*, then* $\mathrm{Var}(\breve{\xi}) = \boldsymbol{I}_p$*. If* $\varepsilon \sim \mathcal{N}(0, \sigma^2\boldsymbol{I}_n)$*, then* $\breve{\xi} \sim \mathcal{N}(0, \boldsymbol{I}_p)$*.*

Finally we present the same result in terms of the original log-likelihood $\mathcal{L}(\boldsymbol{v})$.

Theorem 4.9.8. *Write* $\mathcal{D}^2 = -\nabla^2\mathbb{E}\mathcal{L}(\boldsymbol{v})$ *for the model* (4.39) *in the block form*

$$\mathcal{D}^2 = \begin{pmatrix} D^2 & A \\ A^\top & H^2 \end{pmatrix} \tag{4.56}$$

Let D^2 *and* H^2 *be invertible. Then* $\breve{D}^2$ *and* $\breve{\zeta}$ *in* (4.54) *can be represented as*

$$\breve{D}^2 = D^2 - AH^{-2}A^\top,$$
$$\breve{\zeta} = \nabla_\theta\mathcal{L}(\boldsymbol{v}^*) - AH^{-2}\nabla_\eta\mathcal{L}(\boldsymbol{v}^*).$$

Proof. In view of Theorem 4.9.7, it suffices to check the formulas for $\breve{D}^2$ and $\breve{\zeta}$. One has for $\breve{\Psi} = \Psi(\boldsymbol{I}_n - \Pi_\eta)$ and $A = \sigma^{-2}\Psi\Phi^\top$

$$\breve{D}^2 = \sigma^{-2}\breve{\Psi}\breve{\Psi}^\top = \sigma^{-2}\Psi(I_n - \Pi_\eta)\Psi^\top$$
$$= \sigma^{-2}\Psi\Psi^\top - \sigma^{-2}\Psi\Phi^\top(\Phi\Phi^\top)^{-1}\Phi\Psi^\top = D^2 - AH^{-2}A^\top.$$

Similarly, by $AH^{-2} = \Psi\Phi^\top(\Phi\Phi^\top)^{-1}$, $\nabla_\theta \mathcal{L}(\upsilon^*) = \Psi\varepsilon$, and $\nabla_\eta \mathcal{L}(\upsilon^*) = \Phi\varepsilon$

$$\breve{\zeta} = \breve{\Psi}\varepsilon = \Psi\varepsilon - \Psi\Phi^\top(\Phi\Phi^\top)^{-1}\Phi\varepsilon = \nabla_\theta \mathcal{L}(\upsilon^*) - AH^{-2}\nabla_\eta \mathcal{L}(\upsilon^*).$$

as required.

It is worth stressing again that the result of Theorems 4.9.6 through 4.9.8 is purely geometrical. We only used the condition $\mathbb{E}\varepsilon = 0$ in the model (4.39) and the quadratic structure of the log-likelihood function $\mathcal{L}(\upsilon)$. The distribution of the vector ε does not enter in the results and proofs. However, the representation (4.54) allows for straightforward analysis of the probabilistic properties of the estimate $\tilde{\theta}$.

Theorem 4.9.9. *Consider the model* (4.39) *and let* $\operatorname{Var}(Y) = \operatorname{Var}(\varepsilon) = \Sigma_0$. *Then*

$$\operatorname{Var}(\tilde{\theta}) = \sigma^{-4}\breve{D}^{-2}\breve{\Psi}\Sigma_0\breve{\Psi}^\top\breve{D}^{-2}, \qquad \operatorname{Var}(\breve{\xi}) = \sigma^{-4}\breve{D}^{-1}\breve{\Psi}\Sigma_0\breve{\Psi}^\top\breve{D}^{-1}.$$

In particular, if $\operatorname{Var}(Y) = \sigma^2 I_n$, *this implies that*

$$\operatorname{Var}(\tilde{\theta}) = \breve{D}^{-2}, \qquad \operatorname{Var}(\breve{\xi}) = I_p.$$

Exercise 4.9.10. Check the result of Theorem 4.9.9. Specify this result to the orthogonal case $\Psi\Phi^\top = 0$.

4.9.5 Semiparametric Efficiency Bound

The main goal of this section is to show that the profile method in the semiparametric estimation leads to R-efficient procedures. Remind that the target of estimation is $\theta^* = P\upsilon^*$ for a given linear mapping P. The profile MLE $\tilde{\theta}$ is one natural candidate. The next result claims its optimality.

Theorem 4.9.10 (Gauss–Markov). *Let* Y *follow* $Y = \Upsilon^\top\upsilon^* + \varepsilon$ *for homogeneous errors* ε. *Then the estimate* $\tilde{\theta}$ *of* $\theta^* = P\upsilon^*$ *from* (4.48) *is unbiased and*

$$\operatorname{Var}(\tilde{\theta}) = \sigma^2 P(\Upsilon\Upsilon^\top)^{-1}P^\top$$

yielding

$$\mathbb{E}\|\tilde{\theta} - \theta^*\|^2 = \sigma^2 \operatorname{tr}\{P(\Upsilon\Upsilon^\top)^{-1}P^\top\}.$$

Moreover, this risk is minimal in the class of all unbiased linear estimates of θ^*.

Proof. The statements about the properties of $\tilde{\boldsymbol{\theta}}$ have been already proved. The lower bound can be proved by the same arguments as in the case of the MLE estimation in Sect. 4.4.3. We only outline the main steps. Let $\hat{\boldsymbol{\theta}}$ be any unbiased linear estimate of $\boldsymbol{\theta}^*$. The idea is to show that the difference $\hat{\boldsymbol{\theta}} - \tilde{\boldsymbol{\theta}}$ is orthogonal to $\tilde{\boldsymbol{\theta}}$ in the sense $\mathbb{E}(\hat{\boldsymbol{\theta}} - \tilde{\boldsymbol{\theta}})\tilde{\boldsymbol{\theta}}^\top = 0$. This implies that the variance of $\hat{\boldsymbol{\theta}}$ is the sum of $\operatorname{Var}(\tilde{\boldsymbol{\theta}})$ and $\operatorname{Var}(\hat{\boldsymbol{\theta}} - \tilde{\boldsymbol{\theta}})$ and therefore larger than $\operatorname{Var}(\tilde{\boldsymbol{\theta}})$.

Let $\hat{\boldsymbol{\theta}} = B\boldsymbol{Y}$ for some matrix B. Then $\mathbb{E}\hat{\boldsymbol{\theta}} = B\mathbb{E}\boldsymbol{Y} = B\Upsilon^\top \boldsymbol{\upsilon}^*$. The no-bias property yields the identity $\mathbb{E}\hat{\boldsymbol{\theta}} = \boldsymbol{\theta}^* = P\boldsymbol{\upsilon}^*$ and thus

$$B\Upsilon^\top - P = 0. \tag{4.57}$$

Next, $\mathbb{E}\tilde{\boldsymbol{\theta}} = \mathbb{E}\hat{\boldsymbol{\theta}} = \boldsymbol{\theta}^*$ and thus

$$\begin{aligned}
\mathbb{E}\tilde{\boldsymbol{\theta}}\tilde{\boldsymbol{\theta}}^\top &= \boldsymbol{\theta}^*\boldsymbol{\theta}^{*\top} + \operatorname{Var}(\tilde{\boldsymbol{\theta}}), \\
\mathbb{E}\hat{\boldsymbol{\theta}}\tilde{\boldsymbol{\theta}}^\top &= \boldsymbol{\theta}^*\boldsymbol{\theta}^{*\top} + \mathbb{E}(\hat{\boldsymbol{\theta}} - \mathbb{E}\hat{\boldsymbol{\theta}})(\tilde{\boldsymbol{\theta}} - \mathbb{E}\tilde{\boldsymbol{\theta}})^\top.
\end{aligned}$$

Obviously $\hat{\boldsymbol{\theta}} - \mathbb{E}\hat{\boldsymbol{\theta}} = B\boldsymbol{\varepsilon}$ and $\tilde{\boldsymbol{\theta}} - \mathbb{E}\tilde{\boldsymbol{\theta}} = \mathcal{S}\boldsymbol{\varepsilon}$ yielding $\operatorname{Var}(\tilde{\boldsymbol{\theta}}) = \sigma^2\mathcal{S}\mathcal{S}^\top$ and $\mathbb{E}B\boldsymbol{\varepsilon}(\mathcal{S}\boldsymbol{\varepsilon})^\top = \sigma^2 B\mathcal{S}^\top$. So

$$\mathbb{E}(\hat{\boldsymbol{\theta}} - \tilde{\boldsymbol{\theta}})\tilde{\boldsymbol{\theta}}^\top = \sigma^2(B - \mathcal{S})\mathcal{S}^\top.$$

The identity (4.57) implies

$$\begin{aligned}
(B - \mathcal{S})\mathcal{S}^\top &= \{B - P(\Upsilon\Upsilon^\top)^{-1}\Upsilon\}\Upsilon^\top(\Upsilon\Upsilon^\top)^{-1}P^\top \\
&= (B\Upsilon^\top - P)(\Upsilon\Upsilon^\top)^{-1}P^\top = 0
\end{aligned}$$

and the result follows.

Now we specify the efficiency bound for the $(\boldsymbol{\theta}, \boldsymbol{\eta})$-setup (4.39). In this case P is just the projector onto the $\boldsymbol{\theta}$-coordinates.

4.9.6 *Inference for the Profile Likelihood Approach*

This section discusses the construction of confidence and concentration sets for the profile ML estimation. The key fact behind this construction is the chi-squared result which extends without any change from the parametric to semiparametric framework.

The definition $\tilde{\boldsymbol{\theta}}$ from (4.52) suggests to define a CS for $\boldsymbol{\theta}^*$ as the level set of $\breve{L}(\boldsymbol{\theta}) = \sup_{\boldsymbol{\upsilon}: P\boldsymbol{\upsilon} = \boldsymbol{\theta}} \mathcal{L}(\boldsymbol{\upsilon})$:

$$\mathcal{E}(\mathfrak{z}) \stackrel{\text{def}}{=} \{\boldsymbol{\theta} : \breve{L}(\tilde{\boldsymbol{\theta}}) - \breve{L}(\boldsymbol{\theta}) \le \mathfrak{z}\}.$$

This definition can be rewritten as

$$\mathcal{E}(\mathfrak{z}) \stackrel{\text{def}}{=} \Bigl\{ \boldsymbol{\theta} : \sup_{\boldsymbol{v}} \mathcal{L}(\boldsymbol{v}) - \sup_{\boldsymbol{v}: P\boldsymbol{v}=\boldsymbol{\theta}} \mathcal{L}(\boldsymbol{v}) \le \mathfrak{z} \Bigr\}.$$

It is obvious that the unconstrained optimization of the log-likelihood $\mathcal{L}(\boldsymbol{v})$ w.r.t. $\boldsymbol{v}$ is not smaller than the optimization under the constraint that $P\boldsymbol{v} = \boldsymbol{\theta}$. The point $\boldsymbol{\theta}$ belongs to $\mathcal{E}(\mathfrak{z})$ if the difference between these two values does not exceed $\mathfrak{z}$. As usual, the main question is the choice of a value $\mathfrak{z}$ which ensures the prescribed coverage probability of $\boldsymbol{\theta}^*$. This naturally leads to studying the deviation probability

$$\mathbb{P}\Bigl(\sup_{\boldsymbol{v}} \mathcal{L}(\boldsymbol{v}) - \sup_{\boldsymbol{v}: P\boldsymbol{v}=\boldsymbol{\theta}^*} \mathcal{L}(\boldsymbol{v}) > \mathfrak{z} \Bigr).$$

Such a study is especially simple in the orthogonal case. The answer can be expected: the expression and the value are exactly the same as in the case without any nuisance parameter $\boldsymbol{\eta}$, it simply has no impact. In particular, the chi-squared result still holds.

In this section we follow the line and the notation of Sect. 4.9.4. In particular, we use the block notation (4.56) for the matrix $\mathcal{D}^2 = -\nabla^2 \mathcal{L}(\boldsymbol{v})$.

Theorem 4.9.11. *Consider the model* (4.39). *Let the matrix* $\mathcal{D}^2$ *be non-degenerated. If* $\boldsymbol{\varepsilon} \sim \mathcal{N}(0, \sigma^2 \boldsymbol{I}_n)$*, then*

$$2\bigl\{ \breve{L}(\tilde{\boldsymbol{\theta}}) - \breve{L}(\boldsymbol{\theta}^*) \bigr\} \sim \chi^2_p, \tag{4.58}$$

that is, this $2\bigl\{ \breve{L}(\tilde{\boldsymbol{\theta}}) - \breve{L}(\boldsymbol{\theta}^*) \bigr\}$ *is chi-squared with p degrees of freedom.*

Proof. The result is based on representation (4.55) $2\bigl\{ \breve{L}(\tilde{\boldsymbol{\theta}}) - \breve{L}(\boldsymbol{\theta}^*) \bigr\} = \|\breve{\boldsymbol{\xi}}\|^2$ from Theorem 4.9.7. It remains to note that normality of $\boldsymbol{\varepsilon}$ implies normality of $\breve{\boldsymbol{\xi}}$ and the moment conditions $\mathbb{E}\breve{\boldsymbol{\xi}} = 0$, $\operatorname{Var}(\breve{\boldsymbol{\xi}}) = \boldsymbol{I}_p$ imply (4.58).

This result means that the chi-squared result continues to hold in the general semiparametric framework as well. One possible explanation is as follows: it applies in the orthogonal case, and the general situation can be reduced to the orthogonal case by a change of coordinates which preserves the value of the maximum likelihood.

The statement (4.58) of Theorem 4.9.11 has an interesting geometric interpretation which is often used in analysis of variance. Consider the expansion

$$\breve{L}(\tilde{\boldsymbol{\theta}}) - \breve{L}(\boldsymbol{\theta}^*) = \breve{L}(\tilde{\boldsymbol{\theta}}) - \mathcal{L}(\boldsymbol{\theta}^*, \boldsymbol{\eta}^*) - \bigl\{ \breve{L}(\boldsymbol{\theta}^*) - \mathcal{L}(\boldsymbol{\theta}^*, \boldsymbol{\eta}^*) \bigr\}.$$

The quantity $\mathcal{L}_1 \stackrel{\text{def}}{=} \breve{L}(\tilde{\boldsymbol{\theta}}) - \mathcal{L}(\boldsymbol{v}^*)$ coincides with $\mathcal{L}(\tilde{\boldsymbol{v}}, \boldsymbol{v}^*)$; see (4.47). Thus, $2\mathcal{L}_1$ chi-squared with p^* degrees of freedom by the chi-squared result. Moreover, $2\sigma^2 \mathcal{L}(\tilde{\boldsymbol{v}}, \boldsymbol{v}^*) = \|\Pi_{\upsilon} \boldsymbol{\varepsilon}\|^2$, where $\Pi_{\upsilon} = \Upsilon^\top \bigl(\Upsilon \Upsilon^\top \bigr)^{-1} \Upsilon$ is the projector on the linear subspace spanned by the joint collection of factors $\{\boldsymbol{\psi}_j\}$ and $\{\boldsymbol{\phi}_m\}$. Similarly, the quantity $\mathcal{L}_2 \stackrel{\text{def}}{=} \breve{L}(\boldsymbol{\theta}^*) - \mathcal{L}(\boldsymbol{\theta}^*, \boldsymbol{\eta}^*) = \sup_{\eta} \mathcal{L}(\boldsymbol{\theta}^*, \boldsymbol{\eta}) - \mathcal{L}(\boldsymbol{\theta}^*, \boldsymbol{\eta}^*)$ is the maximum likelihood in the partial $\boldsymbol{\eta}$-model. Therefore, $2\mathcal{L}_2$ is also chi-squared distributed with p_1 degrees of freedom, and $2\sigma^2 \mathcal{L}_2 = \|\Pi_{\eta} \boldsymbol{\varepsilon}\|^2$, where $\Pi_{\eta} = \Phi^{\top} \left(\Phi \Phi^{\top}\right)^{-1} \Phi$ is the projector on the linear subspace spanned by the $\boldsymbol{\eta}$-factors $\{\boldsymbol{\phi}_m\}$. Now we use the decomposition $\Pi_{\upsilon} = \Pi_{\eta} + \Pi_{\upsilon} - \Pi_{\eta}$, in which $\Pi_{\upsilon} - \Pi_{\eta}$ is also a projector on the subspace of dimension p. This explains the result (4.58) that the difference of these two quantities is chi-squared with $p = p^* - p_1$ degrees of freedom. The above consideration leads to the following result.

Theorem 4.9.12. *It holds for the model* (4.39) *with* $\breve{\Pi}_{\theta} = \Pi_{\upsilon} - \Pi_{\eta}$

$$
\begin{aligned}
2\breve{L}(\tilde{\boldsymbol{\theta}}) - 2\breve{L}(\boldsymbol{\theta}^*) &= \sigma^{-2} \left(\|\Pi_{\upsilon} \boldsymbol{\varepsilon}\|^2 - \|\Pi_{\eta} \boldsymbol{\varepsilon}\|^2 \right) \\
&= \sigma^{-2} \|\breve{\Pi}_{\theta} \boldsymbol{\varepsilon}\|^2 = \sigma^{-2} \boldsymbol{\varepsilon}^{\top} \breve{\Pi}_{\theta} \boldsymbol{\varepsilon}.
\end{aligned} \tag{4.59}
$$

Exercise 4.9.11. Check the formula (4.59). Show that it implies (4.58).

4.9.7 Plug-In Method

Although the profile MLE can be represented in a closed form, its computing can be a hard task if the dimensionality p_1 of the nuisance parameter is high. Here we discuss an approach which simplifies the computations but leads to a suboptimal solution.

We start with the approach called *plug-in*. It is based on the assumption that a pilot estimate $\hat{\boldsymbol{\eta}}$ of the nuisance parameter $\boldsymbol{\eta}$ is available. Then one obtains the estimate $\hat{\boldsymbol{\theta}}$ of the target $\boldsymbol{\theta}^*$ from the residuals $\boldsymbol{Y} - \Phi^{\top} \hat{\boldsymbol{\eta}}$.

This means that the residual vector $\hat{\boldsymbol{Y}} = \boldsymbol{Y} - \Phi^{\top} \hat{\boldsymbol{\eta}}$ is used as observations and the estimate $\hat{\boldsymbol{\theta}}$ is defined as the best fit to such observations in the $\boldsymbol{\theta}$-model:

$$
\hat{\boldsymbol{\theta}} = \operatorname*{argmin}_{\boldsymbol{\theta}} \|\hat{\boldsymbol{Y}} - \Psi^{\top} \boldsymbol{\theta}\|^2 = \left(\Psi \Psi^{\top}\right)^{-1} \Psi \hat{\boldsymbol{Y}}. \tag{4.60}
$$

A very particular case of the plug-in method is partial estimation from Sect. 4.9.3 with $\hat{\boldsymbol{\eta}} \equiv \boldsymbol{\eta}^{\circ}$.

The plug-in method can be naturally described in context of partial estimation. We use the following representation of the plug-in method: $\hat{\boldsymbol{\theta}} = \tilde{\boldsymbol{\theta}}(\hat{\boldsymbol{\eta}})$.

Exercise 4.9.12. Check the identity $\hat{\boldsymbol{\theta}} = \tilde{\boldsymbol{\theta}}(\hat{\boldsymbol{\eta}})$ for the plug-in method. Describe the plug-in estimate for $\hat{\boldsymbol{\eta}} \equiv 0$.

The behavior of the $\hat{\boldsymbol{\theta}}$ heavily depends upon the quality of the pilot $\hat{\boldsymbol{\eta}}$. A detailed study is complicated and a closed form solution is only available for the special case of a linear pilot estimate. Let $\hat{\boldsymbol{\eta}} = A\boldsymbol{Y}$. Then (4.60) implies

$$\hat{\boldsymbol{\theta}} = \left(\Psi\Psi^\top\right)^{-1}\Psi(\boldsymbol{Y} - \Phi^\top A\boldsymbol{Y}) = \mathcal{S}\boldsymbol{Y}$$

with $\mathcal{S} = \left(\Psi\Psi^\top\right)^{-1}\Psi(\boldsymbol{I}_n - \Phi^\top A)$. This is a linear estimate whose properties can be studied in a usual way.

4.9.8 Two-Step Procedure

The ideas of partial and plug-in estimation can be combined yielding the so-called *two-step procedures*. One starts with the initial guess $\boldsymbol{\theta}^\circ$ for the target $\boldsymbol{\theta}^*$. A very special choice is $\boldsymbol{\theta}^\circ \equiv 0$. This leads to the partial $\boldsymbol{\eta}$-model $\boldsymbol{Y}(\boldsymbol{\theta}^\circ) = \Phi^\top\boldsymbol{\eta} + \boldsymbol{\varepsilon}$ for the residuals $\boldsymbol{Y}(\boldsymbol{\theta}^\circ) = \boldsymbol{Y} - \Psi^\top\boldsymbol{\theta}^\circ$. Next compute the partial MLE $\tilde{\boldsymbol{\eta}}(\boldsymbol{\theta}^\circ) = \left(\Phi\Phi^\top\right)^{-1}\Phi\boldsymbol{Y}(\boldsymbol{\theta}^\circ)$ in this model and use it as a pilot for the plug-in method: compute the residuals

$$\hat{\boldsymbol{Y}}(\boldsymbol{\theta}^\circ) = \boldsymbol{Y} - \Phi^\top\tilde{\boldsymbol{\eta}}(\boldsymbol{\theta}^\circ) = \boldsymbol{Y} - \Pi_\eta\boldsymbol{Y}(\boldsymbol{\theta}^\circ)$$

with $\Pi_\eta = \Phi^\top\left(\Phi\Phi^\top\right)^{-1}\Phi$, and then estimate the target parameter $\boldsymbol{\theta}$ by fitting $\Psi^\top\boldsymbol{\theta}$ to the residuals $\hat{\boldsymbol{Y}}(\boldsymbol{\theta}^\circ)$. This method results in the estimate

$$\hat{\boldsymbol{\theta}}(\boldsymbol{\theta}^\circ) = \left(\Psi\Psi^\top\right)^{-1}\Psi\hat{\boldsymbol{Y}}(\boldsymbol{\theta}^\circ) \tag{4.61}$$

A simple comparison with the formula (4.53) reveals that the pragmatic two-step approach is sub-optimal: the resulting estimate does not fit the profile MLE $\tilde{\boldsymbol{\theta}}$ unless we have an orthogonal situation with $\Psi\Pi_\eta = 0$. In particular, the estimate $\hat{\boldsymbol{\theta}}(\boldsymbol{\theta}^\circ)$ from (4.61) is biased.

Exercise 4.9.13. Consider the orthogonal case with $\Psi\Phi^\top = 0$. Show that the two-step estimate $\hat{\boldsymbol{\theta}}(\boldsymbol{\theta}^\circ)$ coincides with the partial MLE $\tilde{\boldsymbol{\theta}} = \left(\Psi\Psi^\top\right)^{-1}\Psi\boldsymbol{Y}$.

Exercise 4.9.14. Compute the mean of $\hat{\boldsymbol{\theta}}(\boldsymbol{\theta}^\circ)$. Show that there exists some $\boldsymbol{\theta}^*$ such that $\mathbb{E}\{\hat{\boldsymbol{\theta}}(\boldsymbol{\theta}^\circ)\} \neq \boldsymbol{\theta}^*$ unless the orthogonality condition $\Psi\Phi^\top = 0$ is fulfilled.

Exercise 4.9.15. Compute the variance of $\hat{\boldsymbol{\theta}}(\boldsymbol{\theta}^\circ)$.
Hint: use that $\operatorname{Var}\{\boldsymbol{Y}(\boldsymbol{\theta}^\circ)\} = \operatorname{Var}(\boldsymbol{Y}) = \sigma^2\boldsymbol{I}_n$. Derive that $\operatorname{Var}\{\hat{\boldsymbol{Y}}(\boldsymbol{\theta}^\circ)\} = \sigma^2(\boldsymbol{I}_n - \Pi_\eta)$.

Exercise 4.9.16. Let Ψ be orthogonal, i.e. $\Psi\Psi^\top = \boldsymbol{I}_p$. Show that

$$\operatorname{Var}\{\hat{\boldsymbol{\theta}}(\boldsymbol{\theta}^\circ)\} = \sigma^2(\boldsymbol{I}_p - \Psi\Pi_\eta\Psi^\top).$$

4.9.9 Alternating Method

The idea of partial and two-step estimation can be applied in an iterative way. One starts with some initial value for $\boldsymbol{\theta}^{\circ}$ and sequentially performs the two steps of partial estimation. Set

$$\hat{\boldsymbol{\eta}}_0 = \tilde{\boldsymbol{\eta}}(\boldsymbol{\theta}^{\circ}) = \operatorname*{argmin}_{\boldsymbol{\eta}} \|\boldsymbol{Y} - \Psi^{\top}\boldsymbol{\theta}^{\circ} - \Phi^{\top}\boldsymbol{\eta}\|^2 = \bigl(\Phi\Phi^{\top}\bigr)^{-1}\Phi(\boldsymbol{Y} - \Psi^{\top}\boldsymbol{\theta}^{\circ}).$$

With this estimate fixed, compute $\hat{\boldsymbol{\theta}}_1 = \tilde{\boldsymbol{\theta}}(\hat{\boldsymbol{\eta}}_1)$ and continue in this way. Generically, with $\hat{\boldsymbol{\theta}}_k$ and $\hat{\boldsymbol{\eta}}_k$ computed, one recomputes

$$\hat{\boldsymbol{\theta}}_{k+1} = \tilde{\boldsymbol{\theta}}(\hat{\boldsymbol{\eta}}_k) = \bigl(\Psi\Psi^{\top}\bigr)^{-1}\Psi(\boldsymbol{Y} - \Phi^{\top}\hat{\boldsymbol{\eta}}_k), \tag{4.62}$$

$$\hat{\boldsymbol{\eta}}_{k+1} = \tilde{\boldsymbol{\eta}}(\hat{\boldsymbol{\theta}}_{k+1}) = \bigl(\Phi\Phi^{\top}\bigr)^{-1}\Phi(\boldsymbol{Y} - \Psi^{\top}\hat{\boldsymbol{\theta}}_{k+1}). \tag{4.63}$$

The procedure is especially transparent if the partial design matrices Ψ and Φ are orthonormal: $\Psi\Psi^{\top} = \boldsymbol{I}_p$, $\Phi\Phi^{\top} = I_{p_1}$. Then

$$\begin{aligned}
\hat{\boldsymbol{\theta}}_{k+1} &= \Psi(\boldsymbol{Y} - \Phi^{\top}\hat{\boldsymbol{\eta}}_k),\\
\hat{\boldsymbol{\eta}}_{k+1} &= \Phi(\boldsymbol{Y} - \Psi^{\top}\hat{\boldsymbol{\theta}}_{k+1}).
\end{aligned}$$

In other words, having an estimate $\hat{\boldsymbol{\theta}}$ of the parameter $\boldsymbol{\theta}^*$ one computes the residuals $\hat{\boldsymbol{Y}} = \boldsymbol{Y} - \Psi^{\top}\hat{\boldsymbol{\theta}}$ and then build the estimate $\hat{\boldsymbol{\eta}}$ of the nuisance $\boldsymbol{\eta}^*$ by the empirical coefficients $\Phi\hat{\boldsymbol{Y}}$. Then this estimate $\hat{\boldsymbol{\eta}}$ is used in a similar way to recompute the estimate of $\boldsymbol{\theta}^*$, and so on.

It is worth noting that every doubled step of alternation improves the current value $\mathcal{L}(\hat{\boldsymbol{\theta}}_k, \hat{\boldsymbol{\eta}}_k)$. Indeed, $\hat{\boldsymbol{\theta}}_{k+1}$ is defined by maximizing $\mathcal{L}(\boldsymbol{\theta}, \hat{\boldsymbol{\eta}}_k)$, that is, $\mathcal{L}(\hat{\boldsymbol{\theta}}_{k+1}, \hat{\boldsymbol{\eta}}_k) \ge \mathcal{L}(\hat{\boldsymbol{\theta}}_k, \hat{\boldsymbol{\eta}}_k)$. Similarly, $\mathcal{L}(\hat{\boldsymbol{\theta}}_{k+1}, \hat{\boldsymbol{\eta}}_{k+1}) \ge \mathcal{L}(\hat{\boldsymbol{\theta}}_{k+1}, \hat{\boldsymbol{\eta}}_k)$ yielding

$$\mathcal{L}(\hat{\boldsymbol{\theta}}_{k+1}, \hat{\boldsymbol{\eta}}_{k+1}) \ge \mathcal{L}(\hat{\boldsymbol{\theta}}_k, \hat{\boldsymbol{\eta}}_k). \tag{4.64}$$

A very interesting question is whether the procedure (4.62), (4.63) converges and whether it converges to the maximum likelihood solution. The answer is positive and in the simplest orthogonal case the result is straightforward.

Exercise 4.9.17. Consider the orthogonal situation with $\Psi\Phi^{\top} = 0$. Then the above procedure stabilizes in one step with the solution from Theorem 4.9.4.

In the non-orthogonal case the situation is much more complicated. The idea is to show that the alternating procedure can be represented a sequence of actions of a shrinking linear operator to the data. The key observation behind the result is the following recurrent formula for $\Psi^{\top}\hat{\boldsymbol{\theta}}_k$ and $\Phi^{\top}\hat{\boldsymbol{\eta}}_k$:

$$\Psi^\top \hat{\theta}_{k+1} = \Pi_\theta(Y - \Phi^\top \hat{\eta}_k) = \bigl(\Pi_\theta - \Pi_\theta \Pi_\eta\bigr) Y + \Pi_\theta \Pi_\eta \Psi^\top \hat{\theta}_k, \quad (4.65)$$

$$\Phi^\top \hat{\eta}_{k+1} = \Pi_\eta(Y - \Psi^\top \hat{\theta}_{k+1}) = \bigl(\Pi_\eta - \Pi_\eta \Pi_\theta\bigr) Y + \Pi_\eta \Pi_\theta \Phi^\top \hat{\eta}_k \quad (4.66)$$

with $\Pi_\theta = \Psi^\top \bigl(\Psi\Psi^\top\bigr)^{-1} \Psi$ and $\Pi_\eta = \Phi^\top \bigl(\Phi\Phi^\top\bigr)^{-1} \Phi$.

Exercise 4.9.18. Show (4.65) and (4.66).

This representation explains necessary and sufficient conditions for convergence of the alternating procedure. Namely, the spectral norm $\|\Pi_\eta \Pi_\theta\|_\infty$ (the largest singular value) of the product operator $\Pi_\eta \Pi_\theta$ should be strictly less than one, and similarly for $\Pi_\theta \Pi_\eta$.

Exercise 4.9.19. Show that $\|\Pi_\theta \Pi_\eta\|_\infty = \|\Pi_\eta \Pi_\theta\|_\infty$.

Theorem 4.9.13. *Suppose that $\|\Pi_\eta \Pi_\theta\|_\infty = \lambda < 1$. Then the alternating procedure converges geometrically, the limiting values $\hat{\theta}$ and $\hat{\eta}$ are unique and fulfill*

$$\Psi^\top \hat{\theta} = (I_n - \Pi_\theta \Pi_\eta)^{-1} (\Pi_\theta - \Pi_\theta \Pi_\eta) Y,$$

$$\Phi^\top \hat{\eta} = (I_n - \Pi_\eta \Pi_\theta)^{-1} (\Pi_\eta - \Pi_\eta \Pi_\theta) Y, \quad (4.67)$$

and $\hat{\theta}$ coincides with the profile MLE $\tilde{\theta}$ from (4.52).

Proof. The convergence will be discussed below. Now we comment on the identity $\hat{\theta} = \tilde{\theta}$. A direct comparison of the formulas for these two estimates can be a hard task. Instead we use the monotonicity property (4.64). By definition, $(\tilde{\theta}, \tilde{\eta})$ maximize globally $\mathcal{L}(\theta, \eta)$. If we start the procedure with $\theta^\circ = \tilde{\theta}$, we would improve the value $\mathcal{L}(\tilde{\theta}, \tilde{\eta})$ at every step. By uniqueness, the procedure stabilizes with $\hat{\theta}_k = \tilde{\theta}$ and $\hat{\eta}_k = \tilde{\eta}$ for every k.

Exercise 4.9.20. 1. Show by induction arguments that

$$\Phi^\top \hat{\eta}_{k+1} = A_{k+1} Y + \bigl(\Pi_\eta \Pi_\theta\bigr)^k \Phi^\top \hat{\eta}_1,$$

where the linear operator A_k fulfills $A_1 = 0$ and

$$A_{k+1} = \Pi_\eta - \Pi_\eta \Pi_\theta + \Pi_\eta \Pi_\theta A_k = \sum_{i=0}^{k-1} (\Pi_\eta \Pi_\theta)^i (\Pi_\eta - \Pi_\eta \Pi_\theta).$$

2. Show that A_k converges to $A = (I_n - \Pi_\eta \Pi_\theta)^{-1} (\Pi_\eta - \Pi_\eta \Pi_\theta)$ and evaluate $\|A - A_k\|_\infty$ and $\|\Phi^\top (\hat{\eta}_k - \hat{\eta})\|$.
Hint: use that $\|\Pi_\eta - \Pi_\eta \Pi_\theta\|_\infty \le 1$ and $\|(\Pi_\eta \Pi_\theta)^i\|_\infty \le \|\Pi_\eta \Pi_\theta\|_\infty^i \le \lambda^i$.
3. Prove (4.67) by inserting $\hat{\eta}$ in place of $\hat{\eta}_k$ and $\hat{\eta}_{k+1}$ in (4.66).

4.10 Historical Remarks and Further Reading

The *least squares method* goes back to Carl Gauss (around 1795 but published first 1809) and Adrien Marie Legendre in 1805.

The notion of *linear regression* was introduced by Fransis Galton around 1868 for biological problem and then extended by Karl Pearson and Robert Fisher between 1912 and 1922.

Chi-squared distribution was first described by the German statistician Friedrich Robert Helmert in papers of 1875/1876. The distribution was independently rediscovered by Karl Pearson in the context of goodness of fit.

The *Gauss–Markov theorem* is attributed to Gauß (1995) (originally published in Latin in 1821/1823) and Markoff (1912).

Classical references for the *sandwich formula* (see, e.g., (4.12)) for the variance of the maximum likelihood estimator in a misspecified model are Huber (1967) and White (1982).

It seems that the term *ridge regression* has first been used by Hoerl (1962); see also the original paper by Tikhonov (1963) for what is nowadays referred to as *Tikhonov regularization*. An early reference on *penalized maximum likelihood* is Good and Gaskins (1971) who discussed the usage of a *roughness penalty*.

The original reference for the *Galerkin method* is Galerkin (1915). Its application in the context of regularization of *inverse problems* has been described, e.g., by Donoho (1995). The theory of *shrinkage estimation* started with the fundamental article by Stein (1956).

A systematic treatment of *profile maximum likelihood* is provided by Murphy and van der Vaart (2000), but the origins of this concept can be traced back to Fisher (1956).

The *alternating procedure* has been introduced by Dempster et al. (1977) in the form of the *expectation-maximization algorithm*.

Chapter 5
Bayes Estimation

This chapter discusses the Bayes approach to parameter estimation. This approach differs essentially from classical parametric modeling also called the *frequentist* approach. Classical frequentist modeling assumes that the observed data $\boldsymbol{Y}$ follow a distribution law $\mathbb{P}$ from a given parametric family $(\mathbb{P}_{\boldsymbol{\theta}}, \boldsymbol{\theta} \in \Theta \subset \mathbb{R}^p)$, that is,

$$\mathbb{P} = \mathbb{P}_{\boldsymbol{\theta}^*} \in (\mathbb{P}_{\boldsymbol{\theta}}).$$

Suppose that the family $(\mathbb{P}_{\boldsymbol{\theta}})$ is dominated by a measure μ_0 and denote by $p(y \mid \boldsymbol{\theta})$ the corresponding density:

$$p(y \mid \boldsymbol{\theta}) = \frac{d\mathbb{P}_{\boldsymbol{\theta}}}{d\mu_0}(y).$$

The likelihood is defined as the density at the observed point and the maximum likelihood approach tries to recover the true parameter $\boldsymbol{\theta}^*$ by maximizing this likelihood over $\boldsymbol{\theta} \in \Theta$.

In the Bayes approach, the paradigm is changed and the true data distribution is not assumed to be specified by a single parameter value $\boldsymbol{\theta}^*$. Instead, the unknown parameter is considered to be a random variable $\boldsymbol{\vartheta}$ with a distribution π on the parameter space Θ called a prior. The measure $\mathbb{P}_{\boldsymbol{\theta}}$ can be considered to be the data distribution conditioned that the randomly selected parameter is exactly $\boldsymbol{\theta}$. The target of analysis is not a single value $\boldsymbol{\theta}^*$, this value is no longer defined. Instead one is interested in the *posterior* distribution of the random parameter $\boldsymbol{\vartheta}$ given the observed data:

what is the distribution of $\boldsymbol{\vartheta}$ given the prior π and the data $\boldsymbol{Y}$?

In other words, one aims at inferring on the distribution of $\boldsymbol{\vartheta}$ on the basis of the observed data $\boldsymbol{Y}$ and our prior knowledge π. Below we distinguish between the random variable $\boldsymbol{\vartheta}$ and its particular values $\boldsymbol{\theta}$. However, one often uses the same symbol $\boldsymbol{\theta}$ for denoting the both objects.

V. Spokoiny and T. Dickhaus, *Basics of Modern Mathematical Statistics*, Springer Texts in Statistics, DOI 10.1007/978-3-642-39909-1_5,

5.1 Bayes Formula

The Bayes modeling assumptions can be put together in the form

$$\begin{aligned} \boldsymbol{Y} \mid \boldsymbol{\theta} &\sim p(\cdot \mid \boldsymbol{\theta}), \\ \boldsymbol{\vartheta} &\sim \pi(\cdot). \end{aligned}$$

The first line has to be understood as the conditional distribution of $\boldsymbol{Y}$ given the particular value $\boldsymbol{\theta}$ of the random parameter $\boldsymbol{\vartheta}$: $\boldsymbol{Y} \mid \boldsymbol{\theta}$ means $\boldsymbol{Y} \mid \boldsymbol{\vartheta} = \boldsymbol{\theta}$. This section formalizes and states the Bayes approach in a formal mathematical way. The answer is given by the Bayes formula for the conditional distribution of $\boldsymbol{\vartheta}$ given $\boldsymbol{Y}$. First consider the *joint* distribution $\mathbb{P}$ of $\boldsymbol{Y}$ and $\boldsymbol{\vartheta}$. If B is a Borel set in the space $\mathcal{Y}$ of observations and A is a measurable subset of Θ, then

$$\mathbb{P}(B \times A) = \int_A \left(\int_B \mathbb{P}_{\boldsymbol{\theta}}(d\boldsymbol{y}) \right) \pi(d\boldsymbol{\theta})$$

The *marginal* or *unconditional* distribution of $\boldsymbol{Y}$ is given by averaging the joint probability w.r.t. the distribution of $\boldsymbol{\vartheta}$:

$$\mathbb{P}(B) = \int_{\Theta} \int_B \mathbb{P}_{\boldsymbol{\theta}}(d\boldsymbol{y}) \pi(d\boldsymbol{\theta}) = \int_{\Theta} \mathbb{P}_{\boldsymbol{\theta}}(B) \pi(d\boldsymbol{\theta}).$$

The *posterior* (conditional) distribution of $\boldsymbol{\vartheta}$ given the event $\boldsymbol{Y} \in B$ is defined as the ratio of the joint and marginal probabilities:

$$\mathbb{P}(\boldsymbol{\vartheta} \in A \mid \boldsymbol{Y} \in B) = \frac{\mathbb{P}(B \times A)}{\mathbb{P}(B)}.$$

Equivalently one can write this formula in terms of the related densities. In what follows we denote by the same letter π the prior measure π and its density w.r.t. some other measure λ, e.g. the Lebesgue or uniform measure on Θ. Then the joint measure $\mathbb{P}$ has the density

$$p(\boldsymbol{y}, \boldsymbol{\theta}) = p(\boldsymbol{y} \mid \boldsymbol{\theta}) \pi(\boldsymbol{\theta}),$$

while the *marginal density* $p(\boldsymbol{y})$ is the integral of the joint density w.r.t. the prior π:

$$p(\boldsymbol{y}) = \int_{\Theta} p(\boldsymbol{y}, \boldsymbol{\theta}) d\boldsymbol{\theta} = \int_{\Theta} p(\boldsymbol{y} \mid \boldsymbol{\theta}) \pi(\boldsymbol{\theta}) d\boldsymbol{\theta}.$$

Finally the *posterior (conditional) density* $p(\boldsymbol{\theta} \mid \boldsymbol{y})$ of $\boldsymbol{\vartheta}$ given $\boldsymbol{y}$ is defined as the ratio of the joint density $p(\boldsymbol{y}, \boldsymbol{\theta})$ and the marginal density $p(\boldsymbol{y})$:

$$p(\boldsymbol{\theta} \mid \boldsymbol{y}) = \frac{p(\boldsymbol{y}, \boldsymbol{\theta})}{p(\boldsymbol{y})} = \frac{p(\boldsymbol{y} \mid \boldsymbol{\theta})\pi(\boldsymbol{\theta})}{\int_{\Theta} p(\boldsymbol{y} \mid \boldsymbol{\theta})\pi(\boldsymbol{\theta})d\boldsymbol{\theta}}.$$

Our definitions are summarized in the next lines:

$$\begin{aligned}
\boldsymbol{Y} \mid \boldsymbol{\theta} &\sim p(\boldsymbol{y} \mid \boldsymbol{\theta}), \\
\boldsymbol{\vartheta} &\sim \pi(\boldsymbol{\theta}), \\
\boldsymbol{Y} &\sim p(\boldsymbol{y}) = \int_{\Theta} p(\boldsymbol{y} \mid \boldsymbol{\theta})\pi(\boldsymbol{\theta})d\boldsymbol{\theta}, \\
\boldsymbol{\vartheta} \mid \boldsymbol{Y} &\sim p(\boldsymbol{\theta} \mid \boldsymbol{Y}) = \frac{p(\boldsymbol{Y}, \boldsymbol{\theta})}{p(\boldsymbol{Y})} = \frac{p(\boldsymbol{Y} \mid \boldsymbol{\theta})\pi(\boldsymbol{\theta})}{\int_{\Theta} p(\boldsymbol{Y} \mid \boldsymbol{\theta})\pi(\boldsymbol{\theta})d\boldsymbol{\theta}}. \qquad (5.1)
\end{aligned}$$

Note that given the prior π and the observations $\boldsymbol{Y}$, the posterior density $p(\boldsymbol{\theta} \mid \boldsymbol{Y})$ is uniquely defined and can be viewed as the solution or target of analysis within the Bayes approach. The expression (5.1) for the posterior density is called the *Bayes formula*.

The value $p(\boldsymbol{y})$ of the marginal density of $\boldsymbol{Y}$ at $\boldsymbol{y}$ does not depend on the parameter $\boldsymbol{\theta}$. Given the data $\boldsymbol{Y}$, it is just a numeric normalizing factor. Often one skips this factor writing

$$\boldsymbol{\vartheta} \mid \boldsymbol{Y} \propto p(\boldsymbol{Y} \mid \boldsymbol{\theta})\pi(\boldsymbol{\theta}).$$

Below we consider a couple of examples.

Example 5.1.1. Let $\boldsymbol{Y} = (Y_1, \ldots, Y_n)^\top$ be a sequence of zeros and ones considered to be a realization of a Bernoulli experiment for $n = 10$. Let also the underlying parameter θ be random and let it take the values $1/2$ or 1 each with probability $1/2$, that is,

$$\pi(1/2) = \pi(1) = 1/2.$$

Then the probability of observing $\boldsymbol{y} =$ "10 ones" is

$$\mathbb{P}(\boldsymbol{y}) = \frac{1}{2}\mathbb{P}(\boldsymbol{y} \mid \vartheta = 1/2) + \frac{1}{2}\mathbb{P}(\boldsymbol{y} \mid \vartheta = 1).$$

The first probability is quite small, it is 2^{-10}, while the second one is just one. Therefore, $\mathbb{P}(\boldsymbol{y}) = (2^{-10}+1)/2$. If we observed $\boldsymbol{y} = (1, \ldots, 1)^\top$, then the posterior probability of $\vartheta = 1$ is

$$\mathbb{P}(\vartheta = 1 \mid \boldsymbol{y}) = \frac{\mathbb{P}(\boldsymbol{y} \mid \vartheta = 1)\mathbb{P}(\vartheta = 1)}{\mathbb{P}(\boldsymbol{y})} = \frac{1}{2^{-10} + 1}$$

that is, it is quite close to one.

Exercise 5.1.1. Consider the Bernoulli experiment $\boldsymbol{Y} = (Y_1, \ldots, Y_n)^\top$ with $n = 10$ and let

$$\pi(1/2) = \pi(0.9) = 1/2.$$

Compute the posterior distribution of ϑ if we observe $\boldsymbol{y} = (y_1, \ldots, y_n)^\top$ with

- $\boldsymbol{y} = (1, \ldots, 1)^\top$
- the number of successes $S = y_1 + \ldots + y_n$ is 5.

Show that the posterior density $p(\theta \mid \boldsymbol{y})$ only depends on the numbers of successes S.

5.2 Conjugated Priors

Let $(\mathbb{P}_{\boldsymbol{\theta}})$ be a dominated parametric family with the density function $p(\boldsymbol{y} \mid \boldsymbol{\theta})$. For a prior π with the density $\pi(\boldsymbol{\theta})$, the posterior density is proportional to $p(\boldsymbol{y} \mid \boldsymbol{\theta})\pi(\boldsymbol{\theta})$. Now consider the case when the prior π belongs to some other parametric family indexed by a parameter $\boldsymbol{\alpha}$, that is, $\pi(\boldsymbol{\theta}) = \pi(\boldsymbol{\theta}, \boldsymbol{\alpha})$. A very desirable feature of the Bayes approach is that the posterior density also belongs to this family. Then computing the posterior is equivalent to fixing the related parameter $\boldsymbol{\alpha} = \boldsymbol{\alpha}(\boldsymbol{Y})$. Such priors are usually called *conjugated*.

To illustrate this notion, we present some examples.

Example 5.2.1 (Gaussian Shift). Let $Y_i \sim \mathcal{N}(\theta, \sigma^2)$ with σ known, $i = 1, \ldots, n$. Consider $\vartheta \sim \mathcal{N}(\tau, g^2)$, $\boldsymbol{\alpha} = (\tau, g^2)$. Then for $\boldsymbol{y} = (y_1, \ldots, y_n)^\top \in \mathbb{R}^n$

$$p(\boldsymbol{y} \mid \theta)\pi(\theta, \boldsymbol{\alpha}) = C \exp\{-\sum (y_i - \theta)^2/(2\sigma^2) - (\theta - \tau)^2/(2g^2)\},$$

where the normalizing factor C does not depend on θ and $\boldsymbol{y}$. The expression in the exponent is a quadratic form of θ and the Taylor expansion w.r.t. θ at $\theta = \tau$ implies

$$\vartheta \mid \boldsymbol{Y} \propto \exp\Bigl\{-\sum \frac{(y_i - \tau)^2}{2\sigma^2} + \frac{\theta - \tau}{\sigma^2} \sum (y_i - \tau) - \frac{n(\theta - \tau)^2}{2}(\sigma^{-2} + g^{-2})\Bigr\}.$$

This representation indicates that the conditional distribution of θ given $\boldsymbol{Y}$ is normal. The parameters of the posterior will be computed in the next section.

Example 5.2.2 (Bernoulli). Let Y_i be a Bernoulli r.v. with $\mathbb{P}(Y_i = 1) = \theta$. Then for $\boldsymbol{y} = (y_1, \ldots, y_n)^\top$, it holds $p(\boldsymbol{y} \mid \theta) = \prod_{i=1}^n \theta^{y_i}(1 - \theta)^{1-y_i}$. Consider the family of priors from the Beta-distribution: $\pi(\theta, \boldsymbol{\alpha}) = \pi(\boldsymbol{\alpha})\theta^a(1 - \theta)^b$ for $\boldsymbol{\alpha} = (a, b)$, where $\pi(\boldsymbol{\theta})$ is a normalizing constant. It follows

$$\vartheta \mid \boldsymbol{y} \propto \theta^a(1 - \theta)^b \prod \theta^{y_i}(1 - \theta)^{1-y_i} = \theta^{s+a}(1 - \theta)^{n-s+b}$$

for $s = y_1 + \ldots + y_n$. Obviously given s this is again a distribution from the Beta-family.

Example 5.2.3 (Exponential). Let Y_i be an exponential r.v. with $\mathbb{P}(Y_i \geq y) = \mathrm{e}^{-y\theta}$. Then $p(\boldsymbol{y} \mid \theta) = \prod_{i=1}^{n} \theta \mathrm{e}^{-y_i\theta}$. For the family of priors from the Gamma-distribution with $\pi(\theta, \boldsymbol{\alpha}) = C(\boldsymbol{\alpha})\theta^a \mathrm{e}^{-\theta b}$ for $\boldsymbol{\alpha} = (a, b)$. One has for the vector $\boldsymbol{y} = (y_1, \ldots, y_n)^\top$

$$\vartheta \mid \boldsymbol{y} \propto \theta^a \mathrm{e}^{-\theta b} \prod_{i=1}^{n} \theta \mathrm{e}^{-y_i\theta} = \theta^{n+a} \mathrm{e}^{-(s+b)}.$$

which yields that the posterior is Gamma with the parameters $n + a$ and $s + b$.

All the previous examples can be systematically treated as special case an exponential family. Let $\boldsymbol{Y} = (Y_1, \ldots, Y_n)$ be an i.i.d. sample from a univariate EF (P_θ) with the density

$$p_1(y \mid \theta) = p_1(y) \exp\{yC(\theta) - B(\theta)\}$$

for some fixed functions $C(\theta)$ and $B(\theta)$. For $\boldsymbol{y} = (y_1, \ldots, y_n)^\top$, the joint density at the point $\boldsymbol{y}$ is given by

$$p(\boldsymbol{y} \mid \theta) \propto \exp\{sC(\theta) - nB(\theta)\}.$$

This suggests to take a prior from the family $\pi(\theta, \boldsymbol{\alpha}) = \pi(\boldsymbol{\alpha}) \exp\{aC(\theta) - bB(\theta)\}$ with $\boldsymbol{\alpha} = (a, b)^\top$. This yields for the posterior density

$$\vartheta \mid \boldsymbol{y} \propto p(\boldsymbol{y} \mid \theta)\pi(\theta, \boldsymbol{\alpha}) \propto \exp\{(s + a)C(\theta) - (n + b)B(\theta)\}$$

which is from the same family with the new parameters $\boldsymbol{\alpha}(\boldsymbol{Y}) = (S + a, n + b)$.

Exercise 5.2.1. Build a conjugate prior the Poisson family.

Exercise 5.2.2. Build a conjugate prior the variance of the normal family with the mean zero and unknown variance.

5.3 Linear Gaussian Model and Gaussian Priors

An interesting and important class of prior distributions is given by Gaussian priors. The very nice and desirable feature of this class is that the posterior distribution for the Gaussian model and Gaussian prior is also Gaussian.

5.3.1 *Univariate Case*

We start with the case of a univariate parameter and one observation $Y \sim \mathcal{N}(\theta, \sigma^2)$, where the variance σ^2 is known and only the mean θ is unknown. The Bayes approach suggests to treat θ as a random variable. Suppose that the prior π is also normal with mean τ and variance r^2.

Theorem 5.3.1. *Let $Y \sim \mathcal{N}(\theta, \sigma^2)$, and let the prior π be the normal distribution $\mathcal{N}(\tau, r^2)$:*

$$Y \mid \theta \sim \mathcal{N}(\theta, \sigma^2),$$
$$\vartheta \sim \mathcal{N}(\tau, r^2).$$

Then the joint, marginal, and posterior distributions are normal as well. Moreover, it holds

$$Y \sim \mathcal{N}(\tau, \sigma^2 + r^2),$$
$$\vartheta \mid Y \sim \mathcal{N}\left(\frac{\tau\sigma^2 + Yr^2}{\sigma^2 + r^2}, \frac{\sigma^2 r^2}{\sigma^2 + r^2}\right).$$

Proof. It holds $Y = \vartheta + \varepsilon$ with $\vartheta \sim \mathcal{N}(\tau, r^2)$ and $\varepsilon \sim \mathcal{N}(0, \sigma^2)$ independent of ϑ. Therefore, Y is normal with mean $\mathbb{E}Y = \mathbb{E}\vartheta + \mathbb{E}\varepsilon = \tau$ and the variance is

$$\operatorname{Var}(Y) = \mathbb{E}(Y - \tau)^2 = r^2 + \sigma^2.$$

This implies the formula for the marginal density $p(Y)$. Next, for $\rho = \sigma^2/(r^2 + \sigma^2)$,

$$\mathbb{E}\big[(\vartheta - \tau)(Y - \tau)\big] = \mathbb{E}(\vartheta - \tau)^2 = r^2 = (1 - \rho)\operatorname{Var}(Y).$$

Thus, the random variables $Y - \tau$ and ζ with

$$\zeta = \vartheta - \tau - (1 - \rho)(Y - \tau) = \rho(\vartheta - \tau) - (1 - \rho)\varepsilon$$

are Gaussian and uncorrelated and therefore, independent. The conditional distribution of ζ given Y coincides with the unconditional distribution and hence, it is normal with mean zero and variance

$$\operatorname{Var}(\zeta) = \rho^2 \operatorname{Var}(\vartheta) + (1 - \rho)^2 \operatorname{Var}(\varepsilon) = \rho^2 r^2 + (1 - \rho)^2 \sigma^2 = \frac{\sigma^2 r^2}{\sigma^2 + r^2}.$$

This yields the result because $\vartheta = \zeta + \rho\tau + (1 - \rho)Y$.

Exercise 5.3.1. Check the result of Theorem 5.3.1 by direct calculation using Bayes formula (5.1).

Now consider the i.i.d. model from $\mathcal{N}(\theta, \sigma^2)$ where the variance σ^2 is known.

Theorem 5.3.2. *Let $Y = (Y_1, \ldots, Y_n)^\top$ be i.i.d. and for each Y_i*

$$Y_i \,\big|\, \theta \sim \mathcal{N}(\theta, \sigma^2), \tag{5.2}$$

$$\vartheta \sim \mathcal{N}(\tau, r^2). \tag{5.3}$$

Then for $\overline{Y} = (Y_1 + \ldots + Y_n)/n$

$$\vartheta \,|\, \boldsymbol{Y} \sim \mathcal{N}\left(\frac{\tau\sigma^2/n + \overline{Y} r^2}{r^2 + \sigma^2/n}, \frac{r^2\sigma^2/n}{r^2 + \sigma^2/n} \right).$$

So, the posterior mean of ϑ is a weighted average of the prior mean τ and the sample estimate Y; the sample estimate is pulled back (or *shrunk*) toward the prior mean. Moreover, the weight ρ on the prior mean is close to one if σ^2 is large relative to r^2 (i.e., our prior knowledge is more precise than the data information), producing substantial shrinkage. If σ^2 is small (i.e., our prior knowledge is imprecise relative to the data information), ρ is close to zero and the direct estimate Y is moved very little towards the prior mean.

Exercise 5.3.2. Prove Theorem 5.3.2 using the technique of the proof of Theorem 5.3.1.
Hint: consider $Y_i = \vartheta + \varepsilon_i$, $\overline{Y} = S/n$, and define $\zeta = \vartheta - \tau - (1 - \rho)(\overline{Y} - \tau)$. Check that ζ and $\overline{Y}$ are uncorrelated and hence, independent.

The result of Theorem 5.3.2 can formally be derived from Theorem 5.3.1 by replacing n i.i.d. observations $Y_1, \ldots, Y_n$ with one single observation $\overline{Y}$ with conditional mean θ and variance σ^2/n.

5.3.2 Linear Gaussian Model and Gaussian Prior

Now we consider the general case when both $\boldsymbol{Y}$ and $\boldsymbol{\vartheta}$ are vectors. Namely we consider the linear model $\boldsymbol{Y} = \Psi^\top \boldsymbol{\vartheta} + \boldsymbol{\varepsilon}$ with Gaussian errors $\boldsymbol{\varepsilon}$ in which the random parameter vector $\boldsymbol{\vartheta}$ is multivariate normal as well:

$$\boldsymbol{\vartheta} \sim \mathcal{N}(\boldsymbol{\tau}, \Gamma), \qquad \boldsymbol{Y} \,\big|\, \boldsymbol{\theta} \sim \mathcal{N}(\Psi^\top \boldsymbol{\theta}, \Sigma). \tag{5.4}$$

Here Ψ is a given $p \times n$ design matrix, and Σ is a given error covariance matrix. Below we assume that both Σ and Γ are non-degenerate. The model (5.4) can be represented in the form

$$\boldsymbol{\vartheta} = \boldsymbol{\tau} + \boldsymbol{\xi}, \qquad \boldsymbol{\xi} \sim \mathcal{N}(0, \Gamma), \tag{5.5}$$

$$\boldsymbol{Y} = \Psi^\top \boldsymbol{\tau} + \Psi^\top \boldsymbol{\xi} + \boldsymbol{\varepsilon}, \qquad \boldsymbol{\varepsilon} \sim \mathcal{N}(0, \Sigma), \quad \boldsymbol{\varepsilon} \perp \boldsymbol{\xi}, \tag{5.6}$$

where $\xi \perp \varepsilon$ means independence of the error vectors ξ and ε. This representation makes clear that the vectors ϑ, Y are jointly normal. Now we state the result about the conditional distribution of ϑ given Y.

Theorem 5.3.3. *Assume* (5.4). *Then the joint distribution of* ϑ, Y *is normal with*

$$\mathbb{E} = \begin{pmatrix} \vartheta \\ Y \end{pmatrix} = \begin{pmatrix} \tau \\ \Psi^\top \tau \end{pmatrix} \qquad \operatorname{Var} \begin{pmatrix} \vartheta \\ Y \end{pmatrix} = \begin{pmatrix} \Gamma & \Psi^\top \Gamma \\ \Gamma \Psi & \Psi^\top \Gamma \Psi + \Sigma \end{pmatrix}.$$

Moreover, the posterior $\vartheta \mid Y$ *is also normal. With* $B = \Gamma^{-1} + \Psi \Sigma^{-1} \Psi^\top$,

$$\begin{aligned} \mathbb{E}(\vartheta \mid Y) &= \tau + \Gamma \Psi \bigl(\Psi^\top \Gamma \Psi + \Sigma \bigr)^{-1} (Y - \Psi^\top \tau) \\ &= B^{-1} \Gamma^{-1} \tau + B^{-1} \Psi \Sigma^{-1} Y, && (5.7) \\ \operatorname{Var}(\vartheta \mid Y) &= B^{-1}. && (5.8) \end{aligned}$$

Proof. The following technical lemma explains a very important property of the normal law: conditional normal is again normal.

Lemma 5.3.1. *Let* ξ *and* η *be jointly normal. Denote* $U = \operatorname{Var}(\xi)$, $W = \operatorname{Var}(\eta)$, $C = \operatorname{Cov}(\xi, \eta) = \mathbb{E}(\xi - \mathbb{E}\xi)(\eta - \mathbb{E}\eta)^\top$. *Then the conditional distribution of* ξ *given* η *is also normal with*

$$\begin{aligned} \mathbb{E}\bigl[\xi \mid \eta\bigr] &= \mathbb{E}\xi + C W^{-1} (\eta - \mathbb{E}\eta), \\ \operatorname{Var}\bigl[\xi \mid \eta\bigr] &= U - C W^{-1} C^\top. \end{aligned}$$

Proof. First consider the case when ξ and η are zero-mean. Then the vector

$$\zeta \stackrel{\text{def}}{=} \xi - C W^{-1} \eta$$

is also normal zero mean and fulfills

$$\begin{aligned} \mathbb{E}(\zeta \eta^\top) &= \mathbb{E}\bigl[(\xi - C W^{-1} \eta) \eta^\top\bigr] = \mathbb{E}(\xi \eta^\top) - C W^{-1} \mathbb{E}(\eta \eta^\top) = 0, \\ \operatorname{Var}(\zeta) &= \mathbb{E}\bigl[(\xi - C W^{-1} \eta)(\xi - C W^{-1} \eta)^\top\bigr] \\ &= \mathbb{E}\bigl[(\xi - C W^{-1} \eta) \xi^\top = U - C W^{-1} C^\top. \end{aligned}$$

The vectors ζ and η are jointly normal and uncorrelated, thus, independent. This means that the conditional distribution of ζ given η coincides with the unconditional one. It remains to note that the $\xi = \zeta + C W^{-1} \eta$. Hence, conditionally on η, the vector ξ is just a shift of the normal vector ζ by a fixed vector $C W^{-1} \eta$. Therefore, the conditional distribution of ξ given η is normal with mean $C W^{-1} \eta$ and the variance $\operatorname{Var}(\zeta) = U - C W^{-1} C^\top$.

Exercise 5.3.3. Extend the proof of Lemma 5.3.1 to the case when the vectors ξ and η are not zero mean.

It remains to deduce the desired result about posterior distribution from this lemma. The formulas for the first two moments of $\boldsymbol{\vartheta}$ and $\boldsymbol{Y}$ follow directly from (5.5) and (5.6). Now we apply Lemma 5.3.1 with $U = \Gamma$, $C = \Gamma\Psi$, $W = \Psi^\top\Gamma\Psi + \Sigma$. It follows that the vector $\boldsymbol{\vartheta}$ conditioned on $\boldsymbol{Y}$ is normal with

$$\begin{aligned} \mathbb{E}(\boldsymbol{\vartheta} \mid \boldsymbol{Y}) &= \boldsymbol{\tau} + \Gamma\Psi W^{-1}(\boldsymbol{Y} - \Psi^\top\boldsymbol{\tau}) \\ \operatorname{Var}(\boldsymbol{\vartheta} \mid \boldsymbol{Y}) &= \Gamma - \Gamma\Psi W^{-1}\Psi^\top\Gamma. \end{aligned} \tag{5.9}$$

Now we apply the following identity: for any $p \times n$-matrix A

$$\begin{aligned} A\bigl(\boldsymbol{I}_n + A^\top A\bigr)^{-1} &= \bigl(\boldsymbol{I}_p + AA^\top\bigr)^{-1}A. \\ \boldsymbol{I}_p - A\bigl(\boldsymbol{I}_n + A^\top A\bigr)^{-1}A^\top &= \bigl(\boldsymbol{I}_p + AA^\top\bigr)^{-1}, \end{aligned} \tag{5.10}$$

The latter implies with $A = \Gamma^{1/2}\Psi\Sigma^{-1/2}$ that

$$\begin{aligned} \Gamma - \Gamma\Psi W^{-1}\Psi^\top\Gamma &= \Gamma^{1/2}\bigl\{\boldsymbol{I}_p - A\bigl(\boldsymbol{I}_n + A^\top A\bigr)^{-1}A^\top\bigr\}\Gamma^{1/2} \\ &= \Gamma^{1/2}\bigl(\boldsymbol{I}_p + AA^\top\bigr)^{-1}\Gamma^{1/2} = \bigl(\Gamma^{-1} + \Psi\Sigma^{-1}\Psi^\top\bigr)^{-1} = B^{-1} \end{aligned}$$

Similarly (5.10) yields with the same A

$$\begin{aligned} \Gamma\Psi W^{-1} &= \Gamma^{1/2}A\bigl(\boldsymbol{I}_n + A^\top A\bigr)^{-1}\Sigma^{-1/2} \\ &= \Gamma^{1/2}\bigl(\boldsymbol{I}_p + AA^\top\bigr)^{-1}A\Sigma^{-1/2} = B^{-1}\Psi\Sigma^{-1}. \end{aligned}$$

This implies (5.7) by (5.9).

Exercise 5.3.4. Check the details of the proof of Theorem 5.3.3.

Exercise 5.3.5. Derive the result of Theorem 5.3.3 by direct computation of the density of $\boldsymbol{\vartheta}$ given $\boldsymbol{Y}$.
Hint: use that $\boldsymbol{\vartheta}$ and $\boldsymbol{Y}$ are jointly normal vectors. Consider their joint density $p(\boldsymbol{\theta}, \boldsymbol{Y})$ for $\boldsymbol{Y}$ fixed and obtain the conditional density by analyzing its linear and quadratic terms w.r.t. $\boldsymbol{\theta}$.

Exercise 5.3.6. Show that $\operatorname{Var}(\boldsymbol{\vartheta} \mid \boldsymbol{Y}) < \operatorname{Var}(\boldsymbol{\vartheta}) = \Gamma$.
Hint: use that $\operatorname{Var}\bigl(\boldsymbol{\vartheta} \mid \boldsymbol{Y}\bigr) = B^{-1}$ and $B \stackrel{\text{def}}{=} \Gamma^{-1} + \Psi\Sigma^{-1}\Psi^\top > \Gamma^{-1}$.

The last exercise delivers an important message: *the variance of the posterior is smaller than the variance of the prior*. This is intuitively clear because the posterior utilizes the both sources of information: those contained in the prior and those we get from the data $\boldsymbol{Y}$. However, even in the simple Gaussian case, the proof is quite complicated. Another interpretation of this fact will be given later: the Bayes approach effectively performs a kind of regularization and thus, leads to a reduction of the variance; cf. Sect. 4.7. Another conclusion from the formulas (5.7), (5.8) is that the moments of the posterior distribution approach the moments of the MLE $\tilde{\boldsymbol{\theta}} = \bigl(\Psi\Sigma^{-1}\Psi^\top\bigr)^{-1}\Psi\Sigma^{-1}\boldsymbol{Y}$ as Γ grows.

5.3.3 Homogeneous Errors, Orthogonal Design

Consider a linear model $Y_i = \Psi_i^\top\boldsymbol{\vartheta} + \varepsilon_i$ for $i = 1,\ldots,n$, where Ψ_i are given vectors in $\mathbb{R}^p$ and ε_i are i.i.d. normal $\mathcal{N}(0,\sigma^2)$. This model is a special case of the model (5.4) with $\Psi = (\Psi_1,\ldots,\Psi_n)$ and uncorrelated homogeneous errors $\boldsymbol{\varepsilon}$ yielding $\Sigma = \sigma^2 I_n$. Then $\Sigma^{-1} = \sigma^{-2} I_n$, $B = \Gamma^{-1} + \sigma^{-2}\Psi\Psi^\top$

$$\begin{aligned}\mathbb{E}\bigl(\boldsymbol{\vartheta} \mid \boldsymbol{Y}\bigr) &= B^{-1}\Gamma^{-1}\boldsymbol{\tau} + \sigma^{-2}B^{-1}\Psi\boldsymbol{Y}, \\ \operatorname{Var}\bigl(\boldsymbol{\vartheta} \mid \boldsymbol{Y}\bigr) &= B^{-1},\end{aligned} \tag{5.11}$$

where $\Psi\Psi^\top = \sum_i \Psi_i\Psi_i^\top$. If the prior variance is also homogeneous, that is, $\Gamma = r^2 I_p$, then the formulas can be further simplified. In particular,

$$\operatorname{Var}\bigl(\boldsymbol{\vartheta} \mid \boldsymbol{Y}\bigr) = \bigl(r^{-2}I_p + \sigma^{-2}\Psi\Psi^\top\bigr)^{-1}.$$

The most transparent case corresponds to the orthogonal design with $\Psi\Psi^\top = \eta^2 I_p$ for some $\eta^2 > 0$. Then

$$\mathbb{E}\bigl(\boldsymbol{\vartheta} \mid \boldsymbol{Y}\bigr) = \frac{\sigma^2/r^2}{\eta^2 + \sigma^2/r^2}\boldsymbol{\tau} + \frac{1}{\eta^2 + \sigma^2/r^2}\Psi\boldsymbol{Y}, \tag{5.12}$$

$$\operatorname{Var}\bigl(\boldsymbol{\vartheta} \mid \boldsymbol{Y}\bigr) = \frac{\sigma^2}{\eta^2 + \sigma^2/r^2} I_p. \tag{5.13}$$

Exercise 5.3.7. Derive (5.12) and (5.13) from Theorem 5.3.3 with $\Sigma = \sigma^2 I_n$, $\Gamma = r^2 I_p$, and $\Psi\Psi^\top = I_p$.

Exercise 5.3.8. Show that the posterior mean is the convex combination of the MLE $\tilde{\boldsymbol{\theta}} = \eta^{-2}\Psi\boldsymbol{Y}$ and the prior mean $\boldsymbol{\tau}$:

$$\mathbb{E}\bigl(\boldsymbol{\vartheta} \mid \boldsymbol{Y}\bigr) = \rho\boldsymbol{\tau} + (1-\rho)\tilde{\boldsymbol{\theta}},$$

with $\rho = (\sigma^2/r^2)/(\eta^2 + \sigma^2/r^2)$. Moreover, $\rho \to 0$ as $\eta \to \infty$, that is, the posterior mean approaches the MLE $\tilde{\boldsymbol{\theta}}$.

5.4 Non-informative Priors

The Bayes approach requires to fix a prior distribution on the values of the parameter $\boldsymbol{\vartheta}$. What happens if no such information is available? Is the Bayes approach still applicable? An immediate answer is "no," however it is a bit hasty. Actually one can still apply the Bayes approach with the priors which do not give any preference to one point against the others. Such priors are called *non-informative*. Consider first the case when the set Θ is finite: $\Theta = \{\boldsymbol{\theta}_1, \ldots, \boldsymbol{\theta}_M\}$. Then the non-informative prior is just the uniform measure on Θ giving to every point $\boldsymbol{\theta}_m$ the equal probability $1/M$. Then the joint probability of $\boldsymbol{Y}$ and $\boldsymbol{\vartheta}$ is the average of the measures $\mathbb{P}_{\boldsymbol{\theta}_m}$ and the same holds for the marginal distribution of the data:

$$p(y) = \frac{1}{M} \sum_{m=1}^{M} p(y \mid \boldsymbol{\theta}_m).$$

The posterior distribution is already "informative" and it differs from the uniform prior:

$$p(\boldsymbol{\theta}_k \mid y) = \frac{p(y \mid \boldsymbol{\theta}_k)\pi(\boldsymbol{\theta}_k)}{p(y)} = \frac{p(y \mid \boldsymbol{\theta}_k)}{\sum_{m=1}^{M} p(y \mid \boldsymbol{\theta}_m)}, \qquad k = 1, \ldots, M.$$

Exercise 5.4.1. Check that the posterior measure is non-informative iff all the measures $\mathbb{P}_{\boldsymbol{\theta}_m}$ coincide.

A similar situation arises if the set Θ is a non-discrete bounded subset in $\mathbb{R}^p$. A typical example is given by the case of a univariate parameter restricted to a finite interval $[a, b]$. Define $\pi(\boldsymbol{\theta}) = 1/\pi(\Theta)$, where

$$\pi(\Theta) \stackrel{\text{def}}{=} \int_{\Theta} d\boldsymbol{\theta}.$$

Then

$$p(y) = \frac{1}{\pi(\Theta)} \int_{\Theta} p(y \mid \boldsymbol{\theta}) d\boldsymbol{\theta}.$$

$$p(\boldsymbol{\theta} \mid y) = \frac{p(y \mid \boldsymbol{\theta})\pi(\boldsymbol{\theta})}{p(y)} = \frac{p(y \mid \boldsymbol{\theta})}{\int_{\Theta} p(y \mid \boldsymbol{\theta}) d\boldsymbol{\theta}}. \tag{5.14}$$

In some cases the non-informative uniform prior can be used even for unbounded parameter sets. Indeed, what we really need is that the integrals in the denominator of the last formula are finite:

$$\int_{\Theta} p(y \mid \boldsymbol{\theta}) d\boldsymbol{\theta} < \infty \qquad \forall y.$$

Then we can apply (5.14) even if Θ is unbounded.

Exercise 5.4.2. Consider the Gaussian Shift model (5.2) and (5.3).

(i) Check that for $n = 1$, the value $\int_{-\infty}^{\infty} p(y \mid \theta) d\theta$ is finite for every y and the posterior distribution of ϑ coincides with the distribution of Y.
(ii) Compute the posterior for $n > 1$.

Exercise 5.4.3. Consider the Gaussian regression model $\boldsymbol{Y} = \Psi^{\top}\boldsymbol{\vartheta} + \boldsymbol{\varepsilon}$, $\boldsymbol{\varepsilon} \sim \mathcal{N}(0, \Sigma)$, and the non-informative prior π which is the Lebesgue measure on the space $\mathbb{R}^p$. Show that the posterior for $\boldsymbol{\vartheta}$ is normal with mean $\tilde{\boldsymbol{\theta}} = (\Psi\Sigma^{-1}\Psi^{\top})^{-1}\Psi\Sigma^{-1}\boldsymbol{Y}$ and variance $(\Psi\Sigma^{-1}\Psi^{\top})^{-1}$. Compare with the result of Theorem 5.3.3.

Note that the result of this exercise can be formally derived from Theorem 5.3.3 by replacing Γ^{-1} with 0.

Another way of tackling the case of an unbounded parameter set is to consider a sequence of priors that approaches the uniform distribution on the whole parameter set. In the case of linear Gaussian models and normal priors, a natural way is to let the prior variance tend to infinity. Consider first the univariate case; see Sect. 5.3.1. A non-informative prior can be approximated by the normal distribution with mean zero and variance r^2 tending to infinity. Then

$$\vartheta \mid Y \sim \mathcal{N}\left(\frac{Yr^2}{\sigma^2 + r^2}, \frac{\sigma^2 r^2}{\sigma^2 + r^2}\right) \xrightarrow{w} \mathcal{N}(Y, \sigma^2) \qquad r \to \infty.$$

It is interesting to note that the case of an i.i.d. sample in fact reduces the situation to the case of a non-informative prior. Indeed, the result of Theorem 5.3.3 implies with $r_n^2 = nr^2$

$$\vartheta \mid \boldsymbol{Y} \sim \mathcal{N}\left(\frac{\overline{Y} r_n^2}{\sigma^2 + r_n^2}, \frac{\sigma^2 r_n^2}{\sigma^2 + r_n^2}\right).$$

One says that the prior information "washes out" from the posterior distribution as the sample size n tends to infinity.

5.5 Bayes Estimate and Posterior Mean

Given a loss function $\wp(\boldsymbol{\theta}, \boldsymbol{\theta}')$ on $\Theta \times \Theta$, the *Bayes risk* of an estimate $\hat{\boldsymbol{\theta}} = \hat{\boldsymbol{\theta}}(\boldsymbol{Y})$ is defined as

$$\mathcal{R}_\pi(\hat{\boldsymbol{\theta}}) \stackrel{\text{def}}{=} \mathbb{E}\wp(\hat{\boldsymbol{\theta}}, \boldsymbol{\vartheta}) = \int_\Theta \left(\int_{\mathcal{Y}} \wp(\hat{\boldsymbol{\theta}}(\boldsymbol{y}), \boldsymbol{\theta})\, p(\boldsymbol{y} \mid \boldsymbol{\theta})\, \boldsymbol{\mu}_0(d\boldsymbol{y}) \right) \pi(\boldsymbol{\theta}) d\boldsymbol{\theta}.$$

Note that $\boldsymbol{\vartheta}$ in this formula is treated as a random variable that follows the prior distribution π. One can represent this formula symbolically in the form

$$\mathcal{R}_\pi(\hat{\boldsymbol{\theta}}) = \mathbb{E}\bigl[\mathbb{E}\bigl(\wp(\hat{\boldsymbol{\theta}}, \boldsymbol{\vartheta}) \bigm| \boldsymbol{\theta}\bigr)\bigr] = \mathbb{E}\,\mathcal{R}(\hat{\boldsymbol{\theta}}, \boldsymbol{\vartheta}).$$

Here the external integration averages the pointwise risk $\mathcal{R}(\hat{\boldsymbol{\theta}}, \boldsymbol{\vartheta})$ over all possible values of $\boldsymbol{\vartheta}$ due to the prior distribution.

The Bayes formula $p(\boldsymbol{y} \mid \boldsymbol{\theta})\pi(\boldsymbol{\theta}) = p(\boldsymbol{\theta} \mid \boldsymbol{y})p(\boldsymbol{y})$ and change of order of integration can be used to represent the Bayes risk via the posterior density:

$$\begin{aligned}\mathcal{R}_\pi(\hat{\boldsymbol{\theta}}) &= \int_{\mathcal{Y}} \Bigl(\int_\Theta \wp(\hat{\boldsymbol{\theta}}(\boldsymbol{y}), \boldsymbol{\vartheta})\, p(\boldsymbol{\theta} \mid \boldsymbol{y})\, d\boldsymbol{\theta}\Bigr) p(\boldsymbol{y})\, \mu_0(d\boldsymbol{y}) \\ &= \mathbb{E}\bigl[\mathbb{E}\{\wp(\hat{\boldsymbol{\theta}}, \boldsymbol{\vartheta}) \mid \boldsymbol{Y}\}\bigr].\end{aligned}$$

The estimate $\tilde{\boldsymbol{\theta}}_\pi$ is called *Bayes* or π-Bayes if it minimizes the corresponding risk:

$$\tilde{\boldsymbol{\theta}}_\pi = \operatorname*{argmin}_{\hat{\boldsymbol{\theta}}} \mathcal{R}_\pi(\hat{\boldsymbol{\theta}}),$$

where the infimum is taken over the class of all feasible estimates. The most widespread choice of the loss function is the quadratic one:

$$\wp(\boldsymbol{\theta}, \boldsymbol{\theta}') \stackrel{\text{def}}{=} \|\boldsymbol{\theta} - \boldsymbol{\theta}'\|^2.$$

The great advantage of this choice is that the Bayes solution can be given explicitly; it is the *posterior mean*

$$\tilde{\boldsymbol{\theta}}_\pi \stackrel{\text{def}}{=} \mathbb{E}(\boldsymbol{\vartheta} \mid \boldsymbol{Y}) = \int_\Theta \boldsymbol{\theta}\, p(\boldsymbol{\theta} \mid \boldsymbol{Y})\, d\boldsymbol{\theta}.$$

Note that due to Bayes' formula, this value can be rewritten

$$\begin{aligned}\tilde{\boldsymbol{\theta}}_\pi &= \frac{1}{p(\boldsymbol{Y})} \int_\Theta \boldsymbol{\theta}\, p(\boldsymbol{Y} \mid \boldsymbol{\theta})\, \pi(\boldsymbol{\theta}) d\boldsymbol{\theta} \\ p(\boldsymbol{Y}) &= \int_\Theta p(\boldsymbol{Y} \mid \boldsymbol{\theta})\, \pi(\boldsymbol{\theta}) d\boldsymbol{\theta}.\end{aligned}$$

Theorem 5.5.1. *It holds for any estimate* $\hat{\boldsymbol{\theta}}$

$$\mathcal{R}_\pi(\hat{\boldsymbol{\theta}}) \ge \mathcal{R}_\pi(\tilde{\boldsymbol{\theta}}_\pi).$$

Proof. The main feature of the posterior mean is that it provides a kind of projection of the data. This property can be formalized as follows:

$$\mathbb{E}\bigl(\tilde{\boldsymbol{\theta}}_\pi - \boldsymbol{\vartheta} \bigm| \boldsymbol{Y}\bigr) = \int_\Theta \bigl(\tilde{\boldsymbol{\theta}}_\pi - \boldsymbol{\theta}\bigr)\, p(\boldsymbol{\theta} \mid \boldsymbol{Y})\, d\boldsymbol{\theta} = 0$$

yielding for any estimate $\hat{\boldsymbol{\theta}} = \hat{\boldsymbol{\theta}}(\boldsymbol{Y})$

$$\begin{aligned}
&\mathbb{E}\big(\|\hat{\boldsymbol{\theta}} - \boldsymbol{\vartheta}\|^2 \,|\, \boldsymbol{Y}\big) \\
&\quad = \mathbb{E}\big(\|\tilde{\boldsymbol{\theta}}_\pi - \boldsymbol{\vartheta}\|^2 \,|\, \boldsymbol{Y}\big) + \mathbb{E}\big(\|\tilde{\boldsymbol{\theta}}_\pi - \hat{\boldsymbol{\theta}}\|^2 \,|\, \boldsymbol{Y}\big) + 2(\hat{\boldsymbol{\theta}} - \tilde{\boldsymbol{\theta}}_\pi)\mathbb{E}(\tilde{\boldsymbol{\theta}}_\pi - \boldsymbol{\vartheta} \,|\, \boldsymbol{Y}) \\
&\quad = \mathbb{E}\big(\|\tilde{\boldsymbol{\theta}}_\pi - \boldsymbol{\vartheta}\|^2 \,|\, \boldsymbol{Y}\big) + \mathbb{E}\big(\|\tilde{\boldsymbol{\theta}}_\pi - \hat{\boldsymbol{\theta}}\|^2 \,|\, \boldsymbol{Y}\big) \\
&\quad \geq \mathbb{E}\big(\|\tilde{\boldsymbol{\theta}}_\pi - \boldsymbol{\vartheta}\|^2 \,|\, \boldsymbol{Y}\big).
\end{aligned}$$

Here we have used that both $\hat{\boldsymbol{\theta}}$ and $\tilde{\boldsymbol{\theta}}_\pi$ are functions of $\boldsymbol{Y}$ and can be considered as constants when taking the conditional expectation w.r.t. $\boldsymbol{Y}$. Now

$$\begin{aligned}
\mathcal{R}_\pi(\hat{\boldsymbol{\theta}}) &= \mathbb{E}\|\hat{\boldsymbol{\theta}} - \boldsymbol{\vartheta}\|^2 = \mathbb{E}\big[\mathbb{E}\big(\|\hat{\boldsymbol{\theta}} - \boldsymbol{\vartheta}\|^2 \,|\, \boldsymbol{Y}\big)\big] \\
&\geq \mathbb{E}\big[\mathbb{E}\big(\|\tilde{\boldsymbol{\theta}}_\pi - \boldsymbol{\vartheta}\|^2 \,|\, \boldsymbol{Y}\big)\big] = \mathcal{R}_\pi(\tilde{\boldsymbol{\theta}}_\pi)
\end{aligned}$$

and the result follows.

Exercise 5.5.1. Consider the univariate case with the loss function $|\theta - \theta'|$. Check that the posterior median minimizes the Bayes risk.

5.6 Posterior Mean and Ridge Regression

Here we again consider the case of a linear Gaussian model

$$\boldsymbol{Y} = \Psi^\top \boldsymbol{\vartheta} + \boldsymbol{\varepsilon}, \qquad \boldsymbol{\varepsilon} \sim \mathcal{N}(0, \sigma^2 I_n).$$

(To simplify the presentation, we focus here on the case of homogeneous errors with $\Sigma = \sigma^2 I_n$.) Recall that the maximum likelihood estimate $\tilde{\boldsymbol{\theta}}$ for this model reads as

$$\tilde{\boldsymbol{\theta}} = \big(\Psi\Psi^\top\big)^{-1}\Psi \boldsymbol{Y}.$$

A penalized MLE $\tilde{\boldsymbol{\theta}}_G$ for the roughness penalty $\|G\boldsymbol{\theta}\|^2$ is given by

$$\tilde{\boldsymbol{\theta}}_G = \big(\Psi\Psi^\top + \sigma^2 G^2\big)^{-1}\Psi \boldsymbol{Y};$$

see Sect. 4.7. It turns out that a similar estimate appears in quite a natural way within the Bayes approach. Consider the normal prior distribution $\boldsymbol{\vartheta} \sim \mathcal{N}(0, G^{-2})$. The posterior will be normal as well with the posterior mean:

$$\tilde{\boldsymbol{\theta}}_\pi = \sigma^{-2} B^{-1} \boldsymbol{Y} = \big(\Psi\Psi^\top + \sigma^2 G^2\big)^{-1}\Psi \boldsymbol{Y};$$

see (5.11). It appears that $\tilde{\boldsymbol{\theta}}_\pi = \tilde{\boldsymbol{\theta}}_G$ for the normal prior $\pi = \mathcal{N}(0, G^{-2})$.

One can say that the Bayes approach with a Gaussian prior leads to a regularization of the least squares method which is similar to quadratic penalization. The degree of regularization is inversely proportional to the variance of the prior. The larger the variance, the closer the prior is to the non-informative one and the posterior mean $\tilde{\boldsymbol{\theta}}_{\pi}$ to the MLE $\tilde{\boldsymbol{\theta}}$.

5.7 Bayes and Minimax Risks

Consider the parametric model $\boldsymbol{Y} \sim \mathbb{P} \in (\mathbb{P}_{\boldsymbol{\theta}}, \boldsymbol{\theta} \in \Theta \subset \mathbb{R}^p)$. Let $\tilde{\boldsymbol{\theta}}$ be an estimate of the parameter $\boldsymbol{\vartheta}$ from the available data $\boldsymbol{Y}$. Formally $\tilde{\boldsymbol{\theta}}$ is a measurable function of $\boldsymbol{Y}$ with values in Θ:

$$\tilde{\boldsymbol{\theta}} = \tilde{\boldsymbol{\theta}}(\boldsymbol{Y}) : \boldsymbol{Y} \to \Theta.$$

The quality of estimation is assigned by the loss function $\varrho(\boldsymbol{\theta}, \boldsymbol{\theta}')$. In estimation problem one usually selects this function in the form $\varrho(\boldsymbol{\theta}, \boldsymbol{\theta}') = \varrho_1(\boldsymbol{\theta} - \boldsymbol{\theta}')$ for another function ϱ_1 of one argument. Typical examples are given by quadratic loss $\varrho_1(\boldsymbol{\theta}) = \|\boldsymbol{\theta}\|^2$ or absolute loss $\varrho_1(\boldsymbol{\theta}) = \|\boldsymbol{\theta}\|$. Given such a loss function, the pointwise risk of $\tilde{\boldsymbol{\theta}}$ at $\boldsymbol{\theta}$ is defined as

$$\mathcal{R}(\tilde{\boldsymbol{\theta}}, \boldsymbol{\theta}) \stackrel{\text{def}}{=} \mathbb{E}_{\boldsymbol{\theta}} \varrho(\tilde{\boldsymbol{\theta}}, \boldsymbol{\theta}).$$

The minimax risk is defined as the maximum of pointwise risks over all $\boldsymbol{\theta} \in \Theta$:

$$\mathcal{R}(\tilde{\boldsymbol{\theta}}) \stackrel{\text{def}}{=} \sup_{\boldsymbol{\theta} \in \Theta} \mathcal{R}(\tilde{\boldsymbol{\theta}}, \boldsymbol{\theta}) = \sup_{\boldsymbol{\theta} \in \Theta} \mathbb{E}_{\boldsymbol{\theta}} \varrho(\tilde{\boldsymbol{\theta}}, \boldsymbol{\theta}).$$

Similarly, the Bayes risk for a prior π is defined by weighting the pointwise risks according to the prior distribution:

$$\mathcal{R}_{\pi}(\tilde{\boldsymbol{\theta}}) \stackrel{\text{def}}{=} \mathcal{R}(\tilde{\boldsymbol{\theta}}, \boldsymbol{\theta}) = \int_{\boldsymbol{\theta}} \mathbb{E}_{\boldsymbol{\theta}} \varrho(\tilde{\boldsymbol{\theta}}, \boldsymbol{\theta}) \pi(\boldsymbol{\theta}) d\boldsymbol{\theta}.$$

It is obvious that the Bayes risk is always smaller or equal than the minimax one, whatever the prior measure is:

$$\mathcal{R}_{\pi}(\tilde{\boldsymbol{\theta}}) \le \mathcal{R}(\tilde{\boldsymbol{\theta}})$$

The famous Le Cam theorem states that the minimax risk can be recovered by taking the maximum over all priors:

$$\mathcal{R}(\tilde{\boldsymbol{\theta}}) = \sup_{\pi} \mathcal{R}_{\pi}(\tilde{\boldsymbol{\theta}}).$$

Moreover, the maximum can be taken over all discrete priors with finite supports.

5.8 Van Trees Inequality

The Cramér–Rao inequality yields a low bounds of the risk for any unbiased estimator. However, it appears to be sub-optimal if the condition of no-bias is dropped. Another way to get a general bound on the quadratic risk of any estimator is to bound from below a Bayes risk for any suitable prior and then to maximize this lower bound in a class of all such priors.

Let $\boldsymbol{Y}$ be an observed vector in $\mathbb{R}^n$ and $(\mathbb{P}_{\boldsymbol{\theta}})$ be the corresponding parametric family with the density function $p(\boldsymbol{y} \mid \boldsymbol{\theta})$ w.r.t. a measure $\boldsymbol{\mu}_0$ on $\mathbb{R}^n$. Let also $\hat{\boldsymbol{\theta}} = \hat{\boldsymbol{\theta}}(\boldsymbol{Y})$ be an arbitrary estimator of $\boldsymbol{\theta}$. For any prior π, we aim to lower bound the π-Bayes risk $\mathcal{R}_\pi(\hat{\boldsymbol{\theta}})$ of $\hat{\boldsymbol{\theta}}$. We already know that this risk minimizes by the posterior mean estimator $\tilde{\boldsymbol{\theta}}_\pi$. However, the presented bound does not rely on a particular structure of the considered estimator. Similarly to the Cramér–Rao inequality, it is entirely based on some geometric properties of the log-likelihood function $p(\boldsymbol{y} \mid \boldsymbol{\theta})$.

To simplify the explanation, consider first the case of a univariate parameter $\theta \in \Theta \subseteq \mathbb{R}$. Below we assume that the prior π has a positive continuously differentiable density $\pi(\boldsymbol{\theta})$. In addition we suppose that the parameter set Θ is an interval, probably infinite, and the prior density $\pi(\theta)$ vanishes at the edges of Θ. By $\mathbb{F}_\pi$ we denote the Fisher information for the prior distribution π:

$$\mathbb{F}_\pi \stackrel{\text{def}}{=} \int_\Theta \frac{|\pi'(\theta)|^2}{\pi(\theta)} d\theta$$

Remind also the definition of the full Fisher information for the family $(\mathbb{P}_\theta)$:

$$\mathbb{F}(\theta) \stackrel{\text{def}}{=} \mathbb{E}_\theta \left| \frac{\partial}{\partial \theta} \log p(\boldsymbol{Y} \mid \theta) \right|^2 = \int \frac{|p'(\boldsymbol{y} \mid \theta)|^2}{p(\boldsymbol{y} \mid \theta)} \boldsymbol{\mu}_0(d\boldsymbol{y}).$$

These quantities will enter in the risk bound. In what follows we also use the notation

$$p_\pi(\boldsymbol{y}, \theta) = p(\boldsymbol{y} \mid \theta)\pi(\theta),$$
$$p'_\pi(\boldsymbol{y}, \theta) \stackrel{\text{def}}{=} \frac{dp_\pi(\boldsymbol{y}, \theta)}{d\theta}.$$

The Bayesian analog of the score function is

$$L'_\pi(\boldsymbol{Y}, \theta) \stackrel{\text{def}}{=} \frac{p'_\pi(\boldsymbol{Y}, \theta)}{p_\pi(\boldsymbol{Y}, \theta)} = \frac{p'(\boldsymbol{Y} \mid \theta)}{p(\boldsymbol{Y} \mid \theta)} + \frac{\pi'(\theta)}{\pi(\theta)}. \tag{5.15}$$

The use of $p_\pi(\boldsymbol{y}, \theta) \equiv 0$ at the edges of Θ implies for any $\hat{\theta} = \hat{\theta}(\boldsymbol{y})$ and any $\boldsymbol{y} \in \mathbb{R}^n$

$$\int_{\Theta} \{(\hat{\theta}(y) - \theta)\, p_{\pi}(y, \theta)\}' d\theta = \Big[\{\hat{\theta}(y) - \theta\}\, p_{\pi}(y, \theta) \Big] = 0,$$

and hence

$$\int_{\Theta} \{\hat{\theta}(y) - \theta\}\, p'_{\pi}(y, \theta)\, d\theta = \int_{\Theta} p_{\pi}(y, \theta) d\theta. \tag{5.16}$$

This is an interesting identity. It holds for each y with μ_0-probability one and the estimate $\hat{\theta}$ only appears in its left-hand side. This can be explained by the fact that $\int_{\Theta} p'_{\pi}(y, \theta)\, d\theta = 0$ which follows by the same calculus. Based on (5.16), one can compute the expectation of the product of $\hat{\theta}(Y) - \vartheta$ and $L'_{\pi}(Y, \vartheta) = p'_{\pi}(Y, \vartheta) / p_{\pi}(Y, \vartheta)$:

$$\begin{aligned} \mathbb{E}_{\pi}\Big[\{\hat{\theta}(Y) - \vartheta\} L'_{\pi}(Y, \vartheta) \Big] &= \int_{\mathbb{R}^n} \int_{\Theta} \{\hat{\theta}(y) - \theta\}\, p'_{\pi}(y, \theta)\, d\theta\, d\mu_0(y) \\ &= \int_{\mathbb{R}^n} \int_{\Theta} p_{\pi}(y, \theta)\, d\theta\, d\mu_0(y) = 1. \end{aligned} \tag{5.17}$$

Again, the remarkable feature of this equality is that the estimate $\hat{\theta}$ only appears in the left-hand side. Now the idea of the obtained bound is very simple. We introduce a r.v. $h(Y, \vartheta) = L'_{\pi}(Y, \vartheta) / \mathbb{E}_{\pi} |L'_{\pi}(Y, \vartheta)|^2$ and use orthogonality of $\hat{\theta}(Y) - \vartheta - h(Y, \vartheta)$ and $h(Y, \vartheta)$ and the Pythagoras Theorem to show that the squared risk of $\hat{\theta}$ is not smaller than $\mathbb{E}_{\pi}\big[h^2(Y, \vartheta)\big]$. More precisely, denote

$$\begin{aligned} \mathcal{I}_{\pi} &\stackrel{\text{def}}{=} \mathbb{E}_{\pi}\Big[\{L'_{\pi}(Y, \vartheta)\}^2 \Big], \\ h(Y, \vartheta) &\stackrel{\text{def}}{=} \mathcal{I}_{\pi}^{-1}\, L'_{\pi}(Y, \vartheta). \end{aligned}$$

Then $\mathbb{E}_{\pi}\big[h^2(Y, \vartheta)\big] = \mathcal{I}_{\pi}^{-1}$ and (5.17) implies

$$\begin{aligned} &\mathbb{E}_{\pi}\Big[\{\hat{\theta}(Y) - \vartheta - h(Y, \vartheta)\} h(Y, \vartheta) \Big] \\ &\quad = \mathcal{I}_{\pi}^{-1} \mathbb{E}_{\pi}\Big[\{\hat{\theta}(Y) - \vartheta\} L'_{\pi}(Y, \vartheta) \Big] - \mathbb{E}_{\pi}\big[h^2(Y, \vartheta)\big] = 0 \end{aligned}$$

and

$$\begin{aligned} \mathbb{E}_{\pi}\Big[(\hat{\theta}(Y) - \vartheta)^2 \Big] &= \mathbb{E}_{\pi}\Big[\{\hat{\theta}(Y) - \vartheta - h(Y, \vartheta) + h(Y, \vartheta)\}^2 \Big] \\ &= \mathbb{E}_{\pi}\Big[\{\hat{\theta}(Y) - \vartheta - h(Y, \vartheta)\}^2 \Big] + \mathbb{E}_{\pi}\big[h^2(Y, \vartheta)\big] \\ &\quad + 2\mathbb{E}_{\pi}\Big[\{\hat{\theta}(Y) - \vartheta - h(Y, \vartheta)\} h(Y, \vartheta) \Big] \end{aligned}$$

$$
\begin{aligned}
&= \mathbb{E}_\pi\Big[\big\{\hat\theta(Y) - \vartheta - h(Y,\vartheta)\big\}^2\Big] + \mathbb{E}_\pi\big[h^2(Y,\vartheta)\big] \\
&\ge \mathbb{E}_\pi\big[h^2(Y,\vartheta)\big] = \mathcal{I}_\pi^{-1}.
\end{aligned}
$$

It remains to compute $\mathcal{I}_\pi$. We use the representation (5.15) for $L'_\pi(Y,\theta)$. Further we use the identity $\int p(y \mid \theta) d\mu_0(y) \equiv 1$ which implies by differentiating in θ

$$
\int p'(y \mid \theta)\, d\mu_0(y) \equiv 0.
$$

This yields that two random variables $p'(Y \mid \theta)/p(Y \mid \theta)$ and $\pi'(\theta)/\pi(\theta)$ are orthogonal under the measure $\mathbb{P}_\pi$:

$$
\begin{aligned}
\mathbb{E}_\pi\left\{\frac{p'(Y \mid \theta)}{p(Y \mid \theta)}\,\frac{\pi'(\theta)}{\pi(\theta)}\right\} &= \int_\Theta \int_{\mathbb{R}^n} p'(y \mid \theta)\, \pi'(\theta)\, d\mu_0(y)\, d\theta \\
&= \int_\Theta \left\{\int_{\mathbb{R}^n} p'(y \mid \theta)\, d\mu_0(y)\right\} \pi'(\theta) d\theta = 0.
\end{aligned}
$$

Now by the Pythagoras Theorem

$$
\begin{aligned}
\mathbb{E}_\pi\big\{L'_\pi(Y,\vartheta)\big\}^2 &= \mathbb{E}_\pi\left\{\frac{p'(Y \mid \vartheta)}{p(Y \mid \vartheta)}\right\}^2 + \mathbb{E}_\pi\left\{\frac{\pi'(\vartheta)}{\pi(\vartheta)}\right\}^2 \\
&= \mathbb{E}_\pi \mathbb{E}_\vartheta\left[\left\{\frac{p'(Y \mid \vartheta)}{p(Y \mid \vartheta)}\right\}^2 \,\middle|\, \vartheta\right] + \mathbb{E}_\pi\left\{\frac{\pi'(\vartheta)}{\pi(\vartheta)}\right\}^2 \\
&= \int_\Theta \mathbb{F}(\theta)\, \pi(\theta)\, d\theta + \mathbb{F}_\pi.
\end{aligned}
$$

Now we can summarize the derivations in the form of *van Trees' inequality*.

Theorem 5.8.1 (van Trees). *Let Θ be an interval on $\mathbb{R}$ and let the prior density $\pi(\theta)$ have piecewise continuous first derivative, $\pi(\theta)$ be positive in the interior of Θ and vanish at the edges. Then for any estimator $\hat\theta$ of θ*

$$
\mathbb{E}_\pi(\hat\theta - \vartheta)^2 \ge \left(\int_\Theta \mathbb{F}(\theta)\, \pi(\theta)\, d\theta + \mathbb{F}_\pi\right)^{-1}.
$$

Now we consider a multivariate extension. We will say that a real function $g(\boldsymbol{\theta})$, $\boldsymbol{\theta} \in \Theta$, is *nice* if it is piecewise continuously differentiable in $\boldsymbol{\theta}$ for almost all values of $\boldsymbol{\theta}$. Everywhere $g'(\cdot)$ means the derivative w.r.t. $\boldsymbol{\theta}$, that is, $g'(\boldsymbol{\theta}) = \frac{d}{d\boldsymbol{\theta}} g(\boldsymbol{\theta})$.

We consider the following assumptions:

1. $p(\boldsymbol{y} \mid \boldsymbol{\theta})$ is nice in $\boldsymbol{\theta}$ for almost all $\boldsymbol{y}$;
2. The full Fisher information matrix

$$\mathbb{F}(\boldsymbol{\theta}) \stackrel{\text{def}}{=} \operatorname{Var}_{\boldsymbol{\theta}}\left\{\frac{p'(Y \mid \boldsymbol{\theta})}{p(Y \mid \boldsymbol{\theta})}\right\} = \mathbb{E}_{\boldsymbol{\theta}}\left[\frac{p'(Y \mid \boldsymbol{\theta})}{p(Y \mid \boldsymbol{\theta})}\left\{\frac{p'(Y \mid \boldsymbol{\theta})}{p(Y \mid \boldsymbol{\theta})}\right\}^{\top}\right]$$

exists and continuous in $\boldsymbol{\theta}$;

3. Θ is compact with boundary which is piecewise C_1-smooth;
4. $\pi(\boldsymbol{\theta})$ is nice; $\pi(\boldsymbol{\theta})$ is positive on the interior of Θ and zero on its boundary. The Fisher information matrix $\mathbb{F}_\pi$ of the prior π is positive and finite:

$$\mathbb{F}_\pi \stackrel{\text{def}}{=} \mathbb{E}_\pi\left[\frac{\pi'(\boldsymbol{\vartheta})}{\pi(\boldsymbol{\vartheta})}\left\{\frac{\pi'(\boldsymbol{\vartheta})}{\pi(\boldsymbol{\vartheta})}\right\}^{\top}\right] < \infty.$$

Theorem 5.8.2. *Let the assumptions 1–4 hold. For any estimate $\hat{\boldsymbol{\theta}} = \hat{\boldsymbol{\theta}}(Y)$, it holds*

$$\mathbb{E}_\pi\Bigl[\{\hat{\boldsymbol{\theta}}(Y) - \boldsymbol{\vartheta}\}\{\hat{\boldsymbol{\theta}}(Y) - \boldsymbol{\vartheta}\}^{\top}\Bigr] \geq \mathcal{I}_\pi^{-1}, \tag{5.18}$$

where

$$\mathcal{I}_\pi \stackrel{\text{def}}{=} \int_\Theta \mathbb{F}(\boldsymbol{\theta})\,\pi(\boldsymbol{\theta})\,d\boldsymbol{\theta} + \mathbb{F}_\pi\,.$$

Proof. The use of $p_\pi(\boldsymbol{y}, \boldsymbol{\theta}) = p(\boldsymbol{y} \mid \boldsymbol{\theta})\pi(\boldsymbol{\theta}) \equiv 0$ at the boundary of Θ implies for any $\hat{\boldsymbol{\theta}} = \hat{\boldsymbol{\theta}}(\boldsymbol{y})$ and any $\boldsymbol{y} \in \mathbb{R}^n$ by Stokes' theorem

$$\int_\Theta [\{\hat{\boldsymbol{\theta}}(\boldsymbol{y}) - \boldsymbol{\theta}\}\, p_\pi(\boldsymbol{y}, \boldsymbol{\theta})]'\, d\boldsymbol{\theta} = 0,$$

and hence

$$\int_\Theta p_\pi'(\boldsymbol{y}, \boldsymbol{\theta})\{\hat{\boldsymbol{\theta}}(\boldsymbol{y}) - \boldsymbol{\theta}\}^{\top}\, d\boldsymbol{\theta} = \boldsymbol{I}_p \int_\Theta p_\pi(\boldsymbol{y}, \boldsymbol{\theta})\, d\boldsymbol{\theta},$$

where $\boldsymbol{I}_p$ is the identity matrix. Therefore, the random vector $L_\pi'(\boldsymbol{Y}, \boldsymbol{\theta}) \stackrel{\text{def}}{=} p_\pi'(\boldsymbol{Y}, \boldsymbol{\theta})/p_\pi(\boldsymbol{Y}, \boldsymbol{\theta})$ fulfills

$$\begin{aligned}\mathbb{E}_\pi\Bigl[L_\pi'(\boldsymbol{Y}, \boldsymbol{\vartheta})\{\hat{\boldsymbol{\theta}}(Y) - \boldsymbol{\vartheta}\}^{\top}\Bigr] &= \int_{\mathbb{R}^n}\int_\Theta p_\pi'(\boldsymbol{y}, \boldsymbol{\theta})\,\{\hat{\boldsymbol{\theta}}(\boldsymbol{y}) - \boldsymbol{\theta}\}^{\top}\, d\boldsymbol{\theta}\, d\boldsymbol{\mu}_0(\boldsymbol{y}) \\ &= \boldsymbol{I}_p \int_{\mathbb{R}^n}\int_\Theta p_\pi(\boldsymbol{y}, \boldsymbol{\theta})\, d\boldsymbol{\theta}\, d\boldsymbol{\mu}_0(\boldsymbol{y}) = \boldsymbol{I}_p.\end{aligned} \tag{5.19}$$

Denote

$$\mathcal{I}_\pi \stackrel{\text{def}}{=} \mathbb{E}_\pi\Big[L_\pi'(\boldsymbol{Y},\boldsymbol{\vartheta})\{L_\pi'(\boldsymbol{Y},\boldsymbol{\vartheta})\}^\top\Big],$$
$$h(\boldsymbol{Y},\boldsymbol{\vartheta}) \stackrel{\text{def}}{=} \mathcal{I}_\pi^{-1}\, L_\pi'(\boldsymbol{Y},\boldsymbol{\vartheta}).$$

Then $\mathbb{E}_\pi\big[h(\boldsymbol{Y},\boldsymbol{\vartheta})\{h(\boldsymbol{Y},\boldsymbol{\vartheta})\}^\top\big] = \mathcal{I}_\pi^{-1}$ and (5.19) implies

$$\begin{aligned}
&\mathbb{E}_\pi\Big[h(\boldsymbol{Y},\boldsymbol{\vartheta})\{\hat{\boldsymbol{\theta}}(\boldsymbol{Y}) - \boldsymbol{\vartheta} - h(\boldsymbol{Y},\boldsymbol{\vartheta})\}^\top\Big] \\
&\quad = \mathcal{I}_\pi^{-1}\mathbb{E}_\pi\Big[L_\pi'(\boldsymbol{Y},\boldsymbol{\vartheta})\{\hat{\boldsymbol{\theta}}(\boldsymbol{Y}) - \boldsymbol{\vartheta}\}^\top\Big] - \mathbb{E}_\pi\Big[h(\boldsymbol{Y},\boldsymbol{\vartheta})\{h(\boldsymbol{Y},\boldsymbol{\vartheta})\}^\top\Big] = 0
\end{aligned}$$

and hence

$$\begin{aligned}
&\mathbb{E}_\pi\Big[\{\hat{\boldsymbol{\theta}}(\boldsymbol{Y}) - \boldsymbol{\vartheta}\}\{\hat{\boldsymbol{\theta}}(\boldsymbol{Y}) - \boldsymbol{\vartheta}\}^\top\Big] \\
&\quad = \mathbb{E}_\pi\Big[\{\hat{\boldsymbol{\theta}}(\boldsymbol{Y}) - \boldsymbol{\vartheta} - h(\boldsymbol{Y},\boldsymbol{\vartheta}) + h(\boldsymbol{Y},\boldsymbol{\vartheta})\}\{\hat{\boldsymbol{\theta}}(\boldsymbol{Y}) - \boldsymbol{\vartheta} - h(\boldsymbol{Y},\boldsymbol{\vartheta}) + h(\boldsymbol{Y},\boldsymbol{\vartheta})\}^\top\Big] \\
&\quad = \mathbb{E}_\pi\Big[\{\hat{\boldsymbol{\theta}}(\boldsymbol{Y}) - \boldsymbol{\vartheta} - h(\boldsymbol{Y},\boldsymbol{\vartheta})\}\{\hat{\boldsymbol{\theta}}(\boldsymbol{Y}) - \boldsymbol{\vartheta} - h(\boldsymbol{Y},\boldsymbol{\vartheta})\}^\top\Big] \\
&\qquad + \mathbb{E}_\pi\Big[\{h(\boldsymbol{Y},\boldsymbol{\vartheta})\}\{h(\boldsymbol{Y},\boldsymbol{\vartheta})\}^\top\Big] \\
&\quad \geq \mathbb{E}_\pi\Big[\{h(\boldsymbol{Y},\boldsymbol{\vartheta})\}\{h(\boldsymbol{Y},\boldsymbol{\vartheta})\}^\top\Big] = \mathcal{I}_\pi^{-1}.
\end{aligned}$$

It remains to compute $\mathcal{I}_\pi$. The definition implies

$$L_\pi'(\boldsymbol{Y},\boldsymbol{\theta}) \stackrel{\text{def}}{=} \frac{1}{p_\pi(\boldsymbol{Y},\boldsymbol{\theta})} p_\pi'(\boldsymbol{Y},\boldsymbol{\theta}) = \frac{1}{p(\boldsymbol{Y}\,|\,\boldsymbol{\theta})} p'(\boldsymbol{Y}\,|\,\boldsymbol{\theta}) + \frac{1}{\pi(\boldsymbol{\theta})}\pi'(\boldsymbol{\theta}).$$

Further, the identity $\int p(\boldsymbol{y}\,|\,\boldsymbol{\theta}) d\mu_0(\boldsymbol{y}) \equiv 1$ yields by differentiating in $\boldsymbol{\theta}$

$$\int p'(\boldsymbol{y}\,|\,\boldsymbol{\theta})\, d\mu_0(\boldsymbol{y}) \equiv 0.$$

Using this identity, we show that the random vectors $p'(\boldsymbol{Y}\,|\,\boldsymbol{\theta})/p(\boldsymbol{Y}\,|\,\boldsymbol{\theta})$ and $\pi'(\boldsymbol{\theta})/\pi(\boldsymbol{\theta})$ are orthogonal w.r.t. the measure $\mathbb{P}_\pi$. Indeed,

$$\begin{aligned}
\mathbb{E}_\pi\left[\frac{p'(\boldsymbol{Y}\,|\,\boldsymbol{\theta})}{p(\boldsymbol{Y}\,|\,\boldsymbol{\theta})}\left\{\frac{\pi'(\boldsymbol{\theta})}{\pi(\boldsymbol{\theta})}\right\}^\top\right] &= \int_\Theta \pi'(\boldsymbol{\theta})\left\{\int_{\mathbb{R}^n} p'(\boldsymbol{y}\,|\,\boldsymbol{\theta})\, d\mu_0(\boldsymbol{y})\right\}^\top d\boldsymbol{\theta} \\
&= \int_\Theta \pi'(\boldsymbol{\theta})\left\{\int_{\mathbb{R}^n} p'(\boldsymbol{y}\,|\,\boldsymbol{\theta})\, d\mu_0(\boldsymbol{y})\right\}^\top d\boldsymbol{\theta} = 0.
\end{aligned}$$

Now by usual Pythagorus calculus, we obtain

$$
\begin{aligned}
&\mathbb{E}_\pi\Big[\{L'_\pi(Y,\vartheta)\}\{L'_\pi(Y,\vartheta)\}^\top\Big] \\
&\quad = \mathbb{E}_\pi\left[\frac{p'(Y\,|\,\vartheta)}{p(Y\,|\,\vartheta)}\left\{\frac{p'(Y\,|\,\vartheta)}{p(Y\,|\,\vartheta)}\right\}^\top\right] + \mathbb{E}_\pi\left[\frac{\pi'(\vartheta)}{\pi(\vartheta)}\left\{\frac{\pi'(\vartheta)}{\pi(\vartheta)}\right\}^\top\right] \\
&\quad = \mathbb{E}_\pi\mathbb{E}_\vartheta\left[\left\{\frac{p'(Y\,|\,\vartheta)}{p(Y\,|\,\vartheta)}\right\}\left\{\frac{p'(Y\,|\,\vartheta)}{p(Y\,|\,\vartheta)}\right\}^\top\,\middle|\,\vartheta\right] + \mathbb{E}_\pi\left[\frac{\pi'(\vartheta)}{\pi(\vartheta)}\left\{\frac{\pi'(\vartheta)}{\pi(\vartheta)}\right\}^\top\right] \\
&\quad = \int_\Theta \mathbb{F}(\theta)\,\pi(\theta)\,d\theta + \mathbb{F}_\pi,
\end{aligned}
$$

and the result follows.

This matrix inequality can be used for obtaining a number of L_2 bounds. We present only two bounds for the squared norm $\|\hat{\theta}(Y) - \vartheta\|^2$.

Corollary 5.8.1. *Under the same conditions 1–4, it holds*

$$
\mathbb{E}_\pi \|\hat{\theta}(Y) - \vartheta\|^2 \geq \operatorname{tr}\big(\mathcal{I}_\pi^{-1}\big) \geq p^2/\operatorname{tr}(\mathcal{I}_\pi).
$$

Proof. The first inequality follows directly from the bound (5.18) of Theorem 5.8.2. For the second one, it suffices to note that for any positive symmetric $p \times p$ matrix B, it holds

$$
\operatorname{tr}(B)\operatorname{tr}(B^{-1}) \geq p^2. \tag{5.20}
$$

This fact can be proved by the Cauchy–Schwarz inequality.

Exercise 5.8.1. Prove (5.20).
Hint: use the Cauchy–Schwarz inequality for the scalar product $\operatorname{tr}(B^{1/2}B^{-1/2})$ of two matrices $B^{1/2}$ and $B^{-1/2}$ (considered as vectors in $\mathbb{R}^{p^2}$).

5.9 Historical Remarks and Further Reading

The origin of the *Bayesian approach* to statistics was the article by Bayes (1763). Further theoretical foundations are due to de Finetti (1937), Savage (1954), and Jeffreys (1957).

The theory of *conjugated priors* was developed by Raiffa and Schlaifer (1961). Conjugated priors for *exponential families* have been characterized by Diaconis and Ylvisaker (1979). *Non-informative priors* were considered by Jeffreys (1961).

Bayes optimality of the *posterior mean* estimator under quadratic loss is a classical result which can be found, for instance, in Sect. 4.4.2 of Berger (1985).

The *van Trees inequality* is originally due to Van Trees (1968), p. 72. Gill and Levit (1995) applied it to the problem of establishing a Bayesian version of the Cramér–Rao bound.

For further reading, we recommend the books by Berger (1985), Bernardo and Smith (1994), and Robert (2001).

Chapter 6
Testing a Statistical Hypothesis

Let Y be the observed sample. The hypothesis testing problem assumes that there is some external information (hypothesis) about the distribution of this sample and the target is to check this hypothesis on the basis of the available data.

6.1 Testing Problem

This section specifies the main notions of the theory of hypothesis testing. We start with a simple hypothesis. Afterwards a composite hypothesis will be discussed. We also introduce the notions of the testing error, level, power, etc.

6.1.1 Simple Hypothesis

The classical testing problem consists in checking a specific hypothesis that the available data indeed follow an externally precisely given distribution. We illustrate this notion by several examples.

Example 6.1.1 (Simple Game). Let $\boldsymbol{Y} = (Y_1, \ldots, Y_n)^\top$ be a Bernoulli sequence of zeros and ones. This sequence can be viewed as the sequence of successes, or results of throwing a coin, etc. The hypothesis about this sequence is that wins (associated with one) and losses (associated with zero) are equally frequent in the long run. This hypothesis can be formalized as follows: $\mathbb{P} = (P_{\theta^*})^{\otimes n}$ with $\theta^* = 1/2$, where P_θ describes the Bernoulli experiment with parameter θ.

Example 6.1.2 (No Treatment Effect). Let (Y_i, Ψ_i) be experimental results, $i = 1, \ldots, n$. The linear regression model assumes a certain dependence of the form $Y_i = \Psi_i^\top \boldsymbol{\theta} + \varepsilon_i$ with errors ε_i having zero mean. The "no effect" hypothesis means

V. Spokoiny and T. Dickhaus, *Basics of Modern Mathematical Statistics*, Springer Texts in Statistics, DOI 10.1007/978-3-642-39909-1_6,

that there is no systematic dependence of Y_i on the factors Ψ_i, i.e., $\boldsymbol{\theta} = \boldsymbol{\theta}^* = 0$, and the observations Y_i are just noise.

Example 6.1.3 (Quality Control). Assume that the Y_i are the results of a production process which can be represented in the form $Y_i = \theta^* + \varepsilon_i$, where θ^* is a nominal value and ε_i is a measurement error. The hypothesis is that the observed process indeed follows this model.

The general problem of *testing a simple hypothesis* is stated as follows: to check on the basis of the available observations $\boldsymbol{Y}$ that their distribution is described by a given measure $\mathbb{P}$. The hypothesis is often called a *null hypothesis* or just *null*.

6.1.2 Composite Hypothesis

More generally, one can treat the problem of testing a composite hypothesis. Let $(\mathbb{P}_{\boldsymbol{\theta}}, \boldsymbol{\theta} \in \Theta \subset \mathbb{R}^p)$ be a given parametric family, and let $\Theta_0 \subset \Theta$ be a nonempty subset in Θ. The hypothesis is that the data distribution $\mathbb{P}$ belongs to the set $(\mathbb{P}_{\boldsymbol{\theta}}, \boldsymbol{\theta} \in \Theta_0)$. Often, this hypothesis and the subset Θ_0 are identified with each other and one says that the hypothesis is given by Θ_0.

We give some typical examples where such a formulation is natural.

Example 6.1.4 (Testing a Subvector). Assume that the vector $\boldsymbol{\theta} \in \Theta$ can be decomposed into two parts: $\boldsymbol{\theta} = (\boldsymbol{\gamma}, \boldsymbol{\eta})$. The subvector $\boldsymbol{\gamma}$ is the target of analysis while the subvector $\boldsymbol{\eta}$ matters for the distribution of the data but is not the target of analysis. It is often called the *nuisance* parameter. The hypothesis we want to test is $\boldsymbol{\gamma} = \boldsymbol{\gamma}^*$ for some fixed value $\boldsymbol{\gamma}^*$. A typical situation in factor analysis where such problems arise is to check on "no effect" for one particular factor in the presence of many different, potentially interrelated factors.

Example 6.1.5 (Interval Testing). Let Θ be the real line and Θ_0 be an interval. The hypothesis is that $\mathbb{P} = \mathbb{P}_{\theta^*}$ for $\theta^* \in \Theta_0$. Such problems are typical for quality control or warning (monitoring) systems when the controlled parameter should be in the prescribed range.

Example 6.1.6 (Testing a Hypothesis About the Error Distribution). Consider the regression model $Y_i = \Psi_i^\top \boldsymbol{\theta} + \varepsilon_i$. The typical assumption about the errors ε_i is that they are zero-mean normal. One may be interested in testing this assumption, having in mind that the cases of discrete, or heavy-tailed, or heteroscedastic errors can also occur.

6.1.3 Statistical Tests

A test is a statistical decision on the basis of the available data whether the prespecified hypothesis is rejected or retained. So the decision space consists of only two points, which we denote by zero and one. A decision ϕ is a mapping of the data $\boldsymbol{Y}$ to this space and is called a *test*:

$$\phi : \mathcal{Y} \to \{0, 1\}.$$

The event $\{\phi = 1\}$ means that the null hypothesis is rejected and the opposite event means non-rejection (acceptance) of the null. Usually the testing results are qualified in the following way: rejection of the null hypothesis means that the data are not consistent with the null, or, equivalently, the data contain some evidence against the null hypothesis. Acceptance simply means that the data do not contradict the null. Therefore, the term "non-rejection" is often considered more appropriate.

The region of acceptance is a subset of the observation space $\mathcal{Y}$ on which $\phi = 0$. One also says that this region is the set of values for which we fail to reject the null hypothesis. *The region of rejection* or *critical region* is, on the other hand, the subset of $\mathcal{Y}$ on which $\phi = 1$.

6.1.4 Errors of the First Kind, Test Level

In the hypothesis testing framework one distinguishes between errors of the first and second kind. An *error of the first kind* means that the null hypothesis is falsely rejected when it is correct. We formalize this notion first for the case of a simple hypothesis and then extend it to the general case.

Let $H_0 : \boldsymbol{Y} \sim \mathbb{P}_{\theta^*}$ be a null hypothesis. The error of the first kind is the situation in which the data indeed follow the null, but the decision of the test is to reject this hypothesis: $\phi = 1$. Clearly the probability of such an error is $\mathbb{P}_{\theta^*}(\phi = 1)$. The latter number in $[0, 1]$ is called the *size of the test* ϕ. One says that ϕ is a test of level α for some $\alpha \in (0, 1)$ if

$$\mathbb{P}_{\theta^*}(\phi = 1) \le \alpha.$$

The value α is called *level of the test* or *significance level*. Often, size and level of a test coincide; however, especially in discrete models, it is not always possible to attain the significance level exactly by the chosen test ϕ, meaning that the actual size of ϕ is smaller than α, see Example 6.1.7 below.

If the hypothesis is composite, then the level of the test is the maximum rejection probability over the null subset Θ_0. Here, a test ϕ is of level α if

$$\sup_{\theta \in \Theta_0} \mathbb{P}_{\theta}(\phi = 1) \le \alpha.$$

Example 6.1.7 (One-Sided Binomial Test). Consider again the situation from Example 6.1.1 (an i.i.d. sample $\boldsymbol{Y} = (Y_1, \ldots, Y_n)^\top$ from a Bernoulli distribution is observed). We let $n = 13$ and $\Theta_0 = [0, 1/5]$. For instance, one may want to test if the cancer-related mortality in a subpopulation of individuals which are exposed to some environmental risk factor is not significantly larger than in the general population, in which it is equal to $1/5$. To this end, death causes are assessed for 13 decedents and for every decedent we get the information if s/he died because of cancer or not. From Sect. 2.6, we know that S_n/n efficiently estimates the success probability θ^*, where $S_n = \sum_{i=1}^n Y_i$. Therefore, it appears natural to use S_n also as the basis for a solution of the test problem. Under θ^*, it holds $S_n \sim \text{Bin}(n, \theta^*)$. Since a large value of S_n implies evidence against the null, we choose a test of the form $\phi = \mathbb{1}_{(c_\alpha, n]}(S_n)$, where the constant c_α has to be chosen such that ϕ is of level α. This condition can equivalently be expressed as

$$\inf_{0 \leq \theta \leq 1/5} \mathbb{P}_\theta(S_n \leq c_\alpha) \geq 1 - \alpha.$$

For fixed $k \in \{0, \ldots, n\}$, we have

$$\mathbb{P}_\theta(S_n \leq k) = \sum_{\ell=0}^{k} \binom{n}{\ell} \theta^\ell (1-\theta)^{n-\ell} = F(\theta, k) \text{ (say)}.$$

Exercise 6.1.1. Show that for all $k \in \{0, \ldots, n\}$, the function $F(\cdot, k)$ is decreasing on $\Theta_0 = [0, 1/5]$.

Due to Exercise 6.1.1, we have to calculate c_α under the *least favorable parameter configuration (LFC)* $\theta = 1/5$ and we obtain

$$c_\alpha = \min\left\{k \in \{0, \ldots, n\} : \sum_{\ell=0}^{k} \binom{n}{\ell} \left(\frac{1}{5}\right)^\ell \left(\frac{4}{5}\right)^{n-\ell} \geq 1 - \alpha\right\},$$

because we want to exhaust the significance level α as tightly as possible. For the standard choice of $\alpha = 0.05$, we have

$$\sum_{\ell=0}^{4} \binom{13}{\ell} \left(\frac{1}{5}\right)^\ell \left(\frac{4}{5}\right)^{13-\ell} \approx 0.901, \quad \sum_{\ell=0}^{5} \binom{13}{\ell} \left(\frac{1}{5}\right)^\ell \left(\frac{4}{5}\right)^{13-\ell} \approx 0.9700,$$

hence, we choose $c_\alpha = 5$. However, the size of the test $\phi = \mathbb{1}_{(5,13]}(S_n)$ is strictly smaller than the significance level $\alpha = 0.05$, namely

$$\sup_{0 \leq \theta \leq 1/5} \mathbb{P}_\theta(S_n > 5) = \mathbb{P}_{1/5}(S_n > 5) = 1 - \mathbb{P}_{1/5}(S_n \leq 5) \approx 0.03 < \alpha.$$

6.1.5 *Randomized Tests*

In some situations it is difficult to decide about acceptance or rejection of the hypothesis. A *randomized* test can be viewed as a weighted decision: with a certain probability the hypothesis is rejected, otherwise retained. The decision space for a randomized test ϕ is the unit interval [0, 1], that is, $\phi(Y)$ is a number between zero and one. The hypothesis H_0 is rejected with probability $\phi(Y)$ on the basis of the observed data Y. If $\phi(Y)$ only admits the binary values 0 and 1 for every Y, then we are back at the usual non-randomized test that we have considered before. The probability of an error of the first kind is naturally given by the value $\mathbb{E}\phi(Y)$. For a simple hypothesis $H_0 : \mathbb{P} = \mathbb{P}_{\theta^*}$, a test ϕ is now of level α if

$$\mathbb{E}\phi(Y) = \alpha.$$

For a randomized test ϕ, the significance level α is typically attainable exactly, even for discrete models. In the case of a composite hypothesis $H_0 : \mathbb{P} \in (\mathbb{P}_{\theta}, \theta \in \Theta_0)$, the level condition reads as

$$\sup_{\theta \in \Theta_0} \mathbb{E}\phi(Y) \le \alpha$$

as before. In what follows we mostly consider non-randomized tests and only comment on whether a randomization can be useful. Note that any randomized test can be reduced to a non-randomized test by extending the probability space.

Exercise 6.1.2. Construct for any randomized test ϕ its non-randomized version using a random data generator.

Randomized tests are a satisfactory solution of the test problem from a mathematical point of view, but they are disliked by practitioners, because the test result may not be reproducible, due to randomization.

Example 6.1.8 (Example 6.1.7 Continued). Under the setup of Example 6.1.7, consider the randomized test ϕ, given by

$$\phi(Y) = \begin{Bmatrix} 0, & S_n < 5 \\ 2/7, & S_n = 5 \\ 1, & S_n > 5 \end{Bmatrix}$$

It is easy to show that under the LFC $\theta = 1/5$, the size of ϕ is (up to rounding) exactly equal to $\alpha = 5\,\%$.

6.1.6 Alternative Hypotheses, Error of the Second Kind, Power of a Test

The setup of hypothesis testing is asymmetric in the sense that it focuses on the null hypothesis. However, for a complete analysis, one has to specify the data distribution when the hypothesis is false. Within the parametric framework, one usually makes the assumption that the unknown data distribution belongs to some parametric family $(\mathbb{P}_{\boldsymbol{\theta}}, \boldsymbol{\theta} \in \Theta \subseteq \mathbb{R}^p)$. This assumption has to be fulfilled independently of whether the hypothesis is true or false. In other words, we assume that $\mathbb{P} \in (\mathbb{P}_{\boldsymbol{\theta}}, \boldsymbol{\theta} \in \Theta)$ and there is a subset $\Theta_0 \subset \Theta$ corresponding to the null hypothesis. The measure $\mathbb{P} = \mathbb{P}_{\boldsymbol{\theta}}$ for $\boldsymbol{\theta} \notin \Theta_0$ is called *an alternative*. Furthermore, we call $\Theta_1 \stackrel{\text{def}}{=} \Theta \setminus \Theta_0$ the *alternative hypothesis*.

Now we can consider the performance of a test ϕ when the hypothesis H_0 is false. The decision to retain the hypothesis when it is false is called the *error of the second kind*. The probability of such an error is equal to $\mathbb{P}(\phi = 0)$, whenever $\mathbb{P}$ is an alternative. This value certainly depends on the alternative $\mathbb{P} = \mathbb{P}_{\boldsymbol{\theta}}$ for $\boldsymbol{\theta} \notin \Theta_0$. The value $\beta(\boldsymbol{\theta}) = 1 - \mathbb{P}_{\boldsymbol{\theta}}(\phi = 0)$ is often called *the test power* at $\boldsymbol{\theta} \notin \Theta_0$. The function $\beta(\boldsymbol{\theta})$ of $\boldsymbol{\theta} \in \Theta \setminus \Theta_0$ given by

$$\beta(\boldsymbol{\theta}) \stackrel{\text{def}}{=} 1 - \mathbb{P}_{\boldsymbol{\theta}}(\phi = 0)$$

is called *power function*. Ideally one would desire to build a test which simultaneously and separately minimizes the size and maximizes the power. These two wishes are somehow contradictory. A decrease of the size usually results in a decrease of the power and vice versa. Usually one imposes the level α constraint on the size of the test and tries to optimize its power under this constraint. Under the general framework of statistical decision problems as discussed in Sect. 1.4, one can thus regard $R(\phi, \boldsymbol{\theta}) = 1 - \beta(\boldsymbol{\theta})$, $\boldsymbol{\theta} \in \Theta_1$, as a risk function. If we agree on this risk measure and restrict attention to level α tests, then the test problem, regarded as a statistical decision problem, is already completely specified by $(\mathcal{Y}, \mathcal{B}(\mathcal{Y}), (\mathbb{P}_{\boldsymbol{\theta}} : \boldsymbol{\theta} \in \Theta), \Theta_0)$.

Definition 6.1.1. A test ϕ^* is called *uniformly most powerful* (UMP) test of level α if it is of level α and for any other test of level α, it holds

$$1 - \mathbb{P}_{\boldsymbol{\theta}}(\phi^* = 0) \geq 1 - \mathbb{P}_{\boldsymbol{\theta}}(\phi = 0), \qquad \boldsymbol{\theta} \notin \Theta_0.$$

Unfortunately, such UMP tests exist only in very few special models; otherwise, optimization of the power given the level is a complicated task.

In the case of a univariate parameter $\theta \in \Theta \subset \mathbb{R}^1$ and a simple hypothesis $\theta = \theta^*$, one often considers one-sided alternatives

$$H_1 : \theta \geq \theta^* \qquad \text{or} \qquad H_1 : \theta \leq \theta^*$$

or a two-sided alternative

$$H_1 : \theta \neq \theta^*.$$

6.2 Neyman–Pearson Test for Two Simple Hypotheses

This section discusses one very special case of hypothesis testing when both the null hypothesis and the alternative are simple one-point sets. This special situation by itself can be viewed as a toy problem, but it is very important from the methodological point of view. In particular, it introduces and justifies the so-called likelihood ratio test and demonstrates its efficiency.

For simplicity we write $\mathbb{P}_0$ for the measure corresponding to the null hypothesis and $\mathbb{P}_1$ for the alternative measure. A *test* ϕ is a measurable function of the observations with values in the two-point set $\{0, 1\}$. The event $\phi = 0$ is treated as acceptance of the null hypothesis H_0 while $\phi = 1$ means rejection of the null hypothesis and, consequently, decision in favor of H_1.

For ease of presentation we assume that the measure $\mathbb{P}_1$ is absolutely continuous w.r.t. the measure $\mathbb{P}_0$ and denote by $Z(\boldsymbol{Y})$ the corresponding derivative at the observation point:

$$Z(\boldsymbol{Y}) \stackrel{\text{def}}{=} \frac{d\mathbb{P}_1}{d\mathbb{P}_0}(\boldsymbol{Y}).$$

Similarly $L(\boldsymbol{Y})$ means the log-density:

$$L(\boldsymbol{Y}) \stackrel{\text{def}}{=} \log Z(\boldsymbol{Y}) = \log \frac{d\mathbb{P}_1}{d\mathbb{P}_0}(\boldsymbol{Y}).$$

The solution of the test problem in the case of two simple hypotheses is known as the Neyman–Pearson test: reject the hypothesis H_0 if the log-likelihood ratio $L(\boldsymbol{Y})$ exceeds a specific critical value $\mathfrak{z}$:

$$\phi_{\mathfrak{z}}^{*} \stackrel{\text{def}}{=} \mathbb{1}\bigl(L(\boldsymbol{Y}) > \mathfrak{z}\bigr) = \mathbb{1}\bigl(Z(\boldsymbol{Y}) > \mathrm{e}^{\mathfrak{z}}\bigr).$$

The Neyman–Pearson test is known as the one minimizing the weighted sum of the errors of the first and second kind. For a non-randomized test this sum is equal to

$$\wp_0 \mathbb{P}_0(\phi = 1) + \wp_1 \mathbb{P}_1(\phi = 0),$$

while the weighted error of a randomized test ϕ is

$$\wp_0 \mathbb{E}_0 \phi + \wp_1 \mathbb{E}_1 (1 - \phi). \tag{6.1}$$

Theorem 6.2.1. *For every two positive values* $\wp_0$ *and* $\wp_1$*, the test* $\phi_{\mathfrak{z}}^{*}$ *with* $\mathfrak{z} = \log(\wp_0/\wp_1)$ *minimizes* (6.1) *over all possible (randomized) tests* ϕ*:*

$$\phi_{\mathfrak{z}}^{*} \stackrel{\text{def}}{=} \mathbb{1}\bigl(L(\boldsymbol{Y}) \geq \mathfrak{z}\bigr) = \operatorname*{argmin}_{\phi}\bigl\{\wp_0 \mathbb{E}_0 \phi + \wp_1 \mathbb{E}_1 (1 - \phi)\bigr\}.$$

Proof. We use the formula for a change of measure:

$$\mathbb{E}_1 \xi = \mathbb{E}_0\bigl[\xi Z(\boldsymbol{Y})\bigr]$$

for any r.v. ξ. It holds for any test ϕ with $\mathfrak{z} = \log(\wp_0/\wp_1)$

$$\begin{aligned} \wp_0 \mathbb{E}_0 \phi + \wp_1 \mathbb{E}_1 (1 - \phi) &= \mathbb{E}_0\bigl[\wp_0 \phi - \wp_1 Z(\boldsymbol{Y})\phi\bigr] + \wp_1 \\ &= -\wp_1 \mathbb{E}_0[(Z(\boldsymbol{Y}) - \mathrm{e}^{\mathfrak{z}})\phi] + \wp_1 \\ &\geq -\wp_1 \mathbb{E}_0[Z(\boldsymbol{Y}) - \mathrm{e}^{\mathfrak{z}}]_+ + \wp_1 \end{aligned}$$

with equality for $\phi = \mathbb{1}(L(\boldsymbol{Y}) \geq \mathfrak{z})$.

The Neyman–Pearson test belongs to a large class of tests of the form

$$\phi = \mathbb{1}(T \geq \mathfrak{z}),$$

where T is a function of the observations $\boldsymbol{Y}$. This random variable is usually called a *test statistic* while the threshold $\mathfrak{z}$ is called a *critical value*. The hypothesis is rejected if the test statistic exceeds the critical value. For the Neyman–Pearson test, the test statistic is the log-likelihood ratio $L(\boldsymbol{Y})$ and the critical value is selected as a suitable quantile of this test statistic.

The next result shows that the Neyman–Pearson test $\phi^*_{\mathfrak{z}}$ with a proper critical value $\mathfrak{z}$ can be constructed to maximize the power $\mathbb{E}_1 \phi$ under the level constraint $\mathbb{E}_0 \phi \leq \alpha$.

Theorem 6.2.2. *Given $\alpha \in (0, 1)$, let $\mathfrak{z}_\alpha$ be such that*

$$\mathbb{P}_0(L(\boldsymbol{Y}) \geq \mathfrak{z}_\alpha) = \alpha. \tag{6.2}$$

Then it holds

$$\phi^*_{\mathfrak{z}_\alpha} \stackrel{\mathrm{def}}{=} \mathbb{1}\bigl(L(\boldsymbol{Y}) \geq \mathfrak{z}_\alpha\bigr) = \operatorname*{argmax}_{\phi : \mathbb{E}_0 \phi \leq \alpha} \bigl\{\mathbb{E}_1 \phi\bigr\}.$$

Proof. Let ϕ satisfy $\mathbb{E}_0 \phi \leq \alpha$. Then

$$\begin{aligned} \mathbb{E}_1 \phi - \alpha \mathfrak{z}_\alpha &\leq \mathbb{E}_0\bigl\{Z(\boldsymbol{Y})\phi\bigr\} - \mathrm{e}^{\mathfrak{z}_\alpha} \mathbb{E}_0 \phi \\ &= \mathbb{E}_0\bigl[\bigl\{Z(\boldsymbol{Y}) - \mathrm{e}^{\mathfrak{z}_\alpha}\bigr\}\phi\bigr] \\ &\leq \mathbb{E}_0[Z(\boldsymbol{Y}) - \mathrm{e}^{\mathfrak{z}_\alpha}]_+, \end{aligned}$$

again with equality for $\phi = \mathbb{1}(L(\boldsymbol{Y}) \geq \mathfrak{z}_\alpha)$.

The previous result assumes that for a given α there is a critical value $\mathfrak{z}_\alpha$ such that (6.2) is fulfilled. However, this is not always the case.

Exercise 6.2.1. Let $L(Y) = \log d\mathbb{P}_1(Y)/d\mathbb{P}_0$.

- Show that the relation (6.2) can always be fulfilled with a proper choice of $\mathfrak{z}_\alpha$ if the pdf of $L(Y)$ under $\mathbb{P}_0$ is a continuous function.
- Suppose that the pdf of $L(Y)$ is discontinuous and $\mathfrak{z}_\alpha$ fulfills

$$\mathbb{P}_0(L(Y) \geq \mathfrak{z}_\alpha) > \alpha, \qquad \mathbb{P}_0(L(Y) \leq \mathfrak{z}_\alpha) > 1 - \alpha.$$

Construct a randomized test that fulfills $\mathbb{E}_0\phi = \alpha$ and maximizes the test power $\mathbb{E}_1\phi$ among all such tests.

The Neyman–Pearson test can be viewed as a special case of the general likelihood ratio test. Indeed, it decides in favor of the null or the alternative by looking at the likelihood ratio. Informally one can say: we decide in favor of the alternative if it is significantly more likely at the point of observation $\boldsymbol{Y}$.

An interesting question that arises in relation with the Neyman–Pearson result is how to interpret it when the true distribution $\mathbb{P}$ does not coincide either with $\mathbb{P}_0$ or with $\mathbb{P}_1$ and probably it is not even within the considered parametric family $(\mathbb{P}_{\boldsymbol{\theta}})$. Wald called this situation the third-kind error. It is worth mentioning that the test $\phi_{\mathfrak{z}}^*$ remains meaningful: it decides which of two given measures $\mathbb{P}_0$ and $\mathbb{P}_1$ describes the given data better. However, it is not any more a likelihood ratio test. In analogy with estimation theory, one can call it a *quasi likelihood ratio test*.

6.2.1 Neyman–Pearson Test for an i.i.d. Sample

Let $\boldsymbol{Y} = (Y_1, \ldots, Y_n)^\top$ be an i.i.d. sample from a measure P. Suppose that P belongs to some parametric family $(P_{\boldsymbol{\theta}}, \boldsymbol{\theta} \in \Theta \subset \mathbb{R}^p)$, that is, $P = P_{\boldsymbol{\theta}^*}$ for $\boldsymbol{\theta}^* \in \Theta$. Let also a special point $\boldsymbol{\theta}_0$ be fixed. A simple null hypothesis can be formulated as $\boldsymbol{\theta}^* = \boldsymbol{\theta}_0$. Similarly, a simple alternative is $\boldsymbol{\theta}^* = \boldsymbol{\theta}_1$ for some other point $\boldsymbol{\theta}_1 \in \Theta$. The Neyman–Pearson test situation is a bit artificial: one reduces the whole parameter set Θ to just these two points $\boldsymbol{\theta}_0$ and $\boldsymbol{\theta}_1$ and tests $\boldsymbol{\theta}_0$ against $\boldsymbol{\theta}_1$.

As usual, the distribution of the data $\boldsymbol{Y}$ is described by the product measure $\mathbb{P}_{\boldsymbol{\theta}} = P_{\boldsymbol{\theta}}^{\otimes n}$. If μ_0 is a dominating measure for $(P_{\boldsymbol{\theta}})$ and $\ell(y, \boldsymbol{\theta}) \stackrel{\text{def}}{=} \log[dP_{\boldsymbol{\theta}}(y)/d\mu_0]$, then the log-likelihood $L(\boldsymbol{Y}, \boldsymbol{\theta})$ is

$$L(\boldsymbol{Y}, \boldsymbol{\theta}) \stackrel{\text{def}}{=} \log \frac{d\mathbb{P}_{\boldsymbol{\theta}}}{\boldsymbol{\mu}_0}(\boldsymbol{Y}) = \sum_i \ell(Y_i, \boldsymbol{\theta}),$$

where $\boldsymbol{\mu}_0 = \mu_0^{\otimes n}$. The log-likelihood ratio of $\mathbb{P}_{\boldsymbol{\theta}_1}$ w.r.t. $\mathbb{P}_{\boldsymbol{\theta}_0}$ can be defined as

$$L(\boldsymbol{Y}, \boldsymbol{\theta}_1, \boldsymbol{\theta}_0) \stackrel{\text{def}}{=} L(\boldsymbol{Y}, \boldsymbol{\theta}_1) - L(\boldsymbol{Y}, \boldsymbol{\theta}_0).$$

The related Neyman–Pearson test can be written as

$$\phi_{\mathfrak{z}}^{*} \stackrel{\text{def}}{=} \mathbb{1}\bigl(L(Y,\theta_1,\theta_0) > \mathfrak{z}\bigr).$$

6.3 Likelihood Ratio Test

This section introduces a general likelihood ratio test in the framework of parametric testing theory. Let, as usual, $\boldsymbol{Y}$ be the observed data, and $\mathbb{P}$ be their distribution. The parametric assumption is that $\mathbb{P} \in (\mathbb{P}_{\boldsymbol{\theta}}, \boldsymbol{\theta} \in \Theta)$, that is, $\mathbb{P} = \mathbb{P}_{\boldsymbol{\theta}^*}$ for $\boldsymbol{\theta}^* \in \Theta$. Let now two subsets Θ_0 and Θ_1 of the set Θ be given. The hypothesis H_0 that we would like to test is that $\mathbb{P} \in (\mathbb{P}_{\boldsymbol{\theta}}, \boldsymbol{\theta} \in \Theta_0)$, or equivalently, $\boldsymbol{\theta}^* \in \Theta_0$. The alternative is that $\boldsymbol{\theta}^* \in \Theta_1$.

The general likelihood approach leads to comparing the (maximum) likelihood values $L(\boldsymbol{Y},\boldsymbol{\theta})$ on the hypothesis and alternative sets. Namely, the hypothesis is rejected if there is one alternative point $\boldsymbol{\theta}_1 \in \Theta_1$ such that the value $L(\boldsymbol{Y},\boldsymbol{\theta}_1)$ exceeds all corresponding values for $\boldsymbol{\theta} \in \Theta_0$ by a certain amount which is defined by assigning losses or by fixing a significance level. In other words, rejection takes place if observing the sample $\boldsymbol{Y}$ under alternative $\mathbb{P}_{\boldsymbol{\theta}_1}$ is significantly more likely than under any measure $\mathbb{P}_{\boldsymbol{\theta}}$ from the null. Formally this relation can be written as:

$$\sup_{\boldsymbol{\theta}\in\Theta_0} L(\boldsymbol{Y},\boldsymbol{\theta}) + \mathfrak{z} < \sup_{\boldsymbol{\theta}\in\Theta_1} L(\boldsymbol{Y},\boldsymbol{\theta}),$$

where the constant $\mathfrak{z}$ makes the term "significantly" explicit. In particular, a simple hypothesis means that the set Θ_0 consists of one single point $\boldsymbol{\theta}_0$ and the latter relation takes of the form

$$L(\boldsymbol{Y},\boldsymbol{\theta}_0) + \mathfrak{z} < \sup_{\boldsymbol{\theta}\in\Theta_1} L(\boldsymbol{Y},\boldsymbol{\theta}).$$

In general, the *likelihood ratio (LR)* test corresponds to the test statistic

$$T \stackrel{\text{def}}{=} \sup_{\boldsymbol{\theta}\in\Theta_1} L(\boldsymbol{Y},\boldsymbol{\theta}) - \sup_{\boldsymbol{\theta}\in\Theta_0} L(\boldsymbol{Y},\boldsymbol{\theta}). \tag{6.3}$$

The hypothesis is rejected if this test statistic exceeds some critical value $\mathfrak{z}$. Usually this critical value is selected to ensure the level condition:

$$\mathbb{P}\bigl(T > \mathfrak{z}_\alpha\bigr) \le \alpha$$

for a given level α, whenever $\mathbb{P}$ is a measure under the null hypothesis.

We have already seen that the LR test is optimal for testing a simple hypothesis against a simple alternative. Later we show that this optimality property can be extended to some more general situations. Now and in the following Sect. 6.4 we consider further examples of an LR test.

Example 6.3.1 (Chi-Square Test for Goodness-of-Fit). Let the observation space (which is a subset of $\mathbb{R}^1$) be split into non-overlapping subsets $A_1, \dots, A_d$ and assume one observes indicators $\mathbb{1}(Y_i \in A_j)$ for $1 \le i \le n$ and $1 \le j \le d$. Define $\theta_j = P_0(A_j) = \int_{A_j} P_0(dy)$ for $1 \le j \le d$. Let counting variables N_j, $1 \le j \le d$, be given by $N_j = \sum_{i=1}^n \mathbb{1}(Y_i \in A_j)$. The vector $N = (N_1, \dots, N_d)^\top$ follows the multinomial distribution with parameters n, d, and $\boldsymbol{\theta} = (\theta_1, \dots, \theta_d)^\top$, where we assume n and d as fixed, leading to $\dim(\Theta) = d - 1$. More specifically, it holds

$$\Theta = \Big\{\boldsymbol{\theta} = (\theta_1, \dots, \theta_d)^\top \in [0, 1]^d : \sum_{j=1}^d \theta_j = 1\Big\}.$$

The likelihood statistic for this model with respect to the counting measure is given by

$$Z(N, \boldsymbol{p}) = \frac{n!}{\prod_{j=1}^d N_j!} \prod_{\ell=1}^d \theta_\ell^{N_\ell},$$

and the MLE is given by $\tilde{\boldsymbol{\theta}}_j = N_j/n$, $1 \le j \le d$. Now, consider the point hypothesis $\boldsymbol{\theta} = \boldsymbol{p}$ for a fixed given vector $\boldsymbol{p} \in \Theta$. We obtain the likelihood ratio statistic

$$T = \sup_{\boldsymbol{\theta} \in \Theta} \log \frac{Z(N, \boldsymbol{\theta})}{Z(N, \boldsymbol{p})},$$

leading to

$$T = n \sum_{j=1}^d \tilde{\theta}_j \log \frac{\tilde{\theta}_j}{p_j},$$

In practice, this LR test is often carried out as Pearson's chi-square test. To this end, consider the function $h : \mathbb{R} \to \mathbb{R}$, given by $h(x) = x \log(x/x_0)$ for a fixed real number $x_0 \in (0, 1)$. Then, the Taylor expansion of $h(x)$ around x_0 is given by

$$h(x) = (x - x_0) + \frac{1}{2x_0}(x - x_0)^2 + o[(x - x_0)^2] \text{ as } x \to x_0.$$

Consequently, for $\tilde{\boldsymbol{\theta}}$ close to $\boldsymbol{p}$, the use of $\sum_j \tilde{\theta}_j = \sum_j p_j = 1$ implies

$$2T \approx \sum_{j=1}^{d} \frac{(N_j - np_j)^2}{np_j}$$

in probability under the null hypothesis. The statistic Q, given by

$$Q \stackrel{\text{def}}{=} \sum_{j=1}^{d} \frac{(N_j - np_j)^2}{np_j},$$

is called Pearson's *chi-square statistic*.

6.4 Likelihood Ratio Tests for Parameters of a Normal Distribution

For all examples considered in this section, we assume that the data $\boldsymbol{Y}$ in form of an i.i.d. sample $(Y_1, \ldots, Y_n)^\top$ follow the model $Y_i = \theta^* + \varepsilon_i$ with $\varepsilon_i \sim \mathcal{N}(0, \sigma^2)$ for σ^2 known or unknown. Equivalently $Y_i \sim \mathcal{N}(\theta^*, \sigma^2)$. The log-likelihood $L(\boldsymbol{Y}, \theta)$ (which we also denote by $L(\theta)$) reads as

$$L(\theta) = -\frac{n}{2}\log(2\pi\sigma^2) - \frac{1}{2\sigma^2}\sum_{i=1}^{n}(Y_i - \theta)^2 \tag{6.4}$$

and the log-likelihood ratio $L(\theta, \theta_0) = L(\theta) - L(\theta_0)$ is given by

$$L(\theta, \theta_0) = \sigma^{-2}\big[(S - n\theta_0)(\theta - \theta_0) - n(\theta - \theta_0)^2/2\big] \tag{6.5}$$

with $S \stackrel{\text{def}}{=} Y_1 + \ldots + Y_n$.

6.4.1 *Distributions Related to an i.i.d. Sample from a Normal Distribution*

As a preparation for the subsequent sections, we introduce here some important probability distributions which correspond to functions of an i.i.d. sample $\boldsymbol{Y} = (Y_1, \ldots, Y_n)^\top$ from a normal distribution.

Lemma 6.4.1. *If Y follows the standard normal distribution on $\mathbb{R}$, then Y^2 has the gamma distribution $\Gamma(1/2, 1/2)$.*

Exercise 6.4.1. Prove Lemma 6.4.1.

Corollary 6.4.1. *Let $Y = (Y_1, \ldots, Y_n)^\top$ denote an i.i.d. sample from the standard normal distribution on $\mathbb{R}$. Then, it holds that*

$$\sum_{i=1}^{n} Y_i^2 \sim \Gamma(1/2, n/2).$$

We call $\Gamma(1/2, n/2)$ the chi-square distribution with n degrees of freedom, χ_n^2 for short.

Proof. From Lemma 6.4.1, we have that $Y_1^2 \sim \Gamma(1/2, 1/2)$. Convolution stability of the family of gamma distributions with respect to the second parameter yields the assertion.

Lemma 6.4.2. *Let $\alpha, r, s > 0$ nonnegative constants and X, Y independent random variables with $X \sim \Gamma(\alpha, r)$ and $Y \sim \Gamma(\alpha, s)$. Then $S = X + Y$ and $R = X/(X+Y)$ are independent with $S \sim \Gamma(\alpha, r+s)$ and $R \sim$ Beta(r, s).*

Exercise 6.4.2. Prove Lemma 6.4.2.

Theorem 6.4.1. *Let $X_1, \ldots, X_m$, $Y_1, \ldots, Y_n$ i.i.d., with X_1 following the standard normal distribution on $\mathbb{R}$. Then, the ratio*

$$F_{m,n} \stackrel{\text{def}}{=} m^{-1} \sum_{i=1}^{m} X_i^2 \Big/ (n^{-1} \sum_{j=1}^{n} Y_j^2)$$

has the following pdf with respect to the Lebesgue measure.

$$f_{m,n}(x) = \frac{m^{m/2} n^{n/2}}{B(m/2, n/2)} \frac{x^{m/2-1}}{(n+mx)^{(m+n)/2}} \mathbb{1}_{(0,\infty)}(x).$$

The distribution of $F_{m,n}$ is called Fisher's F-distribution with m and n degrees of freedom (Sir R. A. Fisher, 1890–1962).

Exercise 6.4.3. Prove Theorem 6.4.1.

Corollary 6.4.2. *Let $X, Y_1, \ldots, Y_n$ i.i.d., with X following the standard normal distribution on $\mathbb{R}$. Then, the statistic*

$$T = \frac{X}{\sqrt{n^{-1} \sum_{j=1}^{n} Y_j^2}}$$

has the Lebesgue density

$$t \mapsto \tau_n(t) = \left(1 + \frac{t^2}{n}\right)^{-\frac{n+1}{2}} \left\{\sqrt{n}\, B(1/2, n/2)\right\}^{-1}.$$

The distribution of T is called Student's t-distribution with n degrees of freedom, t_n for short.

Proof. According to Theorem 6.4.1, $T^2 \sim F_{1,n}$. Thus, due to the transformation formula for densities, $|T| = \sqrt{T^2}$ has Lebesgue density $t \mapsto 2t f_{1,n}(t^2)$, $t > 0$. Because of the symmetry of the standard normal density, also the distribution of T is symmetric around 0, i.e., T and $-T$ have the same distribution. Hence, T has the Lebesgue density $t \mapsto |t| f_{1,n}(t^2) = \tau_n(t)$.

Theorem 6.4.2 (Student 1908).
In the Gaussian product model $(\mathbb{R}^n, \mathcal{B}(\mathbb{R}^n), (\mathcal{N}(\mu,\sigma^2)^{\otimes n})_{\theta=(\mu,\sigma^2)\in\Theta})$, *where* $\Theta = \mathbb{R} \times (0,\infty)$, *it holds for all* $\boldsymbol{\theta} \in \Theta$:

(a) *The statistics* $\overline{Y}_n \stackrel{\text{def}}{=} n^{-1}\sum_{j=1}^n Y_j$ *and* $\tilde{\sigma}^2 \stackrel{\text{def}}{=} (n-1)^{-1}\sum_{i=1}^n (Y_i - \overline{Y}_n)^2$ *are independent.*
(b) $\overline{Y}_n \sim \mathcal{N}(\mu, \sigma^2/n)$ *and* $(n-1)\tilde{\sigma}^2/\sigma^2 \sim \chi^2_{n-1}$.
(c) *The statistic* $T_n \stackrel{\text{def}}{=} \sqrt{n}(\overline{Y}_n - \mu)/\tilde{\sigma}$ *is distributed as* t_{n-1}.

6.4.2 Gaussian Shift Model

Under the measure $\mathbb{P}_{\theta_0}$, the variable $S - n\theta_0$ is normal zero-mean with the variance $n\sigma^2$. This particularly implies that $(S - n\theta_0)/\sqrt{n\sigma^2}$ is standard normal under $\mathbb{P}_{\theta_0}$:

$$\mathcal{L}\left(\frac{1}{\sigma\sqrt{n}}(S - n\theta_0) \,\Big|\, \mathbb{P}_{\theta_0}\right) = \mathcal{N}(0,1).$$

We start with the simplest case of a simple null and simple alternative.

6.4.2.1 Simple Null and Simple Alternative

Let the null $H_0 : \theta^* = \theta_0$ be tested against the alternative $H_1 : \theta^* = \theta_1$ for some fixed $\theta_1 \neq \theta_0$. The log-likelihood $L(\theta_1, \theta_0)$ is given by (6.5) leading to the test statistic

$$T = \sigma^{-2}\bigl[(S - n\theta_0)(\theta_1 - \theta_0) - n(\theta_1 - \theta_0)^2/2\bigr].$$

The proper critical value $\mathfrak{z}$ can be selected from the condition of α-level: $\mathbb{P}_{\theta_0}(T > \mathfrak{z}_\alpha) = \alpha$. We use that the sum $S - n\theta_0$ is under the null normal zero-mean with variance $n\sigma^2$, and hence, the random variable

$$\xi = (S - n\theta_0)/\sqrt{n\sigma^2}$$

is under θ_0 standard normal: $\xi \sim \mathcal{N}(0,1)$. The level condition can be rewritten as

$$\mathbb{P}_{\theta_0}\left(\xi > \frac{1}{|\theta_1 - \theta_0|\sigma\sqrt{n}}\left[\sigma^2 \mathfrak{z}_\alpha + n(\theta_1 - \theta_0)^2/2\right]\right) = \alpha.$$

As ξ is standard normal under θ_0, the proper $\mathfrak{z}_\alpha$ can be computed as a quantile of the standard normal law: if z_α is defined by $\mathbb{P}_{\theta_0}(\xi > z_\alpha) = \alpha$, then

$$\frac{1}{|\theta_1 - \theta_0|\sigma\sqrt{n}}\left[\sigma^2 \mathfrak{z}_\alpha + n|\theta_1 - \theta_0|^2/2\right] = z_\alpha$$

or

$$\mathfrak{z}_\alpha = \sigma^{-2}\left[z_\alpha|\theta_1 - \theta_0|\sigma\sqrt{n} - n|\theta_1 - \theta_0|^2/2\right].$$

It is worth noting that this value actually does not depend on θ_0. It only depends on the difference $|\theta_1 - \theta_0|$ between the null and the alternative. This is a very important and useful property of the normal family and is called *pivotality*. Another way of selecting the critical value $\mathfrak{z}$ is given by minimizing the sum of the first and second-kind error probabilities. Theorem 6.2.1 leads to the choice $\mathfrak{z} = 0$, or equivalently, to the test

$$\begin{aligned} \phi &= \mathbb{1}\{S/n \gtrless (\theta_0 + \theta_1)/2\}, \quad \theta_1 \gtrless \theta_0, \\ &= \mathbb{1}\{\tilde{\theta} \gtrless (\theta_0 + \theta_1)/2\}, \quad \theta_1 \gtrless \theta_0. \end{aligned}$$

This test is also called the Fisher discrimination. It naturally appears in classification problems.

6.4.2.2 Two-Sided Test

Now we consider a more general situation when the simple null $\theta^* = \theta_0$ is tested against the alternative $\theta^* \neq \theta_0$. Then the LR test compares the likelihood at θ_0 with the maximum likelihood over $\Theta \setminus \{\theta_0\}$ which clearly coincides with the maximum over the whole parameter set. This leads to the test statistic:

$$T = \max_{\theta} L(\theta, \theta_0) = \frac{n}{2\sigma^2}|\tilde{\theta} - \theta_0|^2.$$

(see Sect. 2.9), where $\tilde{\theta} = S/n$ is the MLE. The LR test rejects the null if $T \geq \mathfrak{z}$ for a critical value $\mathfrak{z}$. The value $\mathfrak{z}$ can be selected from the level condition:

$$\mathbb{P}_{\theta_0}(T > \mathfrak{z}) = \mathbb{P}_{\theta_0}\left(n\sigma^{-2}|\tilde{\theta} - \theta_0|^2 > 2\mathfrak{z}\right) = \alpha.$$

Now we use that $n\sigma^{-2}|\tilde{\theta} - \theta_0|^2$ is χ^2_1-distributed according to Lemma 6.4.1. If $\mathfrak{z}_\alpha$ is defined by $\mathbb{P}(\xi^2 \geq 2\mathfrak{z}_\alpha) = \alpha$ for standard normal ξ, then the test $\phi = \mathbb{1}(T > \mathfrak{z}_\alpha)$ is of level α. Again, this value does not depend on the null point θ_0, and the LR test is pivotal.

Exercise 6.4.4. Compute the power function of the resulting two-sided test

$$\phi = \mathbb{1}(T > \mathfrak{z}_\alpha).$$

6.4.2.3 One-Sided Test

Now we consider the problem of testing the null $\theta^* = \theta_0$ against the one-sided alternative $H_1 : \theta > \theta_0$. To apply the LR test we have to compute the maximum of the log-likelihood ratio $L(\theta, \theta_0)$ over the set $\Theta_1 = \{\theta > \theta_0\}$.

Exercise 6.4.5. Check that

$$\sup_{\theta > \theta_0} L(\theta, \theta_0) = \begin{cases} n\sigma^{-2}|\tilde{\theta} - \theta_0|^2/2 & \text{if } \tilde{\theta} \ge \theta_0, \\ 0 & \text{otherwise.} \end{cases}$$

Hint: if $\tilde{\theta} \ge \theta_0$, then the supremum over Θ_1 coincides with the global maximum, otherwise it is attained at the edge θ_0.

Now the LR test rejects the null if $\tilde{\theta} > \theta_0$ and $n\sigma^{-2}|\tilde{\theta} - \theta_0|^2 > 2\mathfrak{z}$ for a critical value (CV) $\mathfrak{z}$. That is,

$$\phi = \mathbb{1}\bigl(\tilde{\theta} - \theta_0 > \sigma\sqrt{2\mathfrak{z}/n}\bigr).$$

The CV $\mathfrak{z}$ can be again chosen by the level condition. As $\xi = \sqrt{n}(\tilde{\theta} - \theta_0)/\sigma$ is standard normal under $\mathbb{P}_{\theta_0}$, one has to select $\mathfrak{z}$ such that $\mathbb{P}(\xi > \sqrt{2\mathfrak{z}}) = \alpha$, leading to

$$\phi = \mathbb{1}\bigl(\tilde{\theta} > \theta_0 + \sigma z_{1-\alpha}/\sqrt{n}\bigr),$$

where $z_{1-\alpha}$ denotes the $(1-\alpha)$-quantile of the standard normal distribution.

6.4.3 Testing the Mean When the Variance Is Unknown

This section discusses the Gaussian shift model $Y_i = \theta^* + \sigma^* \varepsilon_i$ with standard normal errors ε_i and unknown variance σ^{*2}. The log-likelihood function is still given by (6.4) but now σ^{*2} is a part of the parameter vector.

6.4.3.1 Two-Sided Test Problem

Here, we are considered with the two-sided testing problem $H_0 : \theta^* = \theta_0$ against $H_1 : \theta_1 \neq \theta_0$. Notice that the null hypothesis is composite, because it involves the unknown variance σ^{*2}.

Maximizing the log-likelihood $L(\theta, \sigma^2)$ under the null leads to the value $L(\theta_0, \tilde{\sigma}_0^2)$ with

$$\tilde{\sigma}_0^2 \stackrel{\text{def}}{=} \operatorname*{argmax}_{\sigma^2} L(\theta_0, \sigma^2) = n^{-1} \sum_i (Y_i - \theta_0)^2.$$

As in Sect. 2.9.2 for the problem of variance estimation, it holds for any σ

$$L(\theta_0, \tilde{\sigma}_0^2) - L(\theta_0, \sigma^2) = n\mathcal{K}(\tilde{\sigma}_0^2, \sigma^2).$$

At the same time, maximizing $L(\theta, \sigma^2)$ over the alternative is equivalent to the global maximization leading to the value $L(\tilde{\theta}, \tilde{\sigma}^2)$ with

$$\tilde{\theta} = S/n, \qquad \tilde{\sigma}^2 = \frac{1}{n} \sum_i (Y_i - \tilde{\theta})^2.$$

The LR test statistic reads as

$$T = L(\tilde{\theta}, \tilde{\sigma}^2) - L(\theta_0, \tilde{\sigma}_0^2).$$

This expression can be decomposed in the following way:

$$\begin{aligned} T &= L(\tilde{\theta}, \tilde{\sigma}^2) - L(\theta_0, \tilde{\sigma}^2) + L(\theta_0, \tilde{\sigma}^2) - L(\theta_0, \tilde{\sigma}_0^2) \\ &= \frac{1}{2\tilde{\sigma}^2} (\tilde{\theta} - \theta_0)^2 - n\mathcal{K}(\tilde{\sigma}_0^2, \tilde{\sigma}^2). \end{aligned}$$

In order to derive the CV $\mathfrak{z}$, notice that

$$\exp(T) = \frac{\exp(L(\tilde{\theta}, \tilde{\sigma}^2))}{\exp(L(\theta_0, \tilde{\sigma}_0^2))} = \left(\frac{\tilde{\sigma}_0^2}{\tilde{\sigma}^2} \right)^{n/2}.$$

Consequently, the LR test rejects for large values of

$$\frac{\tilde{\sigma}_0^2}{\tilde{\sigma}^2} = 1 + \frac{n(\tilde{\theta} - \theta_0)^2}{\sum_{i=1}^n (Y_i - \tilde{\theta})^2}.$$

In view of Theorems 6.4.1 and 6.4.2, $\mathfrak{z}$ is therefore a deterministic transformation of the suitable quantile from Fisher's F-distribution with 1 and $n-1$ degrees of freedom or from Student's t-distribution with $n-1$ degrees of freedom.

Exercise 6.4.6. Derive the explicit form of $\mathfrak{z}_\alpha$ for given significance level α in terms of Fisher's F-distribution and in terms of Student's t-distribution.

Often one considers the case in which the variance is only estimated under the alternative, that is, $\tilde{\sigma}$ is used in place of $\tilde{\sigma}_0$. This is quite natural because the null can be viewed as a particular case of the alternative. This leads to the test statistic

$$T^* = L(\tilde{\theta}, \tilde{\sigma}^2) - L(\theta_0, \tilde{\sigma}^2) = \frac{n}{2\tilde{\sigma}^2}(\tilde{\theta} - \theta_0)^2.$$

Since T^* is an isotone transformation of T, both tests are equivalent.

6.4.3.2 One-Sided Test Problem

In analogy to the considerations in Sect. 6.4.2, the LR test for the one-sided problem $H_0 : \theta^* = \theta_0$ against $H_1 : \theta^* > \theta_0$ rejects if $\tilde{\theta} > \theta_0$ and T exceeds a suitable critical value $\mathfrak{z}$.

Exercise 6.4.7. Derive the explicit form of the LR test for the one-sided test problem. Compute $\mathfrak{z}_\alpha$ for given significance level α in terms of Student's t-distribution.

6.4.4 *Testing the Variance*

In this section, we consider the LR test for the hypothesis $H_0 : \sigma^2 = \sigma_0^2$ against $H_1 : \sigma^2 > \sigma_0^2$ or $H_0 : \sigma^2 = \sigma_0^2$ against $H_1 : \sigma^2 \neq \sigma_0^2$, respectively. In this, we assume that θ^* is known. The case of unknown θ^* can be treated similarly, cf. Exercise 6.4.10 below. As discussed before, maximization of the likelihood under the constraint of known mean yields

$$\tilde{\sigma}_*^2 \stackrel{\text{def}}{=} \operatorname*{argmax}_{\sigma^2} L(\theta^*, \sigma^2) = n^{-1} \sum_i (Y_i - \theta^*)^2.$$

The LR test for H_0 against H_1 rejects the null hypothesis if

$$T = L(\theta^*, \tilde{\sigma}_*^2) - L(\theta^*, \sigma_0^2)$$

exceeds a critical value $\mathfrak{z}$. For determining the rejection regions, notice that

$$2T = n(Q - \log(Q) - 1), \quad Q \stackrel{\text{def}}{=} \tilde{\sigma}_*^2 / \sigma_0^2. \tag{6.6}$$

Exercise 6.4.8. (a) Verify representation (6.6).
(b) Show that $x \mapsto x - \log(x) - 1$ is a convex function on $(0, \infty)$ with minimum value 0 at $x = 1$.

Combining Exercise 6.4.8 and Corollary 6.4.1, we conclude that the critical value $\mathfrak{z}$ for the LR test for the one-sided test problem is a deterministic transformation of a suitable quantile of the χ_n^2-distribution. For the two-sided test problem, the acceptance region of the LR test is an interval for Q which is bounded by deterministic transformations of lower and upper quantiles of the χ_n^2-distribution.

Exercise 6.4.9. Derive the rejection regions of the LR tests for the one- and the two-sided test problems explicitly.

Exercise 6.4.10. Derive one- and two-sided LR tests for σ^2 in the case of unknown θ^*. Hint: Use Theorem 6.4.2(b).

6.5 LR Tests: Further Examples

We return to the models investigated in Sect. 2.9 and derive LR tests for the respective parameters.

6.5.1 *Bernoulli or Binomial Model*

Assume that $(Y_i, \ldots, Y_n)$ are i.i.d. with $Y_1 \sim \text{Bernoulli}(\theta^*)$. Letting $S_n = \sum Y_i$, the log-likelihood is given by

$$L(\theta) = \sum \{Y_i \log \theta + (1 - Y_i) \log(1 - \theta)\}$$

$$= S_n \log \frac{\theta}{1 - \theta} + n \log(1 - \theta).$$

6.5.1.1 Simple Null Versus Simple Alternative

The LR statistic for testing the simple null hypothesis $H_0 : \theta^* = \theta_0$ against the simple alternative $H_1 : \theta^* = \theta_1$ reads as

$$T = L(\theta_1) - L(\theta_0) = S_n \log \frac{\theta_1(1 - \theta_0)}{\theta_0(1 - \theta_1)} + n \log \frac{1 - \theta_1}{1 - \theta_0}.$$

For a fixed significance level α, the resulting LR test rejects if

$$\tilde{\theta} \gtrless \left(\frac{\mathfrak{z}_\alpha}{n} - \log \frac{1 - \theta_1}{1 - \theta_0} \right) \Big/ \log \frac{\theta_1(1 - \theta_0)}{\theta_0(1 - \theta_1)}, \quad \theta_1 \gtrless \theta_0,$$

where $\tilde{\theta} = S_n/n$. For the determination of $\mathfrak{z}_\alpha$, it is convenient to notice that for any pair $(\theta_0, \theta_1) \in (0, 1)^2$, the function

$$x \mapsto x \log \frac{\theta_1(1-\theta_0)}{\theta_0(1-\theta_1)} + \log \frac{1-\theta_1}{1-\theta_0}$$

is increasing (decreasing) in $x \in [0, 1]$ if $\theta_1 > \theta_0$ ($\theta_1 < \theta_0$). Hence, the LR statistic T is an isotone (antitone) transformation of $\tilde{\theta}$ if $\theta_1 > \theta_0$ ($\theta_1 < \theta_0$). Since $S_n = n\tilde{\theta}$ is under H_0 binomially distributed with parameters n and θ_0, the LR test ϕ is given by

$$\phi = \begin{cases} \mathbb{1}\{S_n > F^{-1}_{\mathrm{Bin}(n,\theta_0)}(1-\alpha)\}, & \theta_1 > \theta_0, \\ \mathbb{1}\{S_n < F^{-1}_{\mathrm{Bin}(n,\theta_0)}(\alpha)\}, & \theta_1 < \theta_0. \end{cases} \tag{6.7}$$

6.5.1.2 Composite Alternatives

Obviously, the LR test in (6.7) depends on the value of θ_1 only via the sign of $\theta_1 - \theta_0$. Therefore, the LR test for the one-sided test problem $H_0 : \theta^* = \theta_0$ against $H_1 : \theta^* > \theta_0$ rejects if

$$S_n > F^{-1}_{\mathrm{Bin}(n,\theta_0)}(1-\alpha)$$

and the LR test for H_0 against $H_1 : \theta^* < \theta_0$ rejects if

$$S_n < F^{-1}_{\mathrm{Bin}(n,\theta_0)}(\alpha).$$

The LR test for the two-sided test problem H_0 against $H_1 : \theta^* \neq \theta_0$ rejects if

$$S_n \notin [F^{-1}_{\mathrm{Bin}(n,\theta_0)}(\alpha/2), F^{-1}_{\mathrm{Bin}(n,\theta_0)}(1-\alpha/2)].$$

6.5.2 *Uniform Distribution on* $[0, \theta]$

Consider again the model from Sect. 2.9.4, i.e., $Y_1, \ldots, Y_n$ are i.i.d. with $Y_1 \sim \mathrm{UNI}[0, \theta^*]$, where the upper endpoint θ^* of the support is unknown. It holds that

$$Z(\theta) = \theta^{-n} \mathbb{1}(\max_i Y_i \leq \theta)$$

and that the maximum of $Z(\theta)$ over $(0, \infty)$ is obtained for $\tilde{\theta} = \max_i Y_i$. Let us consider the two-sided test problem $H_0 : \theta^* = \theta_0$ against $H_1 : \theta^* \neq \theta_0$ for some given value $\theta_0 > 0$. We get that

$$\exp(T) = \frac{Z(\tilde{\theta})}{Z(\theta_0)} = \begin{cases} (\theta_0/\tilde{\theta})^n, & \max_i Y_i \le \theta_0, \\ \infty, & \max_i Y_i > \theta_0. \end{cases}$$

It follows that $\exp(T) > \mathfrak{z}$ if $\tilde{\theta} > \theta_0$ or $\tilde{\theta} < \theta_0 \mathfrak{z}^{-1/n}$. We compute the critical value $\mathfrak{z}_\alpha$ for a level α test by noticing that

$$\mathbb{P}_{\theta_0}(\{\tilde{\theta} > \theta_0\} \cup \{\tilde{\theta} < \theta_0 \mathfrak{z}^{-1/n}\}) = \mathbb{P}_{\theta_0}(\tilde{\theta} < \theta_0 \mathfrak{z}^{-1/n}) = \left(F_{Y_1}(\theta_0 \mathfrak{z}^{-1/n})\right)^n = 1/\mathfrak{z}.$$

Thus, the LR test at level α for the two-sided problem H_0 against H_1 is given by

$$\phi = \mathbb{1}\{\tilde{\theta} > \theta_0\} + \mathbb{1}\{\tilde{\theta} < \theta_0 \alpha^{1/n}\}.$$

6.5.3 Exponential Model

We return to the model considered in Sect. 2.9.7 and assume that $Y_1, \ldots, Y_n$ are i.i.d. exponential random variables with parameter $\theta^* > 0$. The corresponding log-likelihood can be written as

$$L(\theta) = -n \log \theta - \sum_{i=1}^{n} Y_i/\theta = -S/\theta - n \log \theta,$$

where $S = Y_1 + \ldots + Y_n$.

In order to derive the LR test for the simple hypothesis $H_0 : \theta^* = \theta_0$ against the simple alternative $H_0 : \theta^* = \theta_1$, notice that the LR statistic T is given by

$$T = L(\theta_1) - L(\theta_0) = S\left(\frac{\theta_1 - \theta_0}{\theta_1 \theta_0}\right) + n \log \frac{\theta_0}{\theta_1}.$$

Since the function

$$x \mapsto x\left(\frac{\theta_1 - \theta_0}{\theta_1 \theta_0}\right) + n \log(\theta_0/\theta_1)$$

is increasing (decreasing) in $x > 0$ whenever $\theta_1 > \theta_0$ ($\theta_1 < \theta_0$), the LR test rejects for large values of S in the case that $\theta_1 > \theta_0$ and for small values of S if $\theta_1 < \theta_0$. Due to the facts that the exponential distribution with parameter θ_0 is identical to $\text{Gamma}(\theta_0, 1)$ and the family of gamma distributions is convolution-stable with respect to its second parameter whenever the first parameter is fixed, we obtain that S is under θ_0 distributed as $\text{Gamma}(\theta_0, n)$. This implies that the LR test ϕ for H_0 against H_1 at significance level α is given by

$$\phi = \begin{cases} \mathbb{1}\{S > F^{-1}_{\text{Gamma}(\theta_0,n)}(1-\alpha)\}, & \theta_1 > \theta_0, \\ \mathbb{1}\{S < F^{-1}_{\text{Gamma}(\theta_0,n)}(\alpha)\}, & \theta_1 < \theta_0. \end{cases}$$

Moreover, composite alternatives can be tested in analogy to the considerations in Sect. 6.5.1.

6.5.4 Poisson Model

Let $Y_1, \ldots, Y_n$ be i.i.d. Poisson random variables satisfying $\mathbb{P}(Y_i = m) = |\theta^*|^m e^{-\theta^*}/m!$ for $m = 0, 1, 2, \ldots$. According to Sect. 2.9.8, we have that

$$L(\theta) = S \log \theta - n\theta + R,$$

$$L(\theta_1) - L(\theta_0) = S \log \frac{\theta_1}{\theta_0} + n(\theta_0 - \theta_1),$$

where the remainder term R does not depend on θ. In order to derive the LR test for the simple hypothesis $H_0 : \theta^* = \theta_0$ against the simple alternative $H_0 : \theta^* = \theta_1$, we again check easily that $x \mapsto x \log(\theta_1/\theta_0) + n(\theta_0 - \theta_1)$ is increasing (decreasing) in $x > 0$ if $\theta_1 > \theta_0$ ($\theta_1 < \theta_0$). Convolution stability of the family of Poisson distributions entails that the LR test ϕ for H_0 against H_1 at significance level α is given by

$$\phi = \begin{cases} \mathbb{1}\{S > F^{-1}_{\text{Poisson}(n\theta_0)}(1-\alpha)\}, & \theta_1 > \theta_0, \\ \mathbb{1}\{S < F^{-1}_{\text{Poisson}(n\theta_0)}(\alpha)\}, & \theta_1 < \theta_0. \end{cases}$$

Moreover, composite alternatives can be tested in analogy to Sect. 6.5.1.

6.6 Testing Problem for a Univariate Exponential Family

Let $(P_\theta, \theta \in \Theta \subseteq \mathbb{R}^1)$ be a univariate exponential family. The choice of parametrization is unimportant, any parametrization can be taken. To be specific, we assume the natural parametrization that simplifies the expression for the maximum likelihood estimate.

We assume that the two functions $C(\theta)$ and $B(\theta)$ of θ are fixed, with which the log-density of P_θ can be written in the form:

$$\ell(y, \theta) \stackrel{\text{def}}{=} \log p(y, \theta) = yC(\theta) - B(\theta) - \ell(y)$$

for some other function $\ell(y)$. The function $C(\theta)$ is monotonic in θ and $C(\theta)$ and $B(\theta)$ are related (for the case of an EFn) by the identity $B'(\theta) = \theta C'(\theta)$, see Sect. 2.11.

Let now $Y = (Y_1, \ldots, Y_n)^\top$ be an i.i.d. sample from P_{θ^*} for $\theta^* \in \Theta$. The task is to test a simple hypothesis $\theta^* = \theta_0$ against an alternative $\theta^* \in \Theta_1$ for some subset Θ_1 that does not contain θ_0.

6.6.1 Two-Sided Alternative

We start with the case of a simple hypothesis $H_0 : \theta^* = \theta_0$ against a full two-sided alternative $H_1 : \theta^* \neq \theta_0$. The likelihood ratio approach suggests to compare the likelihood at θ_0 with the maximum of the likelihood over the alternative, which effectively means the maximum over the whole parameter set Θ. In the case of a univariate exponential family, this maximum is computed in Sect. 2.11. For

$$L(\theta, \theta_0) \stackrel{\text{def}}{=} L(\theta) - L(\theta_0) = S\bigl[C(\theta) - C(\theta_0)\bigr] - n\bigl[B(\theta) - B(\theta_0)\bigr]$$

with $S = Y_1 + \ldots + Y_n$, it holds

$$T \stackrel{\text{def}}{=} \sup_{\theta} L(\theta, \theta_0) = n\mathcal{K}(\tilde{\theta}, \theta_0),$$

where $\mathcal{K}(\theta, \theta') = E_\theta \ell(\theta, \theta')$ is the Kullback–Leibler divergence between the measures P_θ and $P_{\theta'}$. For an EFn, the MLE $\tilde{\theta}$ is the empirical mean of the observations Y_i, $\tilde{\theta} = S/n$, and the KL divergence $\mathcal{K}(\theta, \theta_0)$ is of the form

$$\mathcal{K}(\theta, \theta_0) = \theta\bigl[C(\theta) - C(\theta_0)\bigr] - \bigl[B(\theta) - B(\theta_0)\bigr].$$

Therefore, the test statistic T is a function of the empirical mean $\tilde{\theta} = S/n$:

$$T = n\mathcal{K}(\tilde{\theta}, \theta_0) = n\tilde{\theta}\bigl[C(\tilde{\theta}) - C(\theta_0)\bigr] - n\bigl[B(\tilde{\theta}) - B(\theta_0)\bigr]. \tag{6.8}$$

The LR test rejects H_0 if the test statistic T exceeds a critical value $\mathfrak{z}$. Given $\alpha \in (0, 1)$, a proper CV $\mathfrak{z}_\alpha$ can be specified by the level condition

$$\mathbb{P}_{\theta_0}(T > \mathfrak{z}_\alpha) = \alpha.$$

In view of (6.8), the LR test rejects the null if the "distance" $\mathcal{K}(\tilde{\theta}, \theta_0)$ between the estimate $\tilde{\theta}$ and the null θ_0 is significantly larger than zero. In the case of an exponential family, one can simplify the test just by considering the estimate $\tilde{\theta}$ as test statistic. We use the following technical result for the KL divergence $\mathcal{K}(\theta, \theta_0)$:

Lemma 6.6.1. *Let* (P_θ) *be an EFn. Then for every* $\mathfrak{z}$ *there are two positive values* $t^-(\mathfrak{z})$ *and* $t^+(\mathfrak{z})$ *such that*

$$\{\theta : \mathcal{K}(\theta, \theta_0) \le \mathfrak{z}\} = \{\theta : \theta_0 - t^-(\mathfrak{z}) \le \theta \le \theta_0 + t^+(\mathfrak{z})\}. \tag{6.9}$$

In other words, the conditions $\mathcal{K}(\theta, \theta_0) \le \mathfrak{z}$ and $\theta_0 - t^-(\mathfrak{z}) \le \theta \le \theta_0 + t^+(\mathfrak{z})$ are equivalent.

Proof. The function $\mathcal{K}(\theta, \theta_0)$ of the first argument θ fulfills

$$\frac{\partial \mathcal{K}(\theta, \theta_0)}{\partial \theta} = C(\theta) - C(\theta_0), \qquad \frac{\partial^2 \mathcal{K}(\theta, \theta_0)}{\partial \theta^2} = C'(\theta) > 0.$$

Therefore, it is convex in θ with minimum at θ_0, and it can cross the level $\mathfrak{z}$ only once from the left of θ_0 and once from the right. This yields that for any $\mathfrak{z} > 0$, there are two positive values $t^-(\mathfrak{z})$ and $t^+(\mathfrak{z})$ such that (6.9) holds. Note that one or even both of these values can be infinite.

Due to the result of this lemma, the LR test can be rewritten as

$$\begin{aligned} \phi = \mathbb{1}(T > \mathfrak{z}) &= 1 - \mathbb{1}(T \le \mathfrak{z}) \\ &= 1 - \mathbb{1}\bigl(-t^-(\mathfrak{z}) \le \tilde{\theta} - \theta_0 \le t^+(\mathfrak{z})\bigr) \\ &= \mathbb{1}\bigl(\tilde{\theta} > \theta_0 + t^+(\mathfrak{z})\bigr) + \mathbb{1}\bigl(\tilde{\theta} < \theta_0 - t^-(\mathfrak{z})\bigr), \end{aligned}$$

that is, the test rejects the null hypothesis if the estimate $\tilde{\theta}$ deviates significantly from θ_0.

6.6.2 One-Sided Alternative

Now we consider the problem of testing the same null $H_0 : \theta^* = \theta_0$ against the one-sided alternative $H_1 : \theta^* > \theta_0$. Of course, the other one-sided alternative $H_1 : \theta^* < \theta_0$ can be considered analogously.

The LR test requires computing the maximum of the log-likelihood over the alternative set $\{\theta : \theta > \theta_0\}$. This can be done as in the Gaussian shift model. If $\tilde{\theta} > \theta_0$, then this maximum coincides with the global maximum over all θ. Otherwise, it is attained at $\theta = \theta_0$.

Lemma 6.6.2. *Let* (P_θ) *be an EFn. Then*

$$\sup_{\theta > \theta_0} L(\theta, \theta_0) = \begin{cases} n\mathcal{K}(\tilde{\theta}, \theta_0) & \textit{if } \ \tilde{\theta} \ge \theta_0, \\ 0 & \textit{otherwise}. \end{cases}$$

Proof. It is only necessary to consider the case $\tilde{\theta} < \theta_0$. The difference $L(\theta) - L(\theta_0)$ can be represented as $S[C(\theta) - C(\theta_0)] - n[B(\theta) - B(\theta_0)]$. Next, usage of the identities $\tilde{\theta} = S/n$ and $B'(\theta) = \theta C'(\theta)$ yields

$$\frac{\partial L(\theta, \theta_0)}{\partial \theta} = \frac{\partial L(\tilde{\theta}, \theta)}{\partial \theta} = -n \frac{\partial \mathcal{K}(\tilde{\theta}, \theta)}{\partial \theta} = n(\tilde{\theta} - \theta) C'(\theta) < 0$$

for any $\theta > \tilde{\theta}$. This implies that $L(\theta) - L(\theta_0)$ becomes negative as θ grows beyond θ_0 and thus, $L(\theta, \theta_0)$ has its supremum over $\{\theta : \theta > \theta_0\}$ at $\theta = \theta_0$, yielding the assertion.

This fact implies the following representation of the LR test in the case of a one-sided alternative.

Theorem 6.6.1. *Let (P_θ) be an EFn. Then the α-level LR test for the null $H_0 : \theta^* = \theta_0$ against the one-sided alternative $H_1 : \theta^* > \theta_0$ is*

$$\phi = \mathbb{1}(\tilde{\theta} > \theta_0 + t_\alpha), \tag{6.10}$$

where t_α is selected to ensure $\mathbb{P}_{\theta_0}(\tilde{\theta} > \theta_0 + t_\alpha) = \alpha$.

Proof. Let T be the LR test statistic. Due to Lemmas 6.6.2 and 6.6.1, the event $\{T \geq \mathfrak{z}\}$ can be rewritten as $\{\tilde{\theta} > \theta_0 + t(\mathfrak{z})\}$ for some constant $t(\mathfrak{z})$. It remains to select a proper value $t(\mathfrak{z}) = t_\alpha$ to fulfill the level condition.

This result can be extended naturally to the case of a composite null hypothesis $H_0 : \theta^* \leq \theta_0$.

Theorem 6.6.2. *Let (P_θ) be an EFn. Then the α-level LR test for the composite null $H_0 : \theta^* \leq \theta_0$ against the one-sided alternative $H_1 : \theta^* > \theta_0$ is*

$$\phi^*_\alpha = \mathbb{1}(\tilde{\theta} > \theta_0 + t_\alpha), \tag{6.11}$$

where t_α is selected to ensure $\mathbb{P}_{\theta_0}(\tilde{\theta} > \theta_0 + t_\alpha) = \alpha$.

Proof. The same arguments as in the proof of Theorem 6.6.1 lead to exactly the same LR test statistic T and thus to the test of the form (6.10). In particular, the estimate $\tilde{\theta}$ should significantly deviate from the null set. It remains to check that the level condition for the edge point θ_0 ensures the level for all $\theta < \theta_0$. This follows from the next monotonicity property.

Lemma 6.6.3. *Let (P_θ) be an EFn. Then for any $t \geq 0$*

$$\mathbb{P}_\theta(\tilde{\theta} > \theta_0 + t) \leq \mathbb{P}_{\theta_0}(\tilde{\theta} > \theta_0 + t), \qquad \forall \theta < \theta_0 .$$

Proof. Let $\theta < \theta_0$. We apply

$$\mathbb{P}_\theta(\tilde{\theta} > \theta_0 + t) = \mathbb{E}_{\theta_0}\Big[\exp\{L(\theta, \theta_0)\} \mathbb{1}(\tilde{\theta} > \theta_0 + t)\Big].$$

Now the monotonicity of $L(\theta, \theta_0)$ w.r.t. θ (see the proof of Lemma 6.6.2) implies $L(\theta, \theta_0) < 0$ on the set $\{\theta < \theta_0 < \tilde{\theta}\}$. This yields the result.

Therefore, if the level is controlled under $\mathbb{P}_{\theta_0}$, it is well checked for all other points in the null set.

A very nice feature of the LR test is that it can be universally represented in terms of $\tilde{\theta}$ independently of the form of the alternative set. In particular, for the case of a one-sided alternative, this test just compares the estimate $\tilde{\theta}$ with the value $\theta_0 + t_\alpha$. Moreover, the value t_α only depends on the distribution of $\tilde{\theta}$ under $\mathbb{P}_{\theta_0}$ via the level condition. This and the monotonicity of the error probability from Lemma 6.6.3 allow us to state the nice optimality property of this test: ϕ_α^* is uniformly most powerful in the sense of Definition 6.1.1, that is, it maximizes the test power under the level constraint.

Theorem 6.6.3. *Let (P_θ) be an EFn, and let ϕ_α^* be the test from* (6.11) *for testing $H_0 : \theta^* \le \theta_0$ against $H_1 : \theta^* > \theta_0$. For any (randomized) test ϕ satisfying $\mathbb{E}_{\theta_0} \le \alpha$ and any $\theta > \theta_0$, it holds*

$$\mathbb{E}_\theta \phi \le \mathbb{P}_\theta(\phi_\alpha^* = 1).$$

In fact, this theorem repeats the Neyman–Pearson result of Theorem 6.2.2, because the test ϕ_α^* is at the same time the LR α-level test of the simple hypothesis $\theta^* = \theta_0$ against $\theta^* = \theta_1$, for any value $\theta_1 > \theta_0$.

6.6.3 Interval Hypothesis

In some applications, the null hypothesis is naturally formulated in the form that the parameter θ^* belongs to a given interval $[\theta_0, \theta_1]$. The alternative $H_1 : \theta^* \in \Theta \setminus [\theta_0, \theta_1]$ is the complement of this interval. The likelihood ratio test is based on the test statistic T from (6.3) which compares the maximum of the log-likelihood $L(\theta)$ under the null $[\theta_0, \theta_1]$ with the maximum over the alternative set. The special structure of the log-likelihood in the case of an EFn permits representing this test statistics in terms of the estimate $\tilde{\theta}$: the hypothesis is rejected if the estimate $\tilde{\theta}$ significantly deviates from the interval $[\theta_0, \theta_1]$.

Theorem 6.6.4. *Let (P_θ) be an EFn. Then the α-level LR test for the null $H_0 : \theta \in [\theta_0, \theta_1]$ against the alternative $H_1 : \theta \notin [\theta_0, \theta_1]$ can be written as*

$$\phi = \mathbb{1}(\tilde{\theta} > \theta_1 + t_\alpha^+) + \mathbb{1}(\tilde{\theta} < \theta_0 - t_\alpha^-), \tag{6.12}$$

where the non-negative constants t_α^+ and t_α^- are selected to ensure $\mathbb{P}_{\theta_0}(\tilde{\theta} < \theta_0 - t_\alpha^-) + \mathbb{P}_{\theta_1}(\tilde{\theta} > \theta_1 + t_\alpha^+) = \alpha$.

Exercise 6.6.1. Prove the result of Theorem 6.6.4.
Hint: Consider three cases: $\tilde{\theta} \in [\theta_0, \theta_1]$, $\tilde{\theta} > \theta_1$, and $\tilde{\theta} < \theta_0$. For every case, apply the monotonicity of $L(\tilde{\theta}, \theta)$ in θ.

One can consider the alternative of the interval hypothesis as a combination of two one-sided alternatives. The LR test ϕ from (6.12) involves only one critical value $\mathfrak{z}$ and the parameters t_α^- and t_α^+ are related via the structure of this test: they are obtained by transforming the inequality $T > \mathfrak{z}_\alpha$ into $\tilde{\theta} > \theta_1 + t_\alpha^+$ and $\tilde{\theta} < \theta_0 - t_\alpha^-$. However, one can just apply two one-sided tests separately: one for the alternative $H_1^- : \theta^* < \theta_0$ and one for $H_1^+ : \theta^* > \theta_1$. This leads to the two tests

$$\phi^- \stackrel{\text{def}}{=} \mathbb{1}(\tilde{\theta} < \theta_0 - t^-), \qquad \phi^+ \stackrel{\text{def}}{=} \mathbb{1}(\tilde{\theta} > \theta_1 + t^+). \tag{6.13}$$

The values t^-, t^+ can be chosen by the so-called *Bonferroni* rule: just perform each of the two tests at level $\alpha/2$.

Exercise 6.6.2. For fixed $\alpha \in (0, 1)$, let the values t_α^-, t_α^+ be selected to ensure

$$\mathbb{P}_{\theta_0}(\tilde{\theta} < \theta_0 - t_\alpha^-) = \alpha/2, \qquad \mathbb{P}_{\theta_1}(\tilde{\theta} > \theta_1 + t_\alpha^+) = \alpha/2.$$

Then for any $\theta \in [\theta_0, \theta_1]$, the test ϕ, given by $\phi = \phi^- + \phi^+$ (cf. (6.13)) fulfills

$$\mathbb{P}_\theta(\phi = 1) \le \alpha.$$

Hint: Use the monotonicity from Lemma 6.6.3.

6.7 Historical Remarks and Further Reading

The theory of *optimal tests* goes back to Jerzy Neyman (1894–1981) and Egon Sharpe Pearson (1895–1980). Some early considerations with respect to *likelihood ratio tests* can be found in Neyman and Pearson (1928). Neyman and Pearson (1933)'s *fundamental lemma* is core to the derivation of most powerful (likelihood ratio) tests. Fisher (1934) showed that *uniformly best tests* (over the parameter subspace defining a composite alternative) only exist in one-parametric exponential families. More details about the origins of the theory of optimal tests can be found in the book by Lehmann (2011).

Important contributions to the theory of *tests for parameters of a normal distribution* have moreover been made by William Sealy Gosset (1876–1937; under the pen name "Student," see Student (1908)) and Ernst Abbe (1840–1905; cf. Kendall (1971) for an account of Abbe's work).

The χ^2-test for *goodness-of-fit* is due to Pearson (1900). The phenomenon that the limiting distribution of twice the log-likelihood ratio statistic in nested models is under regularity conditions a chi-square distribution, has been discovered by Wilks (1938).

An excellent textbook reference for the theory of testing statistical hypotheses is Lehmann and Romano (2005).

Chapter 7
Testing in Linear Models

This chapter discusses testing problems for linear Gaussian models given by the equation

$$\boldsymbol{Y} = \boldsymbol{f}^* + \boldsymbol{\varepsilon} \tag{7.1}$$

with the vector of observations $\boldsymbol{Y}$, response vector $\boldsymbol{f}^*$, and vector of errors $\boldsymbol{\varepsilon}$ in $\mathbb{R}^n$. The linear parametric assumption (linear PA) means that

$$\boldsymbol{Y} = \Psi^\top \boldsymbol{\theta}^* + \boldsymbol{\varepsilon}, \quad \boldsymbol{\varepsilon} \sim \mathcal{N}(0, \Sigma), \tag{7.2}$$

where Ψ is the $p \times n$ design matrix. By $\boldsymbol{\theta}$ we denote the p-dimensional target parameter vector, $\boldsymbol{\theta} \in \Theta \subseteq \mathbb{R}^p$. Usually we assume that the parameter set coincides with the whole space $\mathbb{R}^p$, i.e. $\Theta = \mathbb{R}^p$. The most general assumption about the vector of errors $\boldsymbol{\varepsilon} = (\varepsilon_1, \ldots, \varepsilon_n)^\top$ is $\operatorname{Var}(\boldsymbol{\varepsilon}) = \Sigma$, which permits for inhomogeneous and correlated errors. However, for most results we assume i.i.d. errors $\varepsilon_i \sim \mathcal{N}(0, \sigma^2)$. The variance σ^2 could be unknown as well. As in previous chapters, $\boldsymbol{\theta}^*$ denotes the true value of the parameter vector (assumed that the model (7.2) is correct).

7.1 Likelihood Ratio Test for a Simple Null

This section discusses the problem of testing a simple hypothesis $H_0\colon \boldsymbol{\theta}^* = \boldsymbol{\theta}_0$ for a given vector $\boldsymbol{\theta}_0$. A natural "non-informative" alternative is $H_1\colon \boldsymbol{\theta}^* \neq \boldsymbol{\theta}_0$.

V. Spokoiny and T. Dickhaus, *Basics of Modern Mathematical Statistics*,
Springer Texts in Statistics, DOI 10.1007/978-3-642-39909-1_7,

7.1.1 General Errors

We start from the case of general errors with known covariance matrix Σ. The results obtained for the estimation problem in Chap. 4 will be heavily used in our study. In particular, the MLE $\tilde{\boldsymbol{\theta}}$ of $\boldsymbol{\theta}^*$ is

$$\tilde{\boldsymbol{\theta}} = \bigl(\Psi\Sigma^{-1}\Psi^\top\bigr)^{-1}\Psi\Sigma^{-1}\boldsymbol{Y}$$

and the corresponding maximum likelihood is

$$L(\tilde{\boldsymbol{\theta}},\boldsymbol{\theta}_0) = \frac{1}{2}(\tilde{\boldsymbol{\theta}} - \boldsymbol{\theta}_0)^\top B(\tilde{\boldsymbol{\theta}} - \boldsymbol{\theta}_0)$$

with a $p \times p$-matrix B given by

$$B = \Psi\Sigma^{-1}\Psi^\top.$$

This immediately leads to the following representation for the likelihood ratio (LR) test in this setup:

$$T \stackrel{\text{def}}{=} \sup_{\boldsymbol{\theta}} L(\boldsymbol{\theta},\boldsymbol{\theta}_0) = \frac{1}{2}(\tilde{\boldsymbol{\theta}} - \boldsymbol{\theta}_0)^\top B(\tilde{\boldsymbol{\theta}} - \boldsymbol{\theta}_0). \tag{7.3}$$

Moreover, Wilks' phenomenon claims that under $\mathbb{P}_{\boldsymbol{\theta}_0}$, the test statistic T has a fixed distribution: namely, $2T$ is χ^2_p-distributed (chi-squared with p degrees of freedom).

Theorem 7.1.1. *Consider the model* (7.2) *with* $\boldsymbol{\varepsilon} \sim \mathcal{N}(0,\Sigma)$ *for a known matrix* Σ. *Then the LR test statistic* T *is given by* (7.3). *Moreover, if* $\mathfrak{z}_\alpha$ *fulfills* $\mathbb{P}(\zeta_p > 2\mathfrak{z}_\alpha) = \alpha$ *with* $\zeta_p \sim \chi^2_p$, *then the LR test* ϕ *with*

$$\phi \stackrel{\text{def}}{=} \mathbf{1}\bigl(T > \mathfrak{z}_\alpha\bigr) \tag{7.4}$$

is of exact level α:

$$\mathbb{P}_{\boldsymbol{\theta}_0}(\phi = 1) = \alpha.$$

This result follows directly from Theorems 4.6.1 and 4.6.2. We see again the important pivotal property of the test: the critical value $\mathfrak{z}_\alpha$ only depends on the dimension of the parameter space Θ. It does not depend on the design matrix Ψ, error covariance Σ, and the null value $\boldsymbol{\theta}_0$.

7.1.2 I.i.d. Errors, Known Variance

We now specify this result for the case of i.i.d. errors. We also focus on the residuals

$$\tilde{\varepsilon} \stackrel{\text{def}}{=} Y - \Psi^\top \tilde{\theta} = Y - \tilde{f},$$

where $\tilde{f} = \Psi^\top \tilde{\theta}$ is the estimated response of the true regression function $f^* = \Psi^\top \theta^*$.

We start with some geometric properties of the residuals $\tilde{\varepsilon}$ and the test statistic T from (7.3).

Theorem 7.1.2. *Consider the model* (7.1)*. Let* T *be the LR test statistic built under the assumptions* $f^* = \Psi^\top \theta^*$ *and* $\operatorname{Var}(\varepsilon) = \sigma^2 I_n$ *with a known value* σ^2*. Then* T *is given by*

$$T = \frac{1}{2\sigma^2} \left\| \Psi^\top (\tilde{\theta} - \theta_0) \right\|^2 = \frac{1}{2\sigma^2} \left\| \tilde{f} - f_0 \right\|^2. \tag{7.5}$$

Moreover, the following decompositions for the vector of observations Y *and for the errors* $\varepsilon = Y - f_0$ *hold:*

$$Y - f_0 = (\tilde{f} - f_0) + \tilde{\varepsilon}, \tag{7.6}$$

$$\|Y - f_0\|^2 = \|\tilde{f} - f_0\|^2 + \|\tilde{\varepsilon}\|^2, \tag{7.7}$$

where $\tilde{f} - f_0$ *is the estimation error and* $\tilde{\varepsilon} = Y - \tilde{f}$ *is the vector of residuals.*

Proof. The key step of the proof is the representation of the estimated response $\tilde{f}$ under the model assumption $Y = f^* + \varepsilon$ as a projection of the data on the p-dimensional linear subspace $\mathcal{L}$ in $\mathbb{R}^n$ spanned by the rows of the matrix Ψ:

$$\tilde{f} = \Pi Y = \Pi(f^* + \varepsilon)$$

where $\Pi = \Psi^\top (\Psi \Psi^\top)^{-1} \Psi$ is the projector onto $\mathcal{L}$; see Sect. 4.3. Note that this decomposition is valid for the general linear model; the parametric form of the response f and the noise normality is not required. The identity $\Psi^\top (\tilde{\theta} - \theta_0) = \tilde{f} - f_0$ follows directly from the definition implying the representation (7.5) for the test statistic T. The identity (7.6) follows from the definition. Next, $\Pi f_0 = f_0$ and thus $\tilde{f} - f_0 = \Pi(Y - f_0)$. Similarly,

$$\tilde{\varepsilon} = Y - \tilde{f} = (I_n - \Pi) Y.$$

As Π and $I_n - \Pi$ are orthogonal projectors, it follows

$$\|Y - f_0\|^2 = \|(I_n - \Pi)Y + \Pi(Y - f_0)\|^2 = \|(I_n - \Pi)Y\|^2 + \|\Pi(Y - f_0)\|^2$$

and the decomposition (7.7) follows.

The decomposition (7.6), although straightforward, is very important for understanding the structure of the residuals under the null and under the alternative. Under the null H_0, the response $\boldsymbol{f}^*$ is assumed to be known and coincides with $\boldsymbol{f}_0$, so the residuals $\tilde{\boldsymbol{\varepsilon}}$ coincide with the errors $\boldsymbol{\varepsilon}$. The sum of squared residuals is usually abbreviated as RSS:

$$\text{RSS}_0 \stackrel{\text{def}}{=} \|\boldsymbol{Y} - \boldsymbol{f}_0\|^2$$

Under the alternative, the response is unknown and is estimated by $\tilde{\boldsymbol{f}}$. The residuals are $\tilde{\boldsymbol{\varepsilon}} = \boldsymbol{Y} - \tilde{\boldsymbol{f}}$ resulting in the RSS

$$\text{RSS} \stackrel{\text{def}}{=} \|\boldsymbol{Y} - \tilde{\boldsymbol{f}}\|^2.$$

The decomposition (7.7) can be rewritten as

$$\text{RSS}_0 = \text{RSS} + \|\tilde{\boldsymbol{f}} - \boldsymbol{f}_0\|^2. \tag{7.8}$$

We see that the RSS under the null and the alternative can be essentially different only if the estimate $\tilde{\boldsymbol{f}}$ significantly deviates from the null assumption $\boldsymbol{f}^* = \boldsymbol{f}_0$. The test statistic T from (7.3) can be written as

$$T = \frac{\text{RSS}_0 - \text{RSS}}{2\sigma^2}.$$

For the proofs in the remainder of this chapter, the following results concerning the distribution of quadratic forms of Gaussians are helpful.

Theorem 7.1.3. *1. Let $X \sim \mathcal{N}_n(\mu, \Sigma)$, where Σ is symmetric and positive definite. Then, $(X - \mu)^\top \Sigma^{-1}(X - \mu) \sim \chi^2_n$.*

2. Let $X \sim \mathcal{N}_n(0, \boldsymbol{I}_n)$, R a symmetric, idempotent $(n \times n)$-matrix with rank $(R) = r$ and B a $(p \times n)$-matrix with $p \le n$. Then it holds

(a) $X^\top R X \sim \chi^2_r$

(b) From $BR = 0$, it follows that $X^\top R X$ is independent of BX.

3. Let $X \sim \mathcal{N}_n(0, \boldsymbol{I}_n)$ and R, S symmetric and idempotent $(n \times n)$-matrices with rank $(R) = r$, rank $(S) = s$ and $RS = 0$. Then it holds

(a) $X^\top R X$ and $X^\top S X$ are independent.

(b) $\frac{X^\top RX/r}{X^\top SX/s} \sim F_{r,s}$.

Proof. For proving the first assertion, let $\Sigma^{1/2}$ denote the uniquely defined, symmetric, and positive definite matrix fulfilling $\Sigma^{1/2} \cdot \Sigma^{1/2} = \Sigma$, with inverse matrix $\Sigma^{-1/2}$. Then, $Z \stackrel{\text{def}}{=} \Sigma^{-1/2}(X - \mu) \sim \mathcal{N}_n(0, \boldsymbol{I}_n)$. The assertion $Z^\top Z \sim \chi^2_n$ follows from the definition of the chi-squared distribution.

For the second part, notice that, due to symmetry and idempotence of R, there exists an orthogonal matrix P with $R = PD_r P^\top$, where $D_r = \begin{pmatrix} I_r & 0 \\ 0 & 0 \end{pmatrix}$. Orthogonality of P implies $W \stackrel{\text{def}}{=} P^\top X \sim \mathcal{N}_n(0, I_n)$. We conclude

$$X^\top RX = X^\top R^2 X = (RX)^\top (RX) = (PD_r W)^\top (PD_r W)$$

$$= W^\top D_r P^\top PD_r W = W^\top D_r W = \sum_{i=1}^{r} W_i^2 \sim \chi_r^2.$$

Furthermore, we have $Z_1 \stackrel{\text{def}}{=} BX \sim \mathcal{N}_n(0, B^\top B)$, $Z_2 \stackrel{\text{def}}{=} RX \sim \mathcal{N}_n(0, R)$, and $\operatorname{Cov}(Z_1, Z_2) = \operatorname{Cov}(BX, RX) = B \operatorname{Cov}(X) R^\top = BR = 0$. As uncorrelation implies independence under Gaussianity, statement 2.(b) follows.

To prove the third part, let $Z_1 \stackrel{\text{def}}{=} SX \sim \mathcal{N}_n(0, S)$ and $Z_2 \stackrel{\text{def}}{=} RX \sim \mathcal{N}_n(0, R)$ and notice that $\operatorname{Cov}(Z_1, Z_2) = S \operatorname{Cov}(X) R = SR = S^\top R^\top = (RS)^\top = 0$. Assertion (a) is then implied by the identities $X^\top SX = Z_1^\top Z_1$ and $X^\top RX = Z_2^\top Z_2$. Assertion (b) immediately follows from 3(a) and 1.

Exercise 7.1.1. Prove Theorem 6.4.2.
Hint: Use Theorem 7.1.3.

Now we show that if the model assumptions are correct, the LR test based on T from (7.3) has the exact level α and is pivotal.

Theorem 7.1.4. *Consider the model* (7.1) *with* $\boldsymbol{\varepsilon} \sim \mathcal{N}(0, \sigma^2 I_n)$ *for a known value* σ^2, *implying that the* ε_i *are i.i.d. normal. The LR test* ϕ *from* (7.4) *is of exact level* α. *Moreover,* $\tilde{f} - f_0$ *and* $\tilde{\boldsymbol{\varepsilon}}$ *are under* $\mathbb{P}_{\theta_0}$ *zero-mean independent Gaussian vectors satisfying*

$$2T = \sigma^{-2}\|\tilde{f} - f_0\|^2 \sim \chi_p^2, \qquad \sigma^{-2}\|\tilde{\boldsymbol{\varepsilon}}\|^2 \sim \chi_{n-p}^2. \tag{7.9}$$

Proof. The null assumption $f^* = f_0$ together with $\Pi f_0 = f_0$ implies now the following decomposition:

$$\tilde{f} - f_0 = \Pi\boldsymbol{\varepsilon}, \qquad \tilde{\boldsymbol{\varepsilon}} = \boldsymbol{\varepsilon} - \Pi\boldsymbol{\varepsilon} = (I_n - \Pi)\boldsymbol{\varepsilon}.$$

Next, Π and $I_n - \Pi$ are orthogonal projectors implying orthogonal and thus uncorrelated vectors $\Pi\boldsymbol{\varepsilon}$ and $(I_n - \Pi)\boldsymbol{\varepsilon}$. Under normality of $\boldsymbol{\varepsilon}$, these vectors are also normal, and uncorrelation implies independence. The property (7.9) for the distribution of $\Pi\boldsymbol{\varepsilon}$ was proved in Theorem 4.6.1. For the distribution of $\tilde{\boldsymbol{\varepsilon}} = (I_n - \Pi)\boldsymbol{\varepsilon}$, we can apply Theorem 7.1.3.

Next we discuss the power of the LR test ϕ defined as the probability of detecting the alternative when the response f^* deviates from the null f_0. In the next result we do not assume that the true response f^* follows the linear PA $f^* = \Psi^\top \theta$ and show that the test power depends on the value $\|\Pi(f^* - f_0)\|^2$.

Theorem 7.1.5. *Consider the model* (7.1) *with* $\operatorname{Var}(\boldsymbol{\varepsilon}) = \sigma^2 \boldsymbol{I}_n$ *for a known value* σ^2. *Define*

$$\Delta = \sigma^{-2} \|\Pi(\boldsymbol{f}^* - \boldsymbol{f}_0)\|^2.$$

Then the power of the LR test ϕ *only depends on* Δ, *i.e. it is the same for all* $\boldsymbol{f}^*$ *with equal* Δ*-value. It holds*

$$\mathbb{P}(\phi = 1) = \mathbb{P}\bigl(\bigl|\xi_1 + \sqrt{\Delta}\bigr|^2 + \xi_2^2 + \ldots + \xi_p^2 > 2\mathfrak{z}_\alpha\bigr)$$

with $\boldsymbol{\xi} = (\xi_1, \ldots, \xi_p)^\top \sim \mathcal{N}(0, \boldsymbol{I}_p)$.

Proof. It follows from $\tilde{\boldsymbol{f}} = \Pi \boldsymbol{Y} = \Pi(\boldsymbol{f}^* + \boldsymbol{\varepsilon})$ and $\boldsymbol{f}_0 = \Pi \boldsymbol{f}_0$ for the test statistic $T = (2\sigma^2)^{-1} \|\tilde{\boldsymbol{f}} - \boldsymbol{f}_0\|^2$ that

$$T = (2\sigma^2)^{-1} \|\Pi(\boldsymbol{f}^* - \boldsymbol{f}_0) + \Pi \boldsymbol{\varepsilon}\|^2.$$

Now we show that the distribution of T depends on the response $\boldsymbol{f}^*$ only via the value Δ. For this we compute the Laplace transform of T.

Lemma 7.1.1. *It holds for* $\mu < 1$

$$g(\mu) \stackrel{\text{def}}{=} \log \mathbb{E} \exp\{\mu T\} = \frac{\mu \Delta}{2(1-\mu)} - \frac{p}{2} \log(1-\mu).$$

Proof. For a standard Gaussian random variable ξ and any a, it holds

$$\begin{aligned}
&\mathbb{E} \exp\{\mu |\xi + a|^2/2\} \\
&= e^{\mu a^2/2} (2\pi)^{-1/2} \int \exp\{\mu a x + \mu x^2/2 - x^2/2\} dx \\
&= \exp\left\{\frac{\mu a^2}{2} + \frac{\mu^2 a^2}{2(1-\mu)}\right\} \frac{1}{\sqrt{2\pi}} \int \exp\left\{-\frac{1-\mu}{2}\left(x - \frac{a\mu}{1-\mu}\right)^2\right\} dx \\
&= \exp\left\{\frac{\mu a^2}{2(1-\mu)}\right\} (1-\mu)^{-1/2}.
\end{aligned}$$

The projector Π can be represented as $\Pi = U^\top \Lambda_p U$ for an orthogonal transform U and the diagonal matrix $\Lambda_p = \operatorname{diag}(1, \ldots, 1, 0, \ldots, 0)$ with only p unit eigenvalues. This permits representing T in the form

$$T = \sum_{j=1}^{p} (\xi_j + a_j)^2/2$$

with i.i.d. standard normal r.v. ξ_j and numbers a_j satisfying $\sum_j a_j^2 = \Delta$. The independence of the ξ_j's implies

$$g(\mu) = \sum_{j=1}^{p} \Bigl[\frac{\mu a_j^2}{2} + \frac{\mu^2 a_j^2}{2(1-\mu)} - \frac{1}{2} \log(1-\mu) \Bigr] = \frac{\mu \Delta}{2(1-\mu)} - \frac{p}{2} \log(1-\mu)$$

as required.

The result of Lemma 7.1.1 claims that the Laplace transform of T depends on $\boldsymbol{f}^*$ only via Δ and so this also holds for the distribution of T.

The distribution of the squared norm $\|\boldsymbol{\xi} + \boldsymbol{h}\|^2$ for $\boldsymbol{\xi} \sim \mathcal{N}(0, \boldsymbol{I}_p)$ and any fixed vector $\boldsymbol{h} \in \mathbb{R}^p$ with $\|\boldsymbol{h}\|^2 = \Delta$ is called non-central chi-squared with the non-centrality parameter Δ. In particular, for each α, α_1 one can define the minimal value Δ providing the prescribed error α_1 of the second kind by the equation under the given level α:

$$\mathbb{P}\bigl(\|\boldsymbol{\xi} + \boldsymbol{h}\|^2 \ge 2\mathfrak{z}_\alpha\bigr) \ge 1 - \alpha_1 \text{ subject to } \mathbb{P}\bigl(\|\boldsymbol{\xi}\|^2 \ge 2\mathfrak{z}_\alpha\bigr) \le \alpha \tag{7.10}$$

with $\|\boldsymbol{h}\|^2 \ge \Delta$. The results from Sect. 4.6 indicate that the value $\mathfrak{z}_\alpha$ can be bounded from above by $p + \sqrt{2p \log \alpha^{-1}}$ for moderate values of α^{-1}. For evaluating the value Δ, the following decomposition is useful:

$$\|\boldsymbol{\xi} + \boldsymbol{h}\|^2 - \|\boldsymbol{h}\|^2 - p = \|\boldsymbol{\xi}\|^2 - p + 2\boldsymbol{h}^\top \boldsymbol{\xi}.$$

The right-hand side of this equality is a sum of centered Gaussian quadratic and linear forms. In particular, the cross term $2\boldsymbol{h}^\top \boldsymbol{\xi}$ is a centered Gaussian r.v. with the variance $4\|\boldsymbol{h}\|^2$, while $\operatorname{Var}\bigl(\|\boldsymbol{\xi}\|^2\bigr) = 2p$. These arguments suggest to take Δ of order p to ensure the prescribed power α_1.

Theorem 7.1.6. *For each $\alpha, \alpha_1 \in (0, 1)$, there are absolute constants $\mathtt{C}$ and $\mathtt{C}_1$ such that* (7.10) *is fulfilled for $\|\boldsymbol{h}\|^2 \ge \Delta$ with*

$$\Delta^{1/2} = \sqrt{\mathtt{C} p \log \alpha^{-1}} + \sqrt{\mathtt{C}_1 p \log \alpha_1^{-1}}.$$

The result of Theorem 7.1.6 reveals some problem with the power of the LR test when the dimensionality of the parameter space grows. Indeed, the test remains insensitive for all alternatives in the zone $\sigma^{-2}\|\Pi(\boldsymbol{f}^* - \boldsymbol{f}_0)\|^2 \le \mathtt{C} p$ and this zone becomes larger and larger with p.

7.1.3 *I.i.d. Errors with Unknown Variance*

This section briefly discusses the case when the errors ε_i are i.i.d. but the variance $\sigma^2 = \operatorname{Var}(\varepsilon_i)$ is unknown. A natural idea in this case is to estimate the variance from the data. The decomposition (7.8) and independence of $\text{RSS} = \|\boldsymbol{Y} - \tilde{\boldsymbol{f}}\|^2$

and $\|\tilde{f} - f_0\|^2$ are particularly helpful. Theorem 7.1.4 suggests to estimate σ^2 from RSS by

$$\tilde{\sigma}^2 = \frac{1}{n-p}\,\text{RSS} = \frac{1}{n-p}\|Y - \tilde{f}\|^2.$$

Indeed, due to the result (7.9), $\sigma^{-2}\,\text{RSS} \sim \chi_p^2$ yielding

$$\mathbb{E}\tilde{\sigma}^2 = \sigma^2, \qquad \operatorname{Var}\tilde{\sigma}^2 = \frac{2}{n-p}\sigma^4 \tag{7.11}$$

and therefore, $\tilde{\sigma}^2$ is an unbiased, root-n consistent estimate of σ^2.

Exercise 7.1.2. Check (7.11). Show that $\tilde{\sigma}^2 - \sigma^2 \xrightarrow{\mathbb{P}} 0$.

Now we consider the LR test (7.5) in which the true variance is replaced by its estimate $\tilde{\sigma}^2$:

$$\tilde{T} \stackrel{\text{def}}{=} \frac{1}{2\tilde{\sigma}^2}\|\tilde{f} - f_0\|^2 = \frac{(n-p)\|\tilde{f} - f_0\|^2}{2\|Y - \tilde{f}\|^2} = \frac{\text{RSS}_0 - \text{RSS}}{2\,\text{RSS}/(n-p)}.$$

The result of Theorem 7.1.4 implies the pivotal property for this test statistic as well.

Theorem 7.1.7. *Consider the model (7.1) with $\boldsymbol{\varepsilon} \sim \mathcal{N}(0, \sigma^2 \boldsymbol{I}_n)$ for an unknown value σ^2. Then the distribution of the test statistic $\tilde{T}$ under $\mathbb{P}_{\boldsymbol{\theta}_0}$ only depends on p and $n-p$:*

$$2p^{-1}\tilde{T} = \frac{n-p}{p}\,\frac{\text{RSS}_0 - \text{RSS}}{\text{RSS}} \sim F_{p,n-p},$$

where $F_{p,n-p}$ denotes the Fisher distribution with parameters $p, n-p$:

$$F_{p,n-p} = \mathcal{L}\left(\frac{\|\boldsymbol{\xi}_p\|^2/p}{\|\boldsymbol{\xi}_{n-p}\|^2/(n-p)}\right)$$

where $\boldsymbol{\xi}_p$ and $\boldsymbol{\xi}_{n-p}$ are two independent standard Gaussian vectors of dimension p and $n-p$, see Theorem 6.4.1. In particular, it does not depend on the design matrix Ψ, the noise variance σ^2, and the true parameter $\boldsymbol{\theta}_0$.

This result suggests to fix the critical value $\mathfrak{z}$ for the test statistic $\tilde{T}$ using the quantiles of the Fisher distribution: If t_α is such that $F_{p,n-p}(t_\alpha) = 1 - \alpha$, then $\mathfrak{z}_\alpha = p\,t_\alpha/2$.

Theorem 7.1.8. *Consider the model (7.1) with $\boldsymbol{\varepsilon} \sim \mathcal{N}(0, \sigma^2 \boldsymbol{I}_n)$ for a unknown value σ^2. If $F_{p,n-p}(t_\alpha) = 1 - \alpha$ and $\tilde{\mathfrak{z}}_\alpha = p t_\alpha/2$, then the test $\tilde{\phi} = \mathbf{1}(\tilde{T} \ge \tilde{\mathfrak{z}}_\alpha)$ is a level-α test:*

$$\mathbb{P}_{\theta_0}(\tilde{\phi} = 1) = \mathbb{P}_{\theta_0}(\tilde{T} \geq \mathfrak{z}_\alpha) = \alpha.$$

Exercise 7.1.3. Prove the result of Theorem 7.1.8.

If the sample size n is sufficiently large, then $\tilde{\sigma}^2$ is very close to σ^2 and one can apply an approximate choice of the critical value $\mathfrak{z}_\alpha$ from the case of σ^2 known:

$$\breve{\phi} = \mathbf{1}(\tilde{T} \geq \mathfrak{z}_\alpha).$$

This test is not of exact level α but it is of asymptotic level α. Its power function is also close to the power function of the test ϕ corresponding to the known variance σ^2.

Theorem 7.1.9. *Consider the model* (7.1) *with* $\boldsymbol{\varepsilon} \sim \mathcal{N}(0, \sigma^2 \boldsymbol{I}_n)$ *for a unknown value* σ^2. *Then*

$$\lim_{n\to\infty} \mathbb{P}_{\theta_0}(\breve{\phi} = 1) = \alpha. \tag{7.12}$$

Moreover,

$$\lim_{n\to\infty} \sup_{f^*} \bigl| \mathbb{P}_{\theta_0}(\tilde{\phi} = 1) - \mathbb{P}_{\theta_0}(\breve{\phi} = 1) \bigr| = 0. \tag{7.13}$$

Exercise 7.1.4. Consider the model (7.1) with $\boldsymbol{\varepsilon} \sim \mathcal{N}(0, \sigma^2 \boldsymbol{I}_n)$ for σ^2 unknown and prove (7.12) and (7.13).
Hints:

- The consistency of $\tilde{\sigma}^2$ permits to restrict to the case $\bigl|\tilde{\sigma}^2/\sigma^2 - 1\bigr| \leq \delta_n$ for $\delta_n \to 0$.
- The independence of $\|\tilde{\boldsymbol{f}} - \boldsymbol{f}_0\|^2$ and $\tilde{\sigma}^2$ permits to consider the distribution of $2\tilde{T} = \|\tilde{\boldsymbol{f}} - \boldsymbol{f}_0\|^2/\tilde{\sigma}^2$ as if $\tilde{\sigma}^2$ were a fixed number close to δ.
- Use that for $\zeta_p \sim \chi^2_p$,

$$\mathbb{P}\bigl(\zeta_p \geq \mathfrak{z}_\alpha(1 + \delta_n)\bigr) - \mathbb{P}\bigl(\zeta_p \geq \mathfrak{z}_\alpha\bigr) \to 0, \qquad n \to \infty.$$

7.2 Likelihood Ratio Test for a Subvector

The previous section dealt with the case of a simple null hypothesis. This section considers a more general situation when the null hypothesis concerns a subvector of $\boldsymbol{\theta}$. This means that the whole model is given by (7.2) but the vector $\boldsymbol{\theta}$ is decomposed into two parts: $\boldsymbol{\theta} = (\boldsymbol{\gamma}, \boldsymbol{\eta})$, where $\boldsymbol{\gamma}$ is of dimension $p_0 < p$. The null hypothesis assumes that $\boldsymbol{\eta} = \boldsymbol{\eta}_0$ for all $\boldsymbol{\gamma}$. Usually $\boldsymbol{\eta}_0 = 0$ but the particular value of $\boldsymbol{\eta}_0$ is not important. To simplify the presentation we assume $\boldsymbol{\eta}_0 = 0$ leading to the subset Θ_0 of Θ, given by

$$\Theta_0 = \{\boldsymbol{\theta} = (\boldsymbol{\gamma}, 0)\}.$$

Under the null hypothesis, the model is still linear:

$$\boldsymbol{Y} = \Psi_{\gamma}^{\top} \boldsymbol{\gamma} + \boldsymbol{\varepsilon},$$

where Ψ_{γ} denotes a submatrix of Ψ composed by the rows of Ψ corresponding to the $\boldsymbol{\gamma}$-components of $\boldsymbol{\theta}$.

Fix any point $\boldsymbol{\theta}_0 \in \Theta_0$, e.g., $\boldsymbol{\theta}_0 = 0$ and define the corresponding response $\boldsymbol{f}_0 = \Psi^{\top} \boldsymbol{\theta}_0$. The LR test T can be written in the form

$$T = \max_{\boldsymbol{\theta} \in \Theta} L(\boldsymbol{\theta}, \boldsymbol{\theta}_0) - \max_{\boldsymbol{\theta} \in \Theta_0} L(\boldsymbol{\theta}, \boldsymbol{\theta}_0). \tag{7.14}$$

The results of both maximization problems are known:

$$\max_{\boldsymbol{\theta} \in \Theta} L(\boldsymbol{\theta}, \boldsymbol{\theta}_0) = \frac{1}{2\sigma^2} \| \tilde{\boldsymbol{f}} - \boldsymbol{f}_0 \|^2,$$
$$\max_{\boldsymbol{\theta} \in \Theta_0} L(\boldsymbol{\theta}, \boldsymbol{\theta}_0) = \frac{1}{2\sigma^2} \| \tilde{\boldsymbol{f}}_0 - \boldsymbol{f}_0 \|^2,$$

where $\tilde{\boldsymbol{f}}$ and $\tilde{\boldsymbol{f}}_0$ are estimates of the response under the null and the alternative, respectively. As in Theorem 7.1.2 we can establish the following geometric decomposition.

Theorem 7.2.1. *Consider the model* (7.1). *Let T be the LR test statistic from* (7.14) *built under the assumptions $\boldsymbol{f}^* = \Psi^{\top} \boldsymbol{\theta}^*$ and $\operatorname{Var}(\boldsymbol{\varepsilon}) = \sigma^2 \boldsymbol{I}_n$ with a known value σ^2. Then T is given by*

$$T = \frac{1}{2\sigma^2} \left\| \tilde{\boldsymbol{f}} - \tilde{\boldsymbol{f}}_0 \right\|^2.$$

Moreover, the following decompositions for the vector of observations $\boldsymbol{Y}$ and for the residuals $\tilde{\boldsymbol{\varepsilon}}_0 = \boldsymbol{Y} - \tilde{\boldsymbol{f}}_0$ from the null hold:

$$\boldsymbol{Y} - \tilde{\boldsymbol{f}}_0 = (\tilde{\boldsymbol{f}} - \tilde{\boldsymbol{f}}_0) + \tilde{\boldsymbol{\varepsilon}},$$
$$\| \boldsymbol{Y} - \tilde{\boldsymbol{f}}_0 \|^2 = \| \tilde{\boldsymbol{f}} - \tilde{\boldsymbol{f}}_0 \|^2 + \| \tilde{\boldsymbol{\varepsilon}} \|^2, \tag{7.15}$$

where $\tilde{\boldsymbol{f}} - \tilde{\boldsymbol{f}}_0$ is the difference between the estimated response under the null and under the alternative, and $\tilde{\boldsymbol{\varepsilon}} = \boldsymbol{Y} - \tilde{\boldsymbol{f}}$ is the vector of residuals from the alternative.

Proof. The proof is similar to the proof of Theorem 7.1.2. We use that $\tilde{\boldsymbol{f}} = \Pi \boldsymbol{Y}$ where $\Pi = \Psi^{\top} \left(\Psi \Psi^{\top} \right)^{-1} \Psi$ is the projector on the space $\mathcal{L}$ spanned by the rows of Ψ. Similarly $\tilde{\boldsymbol{f}}_0 = \Pi_0 \boldsymbol{Y}$ where $\Pi_0 = \Psi_{\gamma}^{\top} \left(\Psi_{\gamma} \Psi_{\gamma}^{\top} \right)^{-1} \Psi_{\gamma}$ is the projector on the subspace $\mathcal{L}_0$ spanned by the rows of Ψ_{γ}. This yields the decomposition $\tilde{\boldsymbol{f}} - \boldsymbol{f}_0 = \Pi(\boldsymbol{Y} - \boldsymbol{f}_0)$. Similarly,

$$\tilde{f} - \tilde{f}_0 = (\Pi - \Pi_0)Y, \qquad \tilde{\varepsilon} = Y - \tilde{f} = (I_n - \Pi)Y.$$

As $\Pi - \Pi_0$ and $I_n - \Pi$ are orthogonal projectors, it follows

$$\|Y - \tilde{f}_0\|^2 = \|(I_n - \Pi)Y + (\Pi - \Pi_0)Y\|^2 = \|(I_n - \Pi)Y\|^2 + \|(\Pi - \Pi_0)Y\|^2$$

and the decomposition (7.15) follows.

The decomposition (7.15) can again be represented as $\mathrm{RSS}_0 = \mathrm{RSS} + 2\sigma^2 T$, where RSS is the sum of squared residuals, while RSS_0 is the same as in the case of a simple null.

Now we show how a pivotal test of exact level α based on T can be constructed if the model assumptions are correct.

Theorem 7.2.2. *Consider the model* (7.1) *with* $\varepsilon \sim \mathcal{N}(0, \sigma^2 I_n)$ *for a known value* σ^2, *i.e., the* ε_i *are i.i.d. normal. Then* $\tilde{f} - \tilde{f}_0$ *and* $\tilde{\varepsilon}$ *are under* $\mathbb{P}_{\theta_0}$ *zero-mean independent Gaussian vectors satisfying*

$$2T = \sigma^{-2}\|\tilde{f} - \tilde{f}_0\|^2 \sim \chi^2_{p-p_0}, \qquad \sigma^{-2}\|\tilde{\varepsilon}\|^2 \sim \chi^2_{n-p}. \tag{7.16}$$

Let $\mathfrak{z}_\alpha$ *fulfill* $\mathbb{P}(\zeta_{p-p_0} \ge \mathfrak{z}_\alpha) = \alpha$. *Then the test* $\phi = \mathbf{1}(T \ge \mathfrak{z}_\alpha/2)$ *is an LR test of exact level* α.

Proof. The null assumption $\theta^* \in \Theta_0$ implies $f^* \in \mathcal{L}_0$. This, together with $\Pi_0 f^* = f^*$ implies now the following decomposition:

$$\tilde{f} - \tilde{f}_0 = (\Pi - \Pi_0)\varepsilon, \qquad \tilde{\varepsilon} = \varepsilon - \Pi\varepsilon = (I_n - \Pi)\varepsilon.$$

Next, $\Pi - \Pi_0$ and $I_n - \Pi$ are orthogonal projectors implying orthogonal and thus uncorrelated vectors $(\Pi - \Pi_0)\varepsilon$ and $(I_n - \Pi)\varepsilon$. Under normality of ε, these vectors are also normal, and uncorrelation implies independence. The property (7.16) is similar to (7.9) and can easily be verified by making use of Theorem 7.1.3.

If the variance σ^2 of the noise is unknown, one can proceed exactly as in the case of a simple null: estimate the variance from the residuals using their independence of the test statistic T. This leads to the estimate

$$\tilde{\sigma}^2 = \frac{1}{n-p}\,\mathrm{RSS} = \frac{1}{n-p}\|Y - \tilde{f}\|^2$$

and to the test statistic

$$\tilde{T} = \frac{\mathrm{RSS}_0 - \mathrm{RSS}}{2\,\mathrm{RSS}/(n-p)} = \frac{(n-p)\|\tilde{f} - \tilde{f}_0\|^2}{2\|Y - \tilde{f}\|^2}.$$

The property of pivotality is preserved here as well: properly scaled, the test statistic $\tilde{T}$ has a Fisher distribution.

Theorem 7.2.3. *Consider the model* (7.1) *with* $\boldsymbol{\varepsilon} \sim \mathcal{N}(0, \sigma^2 \boldsymbol{I}_n)$ *for an unknown value* σ^2. *Then* $2\tilde{T}/(p - p_0)$ *has the Fisher distribution* $F_{p-p_0, n-p}$ *with parameters* $p - p_0$ *and* $n - p$. *If* t_α *is the* $1 - \alpha$ *quantile of this distribution, then the test* $\tilde{\phi} = \mathbf{1}\big(\tilde{T} > (p - p_0)t_\alpha/2\big)$ *is of exact level* α.

If the sample size is sufficiently large, one can proceed as if $\tilde{\sigma}^2$ were the true variance ignoring the error of variance estimation. This would lead to the critical value $\mathfrak{z}_\alpha$ from Theorem 7.2.2 and the corresponding test is of asymptotic level α.

Exercise 7.2.1. Prove Theorem 7.2.3.

The study of the power of the test T does not differ from the case of a simple hypothesis. One only needs to redefine Δ as

$$\Delta \stackrel{\text{def}}{=} \sigma^{-2} \|(\Pi - \Pi_0) \boldsymbol{f}^*\|^2.$$

7.3 Likelihood Ratio Test for a Linear Hypothesis

In this section, we generalize the test problem for the Gaussian linear model further and assume that we want to test the *linear hypothesis* $H_0\colon C\boldsymbol{\theta} = \boldsymbol{d}$ for a given contrast matrix $C \in \mathbb{R}^{r \times p}$ with $\operatorname{rank}(C) = r \le p$ and a right-hand side vector $\boldsymbol{d} \in \mathbb{R}^r$. Notice that the point null hypothesis $H_0\colon \boldsymbol{\theta} = \boldsymbol{\theta}_0$ and the hypothesis $H_0\colon \boldsymbol{\eta} = \boldsymbol{\eta}_0$ regarded in Sect. 7.2 are linear hypotheses. Here, we restrict our attention to the case that the error variance σ^2 is unknown, as it is typically the case in practice.

Theorem 7.3.1. *Assume model* (7.1) *with* $\boldsymbol{\varepsilon} \sim \mathcal{N}(0, \sigma^2 \boldsymbol{I}_n)$ *for an unknown value* σ^2 *and consider the linear hypothesis* $H_0\colon C\boldsymbol{\theta} = \boldsymbol{d}$. *Then it holds*

(a) The likelihood ratio statistic $\tilde{T}$ *is an isotone transformation of* $F \stackrel{\text{def}}{=} \frac{n-p}{r} \frac{\Delta \,\mathrm{RSS}}{\mathrm{RSS}}$, *where* $\Delta\,\mathrm{RSS} \stackrel{\text{def}}{=} \mathrm{RSS}_0 - \mathrm{RSS}$.

(b) The restricted MLE $\tilde{\boldsymbol{\theta}}_0$ *is given by*

$$\tilde{\boldsymbol{\theta}}_0 = \tilde{\boldsymbol{\theta}} - (\Psi\Psi^\top)^{-1} C^\top \big\{C(\Psi\Psi^\top)^{-1} C^\top\big\}^{-1} (C\tilde{\boldsymbol{\theta}} - \boldsymbol{d}).$$

(c) $\Delta\,\mathrm{RSS}$ *and* RSS *are independent.*

(d) Under H_0, $\Delta\,\mathrm{RSS}/\sigma^2 \sim \chi^2_r$ *and* $F \sim F_{r,n-p}$.

Proof. For proving (a), verify that $2\tilde{T} = n \log(\hat{\sigma}_0^2/\hat{\sigma}^2)$, where $\hat{\sigma}^2$ and $\hat{\sigma}_0^2$ are the unrestricted and restricted (i.e., under H_0) MLEs of the error variance σ^2, respectively. Plugging in of the explicit representations for $\hat{\sigma}^2$ and $\hat{\sigma}_0^2$ yields that $2\tilde{T} = n \log(\frac{\Delta\,\mathrm{RSS}}{\mathrm{RSS}} + 1)$, implying the assertion. Part (b) is an application of the following well-known result from quadratic optimization theory.

Lemma 7.3.1. *Let $A \in \mathbb{R}^{r \times p}$ and $\boldsymbol{b} \in \mathbb{R}^r$ fixed and define $\mathcal{M} = \{z \in \mathbb{R}^p : Az = \boldsymbol{b}\}$. Moreover, let $f : \mathbb{R}^p \to \mathbb{R}$, given by $f(z) = z^\top Qz/2 - \boldsymbol{c}^\top z$ for a symmetric, positive semi-definite $(p \times p)$-matrix Q and a vector $\boldsymbol{c} \in \mathbb{R}^p$. Then, the unique minimum of f over the search space $\mathcal{M}$ is characterized by solving the system of linear equations*

$$Qz - A^\top y = c$$
$$Az = b$$

for (y, z). The component z of this solution minimizes f over $\mathcal{M}$.

For part (c), we utilize the explicit form of $\tilde{\boldsymbol{\theta}}_0$ from part (b) and write $\Delta\,\mathrm{RSS}$ in the form

$$\Delta\,\mathrm{RSS} = (C\tilde{\boldsymbol{\theta}} - \boldsymbol{d})^\top \{C(\Psi\Psi^\top)^{-1}C^\top\}^{-1}(C\tilde{\boldsymbol{\theta}} - \boldsymbol{d}).$$

This shows that $\Delta\,\mathrm{RSS}$ is a deterministic transformation of $\tilde{\boldsymbol{\theta}}$. Since $\tilde{\boldsymbol{\theta}}$ and RSS are independent, the assertion follows. This representation of $\Delta\,\mathrm{RSS}$ moreover shows that $\Delta\,\mathrm{RSS}/\sigma^2$ is a quadratic form of a (under H_0) standard normal random vector and part (d) follows from Theorem 7.1.3.

To sum up, Theorem 7.3.1 shows that general linear hypotheses can be tested with F-tests under model (7.1) with $\boldsymbol{\varepsilon} \sim \mathcal{N}(0, \sigma^2 \boldsymbol{I}_n)$ for an unknown value σ^2.

7.4 Wald Test

The drawback of the LR test method for testing a linear hypothesis $H_0 : C\boldsymbol{\theta} = \boldsymbol{d}$ without assuming Gaussian noise is that a constrained maximization of the log-likelihood function under the constraints encoded by C and $\boldsymbol{d}$ has to be performed. This computationally intensive step can be avoided by using Wald statistics. The Wald statistic for testing H_0 is given by

$$W = (C\tilde{\boldsymbol{\theta}} - \boldsymbol{d})^\top \left(C\hat{V}C^\top\right)^{-1}(C\tilde{\boldsymbol{\theta}} - \boldsymbol{d}),$$

where $\tilde{\boldsymbol{\theta}}$ and $\hat{V}$ denote the MLE in the full (unrestricted) model and the estimated covariance matrix of $\tilde{\boldsymbol{\theta}}$, respectively.

Theorem 7.4.1. *Under model (7.2), the statistic W is asymptotically equivalent to the LR test statistic $2T$. In particular, under H_0, the distribution of W converges to χ^2_r: $W \xrightarrow{w} \chi^2_r$ as $n \to \infty$. In the case of normally distributed noise, it holds $W = rF$, where F is as in Theorem 7.3.1.*

Proof. For proving the asymptotic χ_r^2-distribution of W, we use the asymptotic normality of the MLE $\tilde{\boldsymbol{\theta}}$. If the model is regular, it holds

$$\sqrt{n}(\tilde{\boldsymbol{\theta}}_n - \boldsymbol{\theta}_0) \xrightarrow{w} \mathcal{N}(0, \mathbb{F}(\boldsymbol{\theta}_0)^{-1}) \text{ under } \boldsymbol{\theta}_0 \text{ for } n \to \infty,$$

and, consequently,

$$\mathbb{F}(\boldsymbol{\theta}_0)^{1/2} n^{1/2} (\tilde{\boldsymbol{\theta}}_n - \boldsymbol{\theta}_0) \xrightarrow{w} \mathcal{N}(0, \boldsymbol{I}_r), \text{ where } r \stackrel{\text{def}}{=} \dim(\boldsymbol{\theta}_0).$$

Applying the Continuous Mapping Theorem, we get

$$(\tilde{\boldsymbol{\theta}}_n - \boldsymbol{\theta}_0)^\top n \mathbb{F}(\boldsymbol{\theta}_0)(\tilde{\boldsymbol{\theta}}_n - \boldsymbol{\theta}_0) \xrightarrow{w} \chi_r^2.$$

If the Fisher information is continuous and $\hat{\mathbb{F}}(\tilde{\boldsymbol{\theta}}_n)$ is a consistent estimator for $\mathbb{F}(\boldsymbol{\theta}_0)$, it still holds that

$$(\tilde{\boldsymbol{\theta}}_n - \boldsymbol{\theta}_0)^\top n \hat{\mathbb{F}}(\tilde{\boldsymbol{\theta}}_n)(\tilde{\boldsymbol{\theta}}_n - \boldsymbol{\theta}_0) \xrightarrow{w} \chi_r^2.$$

Substituting $\tilde{\boldsymbol{\theta}}_n = C\tilde{\boldsymbol{\theta}} - \boldsymbol{d}$ and $\boldsymbol{\theta}_0 = 0 \in \mathbb{R}^r$, we obtain the assertion concerning the asymptotic distribution of W under H_0. For proving the relationship between W and F in the case of Gaussian noise, we notice that

$$\begin{aligned} F &= \frac{n-p}{r} \frac{\Delta RSS}{RSS} \\ &= \frac{n-p}{r} \frac{(C\tilde{\boldsymbol{\theta}} - \boldsymbol{d})^\top \{C(\Psi\Psi^\top)^{-1}C^\top\}^{-1}(C\tilde{\boldsymbol{\theta}} - \boldsymbol{d})}{(n-p)\tilde{\sigma}^2} \\ &= \frac{(C\tilde{\boldsymbol{\theta}} - \boldsymbol{d})^\top (C\hat{V}C^\top)^{-1}(C\tilde{\boldsymbol{\theta}} - \boldsymbol{d})}{r} = \frac{W}{r} \end{aligned}$$

as required.

7.5 Analysis of Variance

Important special cases of linear models arise when all entries of the design matrix Ψ are binary, i.e., $\Psi_{ij} \in \{0, 1\}$ for all $1 \le i \le p$, $1 \le j \le n$. Every row of Ψ then has the interpretation of a group indicator, where $\Psi_{ij} = 1$ if and only if observational unit j belongs to group i. The target of such an *analysis of variance* is to determine if the mean response differs between groups. In this section, we specialize the theory of testing in linear models to this situation.

7.5.1 Two-Sample Test

First, we consider the case of $p = 2$, meaning that *exactly two samples* corresponding to the two groups A and B are given. We let n_A denote the number of observations in group A and $n_B = n - n_A$ the number of observations in group B. The parameter of interest $\boldsymbol{\theta} \in \mathbb{R}^2$ in this model consists of the two population means η_A and η_B and the model reads

$$Y_{ij} = \eta_i + \varepsilon_{ij}, \quad i \in \{A, B\}, \quad j = 1, \ldots, n_i.$$

Noticing that $(\Psi\Psi^\top)^{-1} = \begin{pmatrix} 1/n_A & 0 \\ 0 & 1/n_B \end{pmatrix}$, we get that $\tilde{\boldsymbol{\theta}} = (\overline{Y}_A, \overline{Y}_B)^\top$, where $\overline{Y}_A$ and $\overline{Y}_B$ denote the two sample means. With this, we immediately obtain that the estimator for the noise variance is in this model given by

$$\tilde{\sigma}^2 = \frac{\text{RSS}}{n-p} = \frac{1}{n_A + n_B - 2}\left(\sum_{j=1}^{n_A}(Y_{A,j} - \overline{Y}_A)^2 + \sum_{j=1}^{n_B}(Y_{B,j} - \overline{Y}_B)^2\right).$$

The quantity $s^2 \stackrel{\text{def}}{=} \tilde{\sigma}^2$ is called the *pooled sample variance*. Denoting the group-specific sample variances by

$$s_A^2 = \frac{1}{n_A}\sum_{j=1}^{n_A}(Y_{A,j} - \overline{Y}_A)^2, \qquad s_B^2 = \frac{1}{n_B}\sum_{j=1}^{n_B}(Y_{B,j} - \overline{Y}_B)^2,$$

we have that

$$s^2 = (n_A s_A^2 + n_B s_B^2)/(n_A + n_B - 2)$$

is a weighted average of s_A^2 and s_B^2.

Under this model, we are interested in testing the null hypothesis $H_0\colon \eta_A = \eta_B$ of equal group means against the two-sided alternative $H_1\colon \eta_A \neq \eta_B$ or the one-sided alternative $H_1^+\colon \eta_A > \eta_B$. The one-sided alternative hypothesis $H_1^-\colon \eta_A < \eta_B$ can be treated by switching the group labels.

The null hypothesis $H_0\colon \eta_A - \eta_B = 0$ is a linear hypothesis in the sense of Sect. 7.3, where the number of restrictions is $r = 1$. Under H_0, the grand average

$$\overline{Y} \stackrel{\text{def}}{=} n^{-1} \sum_{i \in \{A,B\}} \sum_{j=1}^{n_i} Y_{ij}$$

is the MLE for the (common) population mean. Straightforwardly, we calculate that the sum of squares of residuals under H_0 is given by

$$\mathrm{RSS}_0 = \sum_{i\in\{A,B\}} \sum_{j=1}^{n_i} (Y_{ij} - \overline{Y})^2.$$

For computing $\Delta\,\mathrm{RSS} = \mathrm{RSS}_0 - \mathrm{RSS}$ and the statistic F from Theorem 7.3.1(a), we obtain the following results.

Lemma 7.5.1.

$$\mathrm{RSS}_0 = \mathrm{RSS} + \left(n_A(\overline{Y}_A - \overline{Y})^2 + n_B(\overline{Y}_B - \overline{Y})^2\right), \tag{7.17}$$

$$F = \frac{n_A n_B}{n_A + n_B} \frac{(\overline{Y}_A - \overline{Y}_B)^2}{s^2}. \tag{7.18}$$

Exercise 7.5.1. Prove Lemma 7.5.1.

We conclude from Theorem 7.3.1 that, if the individual errors are homogeneous between samples, the test statistic

$$t = \frac{(\overline{Y}_A - \overline{Y}_B)}{s\sqrt{1/n_A + 1/n_B}}$$

is t-distributed with $n_A + n_B - 2$ degrees of freedom under the null hypothesis H_0.

For a given significance level α, define the quantile q_α of Student's t-distribution with $n_A + n_B - 2$ degrees of freedom by the equation

$$\mathbb{P}(t_0 > q_\alpha) = \alpha,$$

where t_0 represents the distribution of t in the case that H_0 is true. Utilizing the symmetry property

$$\mathbb{P}(|t_0| > q_{\alpha/2}) = 2\mathbb{P}(t > q_{\alpha/2}) = \alpha,$$

the one-sided two-sample t-test for H_0 versus H_1^+ rejects if $t > q_\alpha$ and the two-sided two-sample t-test for H_0 versus H_1 rejects if $|t| > q_{\alpha/2}$.

7.5.2 Comparing K Treatment Means

This section deals with the more general one-factorial analysis of variance in presence of $K \geq 2$ groups. We assume that K samples are given, each of size n_k for $k = 1, \ldots, K$. The following quantities are relevant for testing the null hypothesis of equal group (treatment) means.

Sample means in treatment group k:

$$\overline{Y}_k = \frac{1}{n_k} \sum_i Y_{ki}$$

Sum of squares (SS) within treatment group k :

$$S_k = \sum_i (Y_{ki} - \overline{Y}_k)^2$$

Sample variance within treatment group k :

$$s_k^2 = S_k/(n_k - 1)$$

Denoting the pooled sample size by $N = n_1 + \ldots + n_K$, we furthermore define the following pooled measures.
Pooled (grand) mean:

$$\overline{Y} = \frac{1}{N}\sum_k \sum_i Y_{ki}$$

Within-treatment sum of squares:

$$S_R = S_1 + \ldots + S_K$$

Between treatment sum of squares:

$$S_T = \sum_k n_k(\overline{Y}_k - \overline{Y})^2$$

Within- and between-treatment mean square:

$$s_R^2 = \frac{S_R}{N-k} \qquad s_T^2 = \frac{S_T}{K-1}$$

In analogy to Lemma 7.5.1, the following results regarding the *decomposition of spread* holds.

Lemma 7.5.2. *Let the overall variation in the data be defined by*

$$S_D \stackrel{\text{def}}{=} \sum_k \sum_i (Y_{ki} - \overline{Y})^2.$$

Then it holds

$$S_D = \sum_k \sum_i (Y_{ki} - \overline{Y}_k)^2 + \sum_k n_k(\overline{Y}_k - \overline{Y})^2 = S_T + S_R.$$

Exercise 7.5.2. Prove Lemma 7.5.2.

The variance components in a one-factorial analysis of variance can be summarized in an analysis of variance table, cf. Table 7.1.

Table 7.1 Variance components in a one-factorial analysis of variance

Source of variation	Sum of squares	Degrees of freedom	Mean square
Average	$S_A = N\overline{Y}^2$	$\nu_A = 1$	$s_A^2 = S_A/\nu_A$
Between treatments	$S_T = \sum_k n_k(\overline{Y}_k - \overline{Y})^2$	$\nu_T = K - 1$	$s_T^2 = S_T/\nu_T$
Within treatments	$S_R = \sum_k \sum_i (Y_{ki} - \overline{Y}_k)^2$	$\nu_R = N - K$	$s_R^2 = S_R/\nu_R$
Total	$S_D = \sum_k \sum_i (Y_{ki} - \overline{Y})^2$	$N - 1$	

For testing the global hypothesis $H_0\colon \eta_1 = \eta_2 = \ldots = \eta_K$ against the (two-sided) alternative $H_1\colon \exists (i, j)$ with $\eta_i \neq \eta_j$, we again apply Theorem 7.3.1, leading to the following result.

Corollary 7.5.1. *Under the model of the analysis of variance with K groups, assume homogeneous Gaussian noise with noise variance $\sigma^2 > 0$. Then it holds:*

(i) $S_R/\sigma^2 \sim \chi^2_{N-K}$.
(ii) Under H_0, $S_T/\sigma^2 \sim \chi^2_{K-1}$.
(iii) S_R *and* S_T *are stochastically independent.*
(iv) Under H_0, *the statistic* $F = \frac{S_T/(K-1)}{S_R/(N-K)}$ *is distributed as* $F_{K-1,N-K}$:

$$F \stackrel{\text{def}}{=} \frac{S_T/(K-1)}{S_R/(N-K)} \sim F_{K-1,N-K}.$$

Therefore, H_0 *is rejected by a level* α F*-test if the value of* F *exceeds the quantile* $F_{K-1,N-K;1-\alpha}$ *of Fisher's* F*-distribution with* $K-1$ *and* $N-K$ *degrees of freedom.*

7.5.2.1 Treatment Effects

For estimating the amount of shift in the mean response caused by treatment k (the k-th *treatment effect*), it is convenient to re-parametrize the model as follows. We start with the basic model, given by

$$Y_{ki} = \eta_k + \varepsilon_{ki}, \qquad \varepsilon_{ki} \sim \mathcal{N}(0, \sigma^2) \text{ i.i.d.},$$

where η_k is the true *treatment mean*.

Now, we introduce the averaged treatment mean η and the k-th treatment effect τ_k, given by

$$\eta \stackrel{\text{def}}{=} \frac{1}{N} \sum n_k \eta_k, \quad \tau_k \stackrel{\text{def}}{=} \eta_k - \eta.$$

The re-parametrized model is then given by

$$Y_{ki} = \eta + \tau_k + \varepsilon_{ki}. \tag{7.19}$$

It is important to notice that this new model representation involves $K + 1$ unknowns. In order to achieve maximum rank of the design matrix, one therefore has to take the constraint $\sum_{k=1}^{K} n_k \tau_k = 0$ into account when building Ψ, for instance by coding

$$\tau_K = -n_K^{-1} \sum_{k=1}^{K-1} n_k \tau_k.$$

For the data analysis, it is helpful to consider the decomposition of the observed data points:

$$Y_{ki} = \overline{Y} + (\overline{Y}_k - \overline{Y}) + (Y_{ki} - \overline{Y}_k).$$

Routine algebra then leads to the following results concerning inference in the treatment effect representation of the analysis of variance model.

Theorem 7.5.1. *Under Model* (7.19) *with* $\sum_{k=1}^{K} n_k \tau_k = 0$, *it holds:*

(i) The MLEs for the unknown model parameters are given by

$$\tilde{\eta} = \overline{Y}, \quad \tilde{\tau}_k = \overline{Y}_k - \overline{Y}, \ 1 \le k \le K.$$

(ii) The F-statistic for testing the global hypothesis $H_0\colon \tau_1 = \tau_2 = \ldots = \tau_{K-1} = 0$ is identical to the one given in Theorem 7.5.1(iv), i.e., $F = \frac{S_T/(K-1)}{S_R/(N-K)}$.

7.5.3 Randomized Blocks

This section deals with a special case of the two-factorial analysis of variance. We assume that the observational units are grouped into n blocks, where in each block the K treatments under investigation are applied exactly once. The total sample size is then given by $N = nK$. Since there may be a "block effect" on the mean response, we consider the model

$$Y_{ki} = \eta + \beta_i + \tau_k + \varepsilon_{ki}, \quad 1 \le k \le K, \quad 1 \le i \le n, \tag{7.20}$$

where η is the *general mean*, β_i the *block effect*, and τ_k the *treatment effect*.

The data can be decomposed as

$$Y_{ki} = \overline{Y} + (\overline{Y}_i - \overline{Y}) + (\overline{Y}_k - \overline{Y}) + (Y_{ki} - \overline{Y}_i - \overline{Y}_k + \overline{Y}),$$

Table 7.2 Analysis of variance table for a randomized block design

Source of variation	Sum of squares	Degrees of freedom
Average (correction factor)	$S = nK\overline{Y}^2$	1
Between blocks	$S_B = K\sum_i(\overline{Y}_i - \overline{Y})^2$	$n-1$
Between treatments	$S_T = n\sum_k(\overline{Y}_k - \overline{Y})^2$	$K-1$
Residuals	$S_R = \sum_k\sum_i(Y_{ki} - \overline{Y}_i - \overline{Y}_k + \overline{Y})^2$	$(n-1)(K-1)$
Total	$S_D = \sum_k\sum_i(Y_{ki} - \overline{Y})^2$	$N-1 = nK-1$

where $\overline{Y}$ is the *grand average*, $\overline{Y}_i$ the *block average*, $\overline{Y}_k$ the *treatment average*, and $Y_{ki} - \overline{Y}_i - \overline{Y}_k + \overline{Y}$ the *residual*.

Applying decomposition of spread to the model given by (7.20), we arrive at Table 7.2. Based on this, the following corollary summarizes the inference techniques under the model defined by (7.20).

Corollary 7.5.2. *Under the model defined by* (7.20)*, it holds:*

(i) The MLEs for the unknown model parameters are given by

$$\tilde{\eta} = \overline{Y}, \quad \tilde{\beta}_i = \overline{Y}_i - \overline{Y}, \quad \tilde{\tau}_k = \overline{Y}_k - \overline{Y}.$$

(ii) The F-statistic for testing the hypothesis H_B of no block effects is given by

$$F_B = \frac{S_B/(n-1)}{S_R/[(n-1)(K-1)]}.$$

Under H_B, the distribution of F_B is Fisher's F-distribution with $(n-1)$ and $(n-1)(K-1)$ degrees of freedom.

(iii) The F-statistic for testing the null hypothesis H_T of no treatment effects is given by

$$F_T = \frac{S_T/(K-1)}{S_R/[(n-1)(K-1)]}.$$

Under H_T, the distribution of F_T is Fisher's F-distribution with $(K-1)$ and $(n-1)(K-1)$ degrees of freedom.

7.6 Historical Remarks and Further Reading

Hotelling (1931) derived a deterministic transformation of *Fisher's F -distribution* and demonstrated its usage in the context of testing for differences among several Gaussian means with a *likelihood ratio test*. The general idea of the *Wald test* goes back to Wald (1943).

A classical textbook on the *analysis of variance* is that of Scheffé (1959). The general theory of *testing linear hypotheses* in linear models is described, e.g., in the textbook by Searle (1971).

Chapter 8
Some Other Testing Methods

This chapter discusses some nonparametric testing methods. First, we treat classical testing procedures such as the Kolmogorov–Smirnov and the Cramér–Smirnov–von Mises test as particular cases of the substitution approach. Then, we are considered with Bayesian approaches towards hypothesis testing. Finally, Sect. 8.4 deals with locally best tests. It is demonstrated that the score function is the natural equivalent to the LR statistic if no uniformly best tests exist, but locally best tests are aimed at, assuming that the model is differentiable in the mean. Conditioning on the ranks of the observations leads to the theory of rank tests. Due to the close connection of rank tests and permutation tests (the null distribution of a rank test is a permutation distribution), we end the chapter with some general remarks on permutation tests.

Let $\boldsymbol{Y} = (Y_1, \ldots, Y_n)^\top$ be an i.i.d. sample from a distribution P. The joint distribution $\mathbb{P}$ of $\boldsymbol{Y}$ is the n-fold product of P, so a hypothesis about $\mathbb{P}$ can be formulated as a hypothesis about the marginal measure P. A simple hypothesis H_0 means the assumption that $P = P_0$ for a given measure P_0. The empirical measure P_n is a natural empirical counterpart of P leading to the idea of testing the hypothesis by checking whether P_n significantly deviates from P_0. As in the estimation problem, this substitution idea can be realized in several different ways. We briefly discuss below the method of moments and the minimal distance method.

8.1 Method of Moments for an i.i.d. Sample

Let $\boldsymbol{g}(\cdot)$ be any d-vector function on $\mathbb{R}^1$. The assumption $P = P_0$ leads to the population moment

$$\boldsymbol{m}_0 = \mathbb{E}_0 \boldsymbol{g}(Y_1).$$

V. Spokoiny and T. Dickhaus, *Basics of Modern Mathematical Statistics*, Springer Texts in Statistics, DOI 10.1007/978-3-642-39909-1_8,

The empirical counterpart of this quantity is given by

$$\boldsymbol{M}_n = \mathbb{E}_n \boldsymbol{g}(Y) = \frac{1}{n} \sum_i \boldsymbol{g}(Y_i).$$

The method of moments (MOM) suggests to consider the difference $\boldsymbol{M}_n - \boldsymbol{m}_0$ for building a reasonable test. The properties of $\boldsymbol{M}_n$ were stated in Sect. 2.4. In particular, under the null $P = P_0$, the first two moments of the vector $\boldsymbol{M}_n - \boldsymbol{m}_0$ can be easily computed: $\mathbb{E}_0(\boldsymbol{M}_n - \boldsymbol{m}_0) = 0$ and

$$\begin{aligned} \operatorname{Var}_0(\boldsymbol{M}) &= \mathbb{E}_0\bigl[(\boldsymbol{M}_n - \boldsymbol{m}_0)\,(\boldsymbol{M}_n - \boldsymbol{m}_0)^\top\bigr] = n^{-1} V, \\ V &\stackrel{\text{def}}{=} E_0\bigl[\bigl(\boldsymbol{g}(Y) - \boldsymbol{m}_0\bigr)\bigl(\boldsymbol{g}(Y) - \boldsymbol{m}_0\bigr)^\top\bigr]. \end{aligned}$$

For simplicity of presentation we assume that the moment function $\boldsymbol{g}$ is selected to ensure a non-degenerate matrix V. Standardization by the covariance matrix leads to the vector

$$\boldsymbol{\xi}_n = n^{1/2} V^{-1/2} (\boldsymbol{M}_n - \boldsymbol{m}_0),$$

which has under the null measure zero mean and a unit covariance matrix. Moreover, $\boldsymbol{\xi}_n$ is, under the null hypothesis, asymptotically standard normal, i.e., its distribution is approximately standard normal if the sample size n is sufficiently large; see Theorem 2.4.4. The MOM test rejects the null hypothesis if the vector $\boldsymbol{\xi}_n$ computed from the available data $\boldsymbol{Y}$ is very unlikely standard normal, that is, if it deviates significantly from zero. We specify the procedure separately for the univariate and multivariate cases.

8.1.1 Univariate Case

Let $g(\cdot)$ be a univariate function with $E_0 g(Y) = m_0$ and $E_0\bigl[g(Y) - m_0\bigr]^2 = \sigma^2$. Define the linear test statistic

$$T_n = \frac{1}{\sqrt{n\sigma^2}} \sum_i \bigl[g(Y_i) - m_0\bigr] = n^{1/2} \sigma^{-1} (M_n - m_0)$$

leading to the test

$$\phi = \mathbf{1}\bigl(|T_n| > z_{\alpha/2}\bigr), \tag{8.1}$$

where z_α denotes the upper α-quantile of the standard normal law.

Theorem 8.1.1. *Let Y be an i.i.d. sample from P. Then the test statistic T_n is asymptotically standard normal under the null and the test ϕ from* (8.1) *for $H_0 : P = P_0$ is of asymptotic level α, that is,*

$$\mathbb{P}_0(\phi = 1) \to \alpha, \qquad n \to \infty.$$

Similarly one can consider a one-sided alternative $H_1^+ : m > m_0$ or $H_1^- : m < m_0$ about the moment $m = Eg(Y)$ of the distribution P and the corresponding one-sided tests

$$\phi^+ = \mathbf{1}(T_n > z_\alpha), \qquad \phi^- = \mathbf{1}(T_n < -z_\alpha).$$

As in Theorem 8.1.1, both tests ϕ^+ and ϕ^- are of asymptotic level α.

8.1.2 Multivariate Case

The components of the vector function $\boldsymbol{g}(\cdot) \in \mathbb{R}^d$ are usually associated with "directions" in which the null hypothesis is tested. The multivariate situation means that we test simultaneously in $d > 1$ directions. The most natural test statistic is the squared Euclidean norm of the standardized vector $\boldsymbol{\xi}_n$:

$$T_n \stackrel{\text{def}}{=} \|\boldsymbol{\xi}_n\|^2 = n\|V^{-1/2}(\boldsymbol{M}_n - \boldsymbol{m}_0)\|^2. \tag{8.2}$$

By Theorem 2.4.4 the vector $\boldsymbol{\xi}_n$ is asymptotically standard normal so that T_n is asymptotically chi-squared with d degrees of freedom. This yields the natural definition of the test ϕ using quantiles of χ_d^2, i.e.,

$$\phi = \mathbf{1}(T_n > \mathfrak{z}_\alpha) \tag{8.3}$$

with $\mathfrak{z}_\alpha$ denoting the upper α-quantile of the χ_d^2 distribution.

Theorem 8.1.2. *Let Y be an i.i.d. sample from P. If $\mathfrak{z}_\alpha$ fulfills $\mathbb{P}(\chi_d^2 > \mathfrak{z}_\alpha) = \alpha$, then the test statistic T_n from* (8.2) *is asymptotically χ_d^2-distributed under the null and the test ϕ from* (8.3) *for $H_0 : P = P_0$ is of asymptotic level α.*

8.1.3 Series Expansion

A standard method of building the moment tests or, alternatively, of choosing the directions $\boldsymbol{g}(\cdot)$ is based on some series expansion. Let $\psi_1, \psi_2, \ldots,$ be a given set of basis functions in the related functional space. It is especially useful to select these basis functions to be orthonormal under the measure P_0:

$$\int \psi_j(y)P_0(dy) = 0, \qquad \int \psi_j(y)\psi_{j'}(y)P_0(dy) = \delta_{j,j'}, \qquad \forall j, j'. \tag{8.4}$$

Select a fixed index d and take the first d basis functions $\psi_1, \dots, \psi_d$ as "directions" or components of g. Then

$$m_{j,0} \stackrel{\text{def}}{=} \int \psi_j(y)P_0(dy) = 0$$

is the j th population moment under the null hypothesis H_0 and it is tested by checking whether the empirical moments $M_{j,n}$ with

$$M_{j,n} \stackrel{\text{def}}{=} \frac{1}{n}\sum_i \psi_j(Y_i)$$

do not deviate significantly from zero. The condition (8.4) effectively permits to test each direction ψ_j independently of the others.

For each d one obtains a test statistic $T_{n,d}$ with

$$T_{n,d} \stackrel{\text{def}}{=} n\bigl(M_{1,n}^2 + \ldots + M_{d,n}^2\bigr)$$

leading to the test

$$\phi_d = \mathbf{1}\bigl(T_{n,d} > \mathfrak{z}_{\alpha,d}\bigr),$$

where $\mathfrak{z}_{\alpha,d}$ is the upper α-quantile of χ^2_d. In practical applications the choice of d is particularly relevant and is subject of various studies.

8.1.4 Testing a Parametric Hypothesis

The method of moments can be extended to the situation when the null hypothesis is parametric: $H_0 : P \in (P_{\boldsymbol{\theta}}, \boldsymbol{\theta} \in \Theta_0)$. It is natural to apply the method of moments both to estimate the parameter $\boldsymbol{\theta}$ under the null and to test the null. So, we assume two different moment vector functions $\boldsymbol{g}_0$ and $\boldsymbol{g}_1$ to be given. The first one is selected to fulfill

$$\boldsymbol{\theta} \equiv E_{\boldsymbol{\theta}}\boldsymbol{g}_0(Y_1), \qquad \boldsymbol{\theta} \in \Theta_0 .$$

This permits estimating the parameter $\boldsymbol{\theta}$ directly by the empirical moment:

$$\tilde{\boldsymbol{\theta}} = \frac{1}{n} \sum_i g_0(Y_i).$$

The second vector of moment functions is composed by directional alternatives. An identifiability condition suggests to select the directional alternative functions orthogonal to g_0 in the following sense. We choose $\boldsymbol{g}_1 = (\boldsymbol{g}_1^{(1)}, \ldots, \boldsymbol{g}_1^{(k)})^\top : \mathbb{R}^r \to \mathbb{R}^k$ such that for all $\boldsymbol{\theta} \in \Theta_0$ it holds $\boldsymbol{g}_1(m_1, \ldots, m_r) = 0 \in \mathbb{R}^k$, where $(m_\ell : 1 \le \ell \le r)$ denote the first r (population) moments of the distribution $P_{\boldsymbol{\theta}}$.

Theorem 8.1.3. *Let $\tilde{m}_\ell = n^{-1} \sum_{i=1}^n Y_i^\ell$ denote the ℓ-th sample moment for $1 \le \ell \le r$. Then, under regularity assumptions discussed in Sect. 2.4 and assuming that each $\boldsymbol{g}_1^{(j)}$ is continuously differentiable and that all $(m_\ell : 1 \le \ell \le 2r)$ are continuous functions of $\boldsymbol{\theta}$, it holds that the distribution of*

$$T_n \stackrel{\text{def}}{=} n\, \boldsymbol{g}_1^\top(\tilde{m}_1, \ldots, \tilde{m}_r)\, V^{-1}(\tilde{\boldsymbol{\theta}})\, \boldsymbol{g}_1(\tilde{m}_1, \ldots, \tilde{m}_r)$$

converges under H_0 weakly to χ_k^2, where

$$V(\boldsymbol{\theta}) \stackrel{\text{def}}{=} J(\boldsymbol{g}_1) \Sigma J(\boldsymbol{g}_1)^\top \in \mathbb{R}^{k \times k},$$

$$J(\boldsymbol{g}_1) \stackrel{\text{def}}{=} \left(\frac{\partial \boldsymbol{g}_1^{(j)}(m_1, \ldots, m_r)}{\partial m_\ell} \right)_{\substack{1 \le j \le k \\ 1 \le \ell \le r}} \in \mathbb{R}^{k \times r}$$

and $\Sigma = (\sigma_{ij}) \in \mathbb{R}^{r \times r}$ with $\sigma_{ij} = m_{i+j} - m_i m_j$.

Theorem 8.1.3, which is an application of the Delta method in connection with the asymptotic normality of MOM estimators, leads to the goodness-of-fit test

$$\phi = \mathbf{1}(T_n > \mathfrak{z}_\alpha),$$

where $\mathfrak{z}_\alpha$ is the upper α-quantile of χ_k^2, for testing H_0.

8.2 Minimum Distance Method for an i.i.d. Sample

The method of moments is especially useful for the case of a simple hypothesis because it compares the population moments computed under the null with their empirical counterpart. However, if a more complicated composite hypothesis is tested, the population moments cannot be computed directly: the null measure is not specified precisely. In this case, the minimum distance idea appears to be useful. Let $(P_{\boldsymbol{\theta}}, \boldsymbol{\theta} \in \Theta \subset \mathbb{R}^p)$ be a parametric family and Θ_0 be a subset of Θ. The null hypothesis about an i.i.d. sample Y from P is that $P \in (P_{\boldsymbol{\theta}}, \boldsymbol{\theta} \in \Theta_0)$. Let $\rho(P, P')$ denote some functional (distance) defined for measures P, P' on the real line. We assume that ρ satisfies the following conditions: $\rho(P_{\boldsymbol{\theta}_1}, P_{\boldsymbol{\theta}_2}) \geq 0$ and $\rho(P_{\boldsymbol{\theta}_1}, P_{\boldsymbol{\theta}_2}) = 0$ iff $\boldsymbol{\theta}_1 = \boldsymbol{\theta}_2$. The condition $P \in (P_{\boldsymbol{\theta}}, \boldsymbol{\theta} \in \Theta_0)$ can be rewritten in the form

$$\inf_{\boldsymbol{\theta} \in \Theta_0} \rho(P, P_{\boldsymbol{\theta}}) = 0.$$

Now we can apply the substitution principle: use P_n in place of P. Define the value T by

$$T \stackrel{\text{def}}{=} \inf_{\boldsymbol{\theta} \in \Theta_0} \rho(P_n, P_{\boldsymbol{\theta}}). \tag{8.5}$$

Large values of the test statistic T indicate a possible violation of the null hypothesis.

In particular, if H_0 is a simple hypothesis, that is, if the set Θ_0 consists of one point $\boldsymbol{\theta}_0$, the test statistic reads as $T = \rho(P_n, P_{\boldsymbol{\theta}_0})$. The critical value for this test is usually selected by the level condition:

$$\mathbb{P}_{\boldsymbol{\theta}_0}\bigl(\rho(P_n, P_{\boldsymbol{\theta}_0}) > t_\alpha\bigr) \leq \alpha.$$

Note that the test statistic (8.5) can be viewed as a combination of two different steps. First we estimate under the null the parameter $\boldsymbol{\theta} \in \Theta_0$ which provides the best possible parametric fit under the assumption $P \in (P_{\boldsymbol{\theta}}, \boldsymbol{\theta} \in \Theta_0)$:

$$\tilde{\boldsymbol{\theta}}_0 = \operatorname*{arginf}_{\boldsymbol{\theta} \in \Theta_0} \rho(P_n, P_{\boldsymbol{\theta}}).$$

Next we formally apply the minimum distance test with the simple hypothesis given by $\boldsymbol{\theta}_0 = \tilde{\boldsymbol{\theta}}_0$.

Below we discuss some standard choices of the distance ρ.

8.2.1 Kolmogorov–Smirnov Test

Let P_0, P_1 be two distributions on the real line with distribution functions F_0, F_1: $F_j(y) = P_j(Y \leq y)$ for $j = 0, 1$. Define

$$\rho(P_0, P_1) = \rho(F_0, F_1) = \sup_y \bigl| F_0(y) - F_1(y) \bigr|. \tag{8.6}$$

Now consider the related test starting from the case of a simple null hypothesis $P = P_0$ with corresponding c.d.f. F_0. Then the distance ρ from (8.6) (properly scaled) leads to the *Kolmogorov–Smirnov test statistic*

$$T_n \stackrel{\text{def}}{=} \sup_y n^{1/2} \bigl| F_0(y) - F_n(y) \bigr|.$$

A nice feature of this test is the property of asymptotic pivotality.

Theorem 8.2.1 (Kolmogorov). *Let F_0 be a continuous c.d.f. Then*

$$T_n = \sup_y n^{1/2} \bigl| F_0(y) - F_n(y) \bigr| \xrightarrow{\mathcal{D}} \eta,$$

where η is a fixed random variable (maximum of a Brownian bridge on $[0,1]$).

Proof. Idea of the proof: The c.d.f. F_0 is monotonic and continuous. Therefore, its inverse function F_0^{-1} is uniquely defined. Consider the r.v.'s

$$U_i \stackrel{\text{def}}{=} F_0(Y_i).$$

The basic fact about this transformation is that the U_i 's are i.i.d. uniform on the interval $[0,1]$.

Lemma 8.2.1. *The r.v.'s U_i are i.i.d. with values in $[0,1]$ and for any $u \in [0,1]$ it holds*

$$\mathbb{P}\bigl(U_i \le u\bigr) = u.$$

By definition of F_0^{-1}, it holds for any $u \in [0,1]$

$$F_0\bigl(F_0^{-1}(u)\bigr) \equiv u.$$

Moreover, if G_n is the c.d.f. of the U_i 's, that is, if

$$G_n(u) \stackrel{\text{def}}{=} \frac{1}{n} \sum_i \mathbf{1}(U_i \le u),$$

then

$$G_n(u) \equiv F_n\bigl[F_0^{-1}(u)\bigr]. \tag{8.7}$$

Exercise 8.2.1. Check Lemma 8.2.1 and (8.7).

Now by the change of variable $y = F_0^{-1}(u)$ we obtain

$$T_n = \sup_{u \in [0,1]} n^{1/2} \bigl| F_0(F_0^{-1}(u)) - F_n(F_0^{-1}(u)) \bigr| = \sup_{u \in [0,1]} n^{1/2} \bigl| u - G_n(u) \bigr|.$$

It is obvious that the right-hand side of this expression does not depend on the original model. Actually, it is for fixed n a precisely described random variable, and so its distribution only depends on n. It only remains to show that this distribution for large n is close to some fixed limit distribution with a continuous c.d.f. allowing for a choice of a proper critical value. We indicate the main steps of the proof.

Given a sample $U_1, \ldots, U_n$, define the random function

$$\xi_n(u) \stackrel{\text{def}}{=} n^{1/2}\big[u - G_n(u)\big].$$

Clearly $T_n = \sup_{u\in[0,1]} \xi_n(u)$. Next, convergence of the random functions $\xi_n(\cdot)$ would imply the convergence of their maximum over $u \in [0,1]$, because the maximum is a continuous functional of a function. Finally, the weak convergence of $\xi_n(\cdot) \xrightarrow{w} \xi(\cdot)$ can be checked if for any continuous function $h(u)$, it holds

$$\langle \xi_n, h\rangle \stackrel{\text{def}}{=} n^{1/2}\int_0^1 h(u)\big[u - G_n(u)\big]du \xrightarrow{w} \langle \xi, h\rangle \stackrel{\text{def}}{=} \int_0^1 h(u)\xi(u)du.$$

Now the result can be derived from the representation

$$\langle \xi_n, h\rangle = n^{1/2}\int_0^1 \big[h(u)G_n(u) - m(h)\big]du = n^{-1/2}\sum_{i=1}^n \big[U_i h(U_i) - m(h)\big]$$

with $m(h) = \int_0^1 h(u)\xi(u)du$ and from the central limit theorem for a sum of i.i.d. random variables.

8.2.1.1 The Case of a Composite Hypothesis

If $H_0 : P \in (P_{\boldsymbol{\theta}}, \boldsymbol{\theta} \in \Theta_0)$ is considered, then the test statistic is described by (8.5). As we already mentioned, testing of a composite hypothesis can be viewed as a two-step procedure. In the first step, $\boldsymbol{\theta}$ is estimated by $\tilde{\boldsymbol{\theta}}_0$ and in the second step, the goodness-of-fit test based on T_n is carried out, where F_0 is replaced by the c.d.f. corresponding to $P_{\tilde{\boldsymbol{\theta}}_0}$. It turns out that pivotality of the distribution of T_n is preserved if $\boldsymbol{\theta}$ is a location and/or scale parameter, but a general (asymptotic) theory allowing to derive t_α analytically is not available. Therefore, computer simulations are typically employed to approximate t_α.

8.2.2 ω^2 Test (Cramér–Smirnov–von Mises)

Here we briefly discuss another distance also based on the c.d.f. of the null measure. Namely, define for a measure P on the real line with c.d.f. F

$$\rho(P_n, P) = \rho(F_n, F) = n\int \big[F_n(y) - F(y)\big]^2 dF(x). \tag{8.8}$$

For the case of a simple hypothesis $P = P_0$, the *Cramér–Smirnov–von Mises* (CSvM) test statistic is given by (8.8) with $F = F_0$. This is another functional of the path of the random function $n^{1/2}\bigl[F_n(y) - F_0(y)\bigr]$. The Kolmogorov test uses the maximum of this function while the CSvM test uses the integral of this function squared. The property of pivotality is preserved for the CSvM test statistic as well.

Theorem 8.2.2. *Let F_0 be a continuous c.d.f. Then*

$$T_n = n \int \bigl[F_n(y) - F_0(y)\bigr]^2 dF(x) \xrightarrow{\mathcal{D}} \eta,$$

where η is a fixed random variable (integral of a Brownian bridge squared on $[0, 1]$).

Proof. The idea of the proof is the same as in the case of the Kolmogorov–Smirnov test. First the transformation by F_0^{-1} translates the general case to the case of the uniform distribution on $[0, 1]$. Next one can again use the functional convergence of the process $\xi_n(u)$.

8.3 Partially Bayes Tests and Bayes Testing

In the above sections we mostly focused on the likelihood ratio testing approach. As in estimation theory, the LR approach is very general and possesses some nice properties. This section briefly discusses some possible alternative approaches including partially Bayes and Bayes approaches.

8.3.1 Partial Bayes Approach and Bayes Tests

Let Θ_0 and Θ_1 be two subsets of the parameter set Θ. We test the null hypothesis $H_0 : \boldsymbol{\theta}^* \in \Theta_0$ against the alternative $H_1 : \boldsymbol{\theta}^* \in \Theta_1$. The LR approach compares the maximum of the likelihood process over Θ_0 with the similar maximum over Θ_1. Let now two measures π_0 on Θ_0 and π_1 on Θ_1 be given. Now instead of the maximum of $L(\boldsymbol{Y}, \boldsymbol{\theta})$ we consider its weighted sum (integral) over Θ_0 (resp. Θ_1) with weights $\pi_0(\theta)$ resp. $\pi_1(\theta)$. More precisely, we consider the value

$$T_{\pi_0,\pi_1} = \int_{\Theta_1} L(\boldsymbol{Y}, \boldsymbol{\theta})\pi_1(\boldsymbol{\theta})\lambda(d\boldsymbol{\theta}) - \int_{\Theta_0} L(\boldsymbol{Y}, \boldsymbol{\theta})\pi_0(\boldsymbol{\theta})\lambda(d\boldsymbol{\theta}).$$

Significantly positive values of this expression indicate that the null hypothesis is likely to be false.

Similarly and more commonly used, we may define measures g_0 and g_1 such that

$$\pi(\boldsymbol{\theta}) = \begin{cases} \pi_0 g_0(\boldsymbol{\theta}), & \boldsymbol{\theta} \in \Theta_0, \\ \pi_1 g_1(\boldsymbol{\theta}), & \boldsymbol{\theta} \in \Theta_1, \end{cases}$$

where π is a prior on the entire parameter space Θ and $\pi_i \stackrel{\text{def}}{=} \int_{\Theta_i} \pi(\boldsymbol{\theta})\lambda(d\boldsymbol{\theta}) = \mathbb{P}(\Theta_i)$ for $i = 0, 1$. Then, the *Bayes factor* for comparing H_0 and H_1 is given by

$$B_{0|1} = \frac{\int_{\Theta_0} L(\boldsymbol{Y}, \boldsymbol{\theta}) g_0(\boldsymbol{\theta}) \lambda(d\boldsymbol{\theta})}{\int_{\Theta_1} L(\boldsymbol{Y}, \boldsymbol{\theta}) g_1(\boldsymbol{\theta}) \lambda(d\boldsymbol{\theta})} = \frac{\mathbb{P}(\Theta_0 \,|\, \boldsymbol{Y}) / \mathbb{P}(\Theta_1 \,|\, \boldsymbol{Y})}{\mathbb{P}(\Theta_0) / \mathbb{P}(\Theta_1)}. \tag{8.9}$$

The representation of the Bayes factor on the right-hand side of (8.9) shows that it can be interpreted as the ratio of the posterior odds for H_0 and the prior odds for H_0. The resulting test rejects the null hypothesis for significantly small values of $B_{0|1}$, or, equivalently, for significantly large values of $B_{1|0} = 1/B_{0|1}$. In the special case that H_0 and H_1 are two simple hypotheses, i.e., $\Theta_0 = \{\boldsymbol{\theta}_0\}$ and $\Theta_1 = \{\boldsymbol{\theta}_1\}$, the Bayes factor is simply given by

$$B_{0|1} = \frac{L(\boldsymbol{Y}, \boldsymbol{\theta}_0)}{L(\boldsymbol{Y}, \boldsymbol{\theta}_1)},$$

hence, in such a case the testing approach based on the Bayes factor is equivalent to the LR approach.

8.3.2 Bayes Approach

Within the Bayes approach the true data distribution and the true parameter value are not defined. Instead one considers the prior and posterior distribution of the parameter. The parametric Bayes model can be represented as

$$\boldsymbol{Y} \mid \boldsymbol{\theta} \sim p(\boldsymbol{y} \,|\, \boldsymbol{\theta}), \qquad \boldsymbol{\theta} \sim \pi(\boldsymbol{\theta}).$$

The posterior density $p(\boldsymbol{\theta} \,|\, \boldsymbol{Y})$ can be computed via the Bayes formula:

$$p(\boldsymbol{\theta} \,|\, \boldsymbol{Y}) = \frac{p(\boldsymbol{Y} \,|\, \boldsymbol{\theta}) \pi(\boldsymbol{\theta})}{p(\boldsymbol{Y})}$$

with the marginal density $p(\boldsymbol{Y}) = \int_{\Theta} p(\boldsymbol{Y} \,|\, \boldsymbol{\theta}) \pi(\boldsymbol{\theta}) \lambda(d\boldsymbol{\theta})$. The Bayes approach suggests instead of checking the hypothesis about the location of the parameter $\boldsymbol{\theta}$ to look directly at the posterior distribution. Namely, one can construct the so-called *credible sets* which contain a prespecified fraction, say $1 - \alpha$, of the mass of the whole posterior distribution. Then one can say that the probability for the parameter

θ to lie outside of this credible set is at most α. So, the testing problem in the frequentist approach is replaced by the problem of confidence estimation for the Bayes method.

Example 8.3.1 (Example 5.2.2 Continued). Consider again the situation of a Bernoulli product likelihood for $Y = (Y_1, \ldots, Y_n)^\top$ with unknown success probability θ. In example 5.2.2 we saw that this family of likelihoods is conjugated to the family of beta distributions as priors on $[0, 1]$. More specifically, if $\theta \sim \text{Beta}(a, b)$, then $\theta \mid Y = y \sim \text{Beta}(a + s, b + n - s)$, where $s = \sum_{i=1}^n y_i$ denotes the observed number of successes. Under quadratic risk, the Bayes-optimal point estimate for θ is given by $\mathbb{E}[\theta \mid Y = y] = (a + s)/(a + b + n)$, and a credible interval can be constructed around this value by utilizing quantiles of the posterior $\text{Beta}(a + s, b + n - s)$-distribution.

8.4 Score, Rank, and Permutation Tests

8.4.1 Score Tests

Testing a composite null hypothesis H_0 against a composite alternative H_1 is in general a challenging problem, because only in some special cases uniformly (over $\theta \in H_1$) most powerful level α-tests exist. In all other cases, one has to decide against which regions in H_1 optimal power is targeted. One class of procedures is given by locally best tests, optimizing power in regions close to H_0. To formalize this class mathematically, one needs the concept of differentiability in the mean.

Definition 8.4.1 (Differentiability in the Mean). Let $(\mathcal{Y}, \mathcal{B}(\mathcal{Y}), (\mathbb{P}_\theta)_{\theta \in \Theta})$ denote a statistical model and assume that $(\mathbb{P}_\theta)_{\theta \in \Theta}$ is a dominated (by μ) family of measures, where $\Theta \subseteq \mathbb{R}$. Then, $(\mathcal{Y}, \mathcal{B}(\mathcal{Y}), (\mathbb{P}_\theta)_{\theta \in \Theta})$ is called differentiable in the mean in $\theta_0 \in \mathring{\Theta}$, if a function $g \in L_1(\mu)$ exists with

$$\left\| t^{-1} \left(\frac{d\mathbb{P}_{\theta_0 + t}}{d\mu} - \frac{d\mathbb{P}_{\theta_0}}{d\mu} \right) - g \right\|_{L_1(\mu)} \to 0 \quad \text{as } t \to 0.$$

The function g is called $L_1(\mu)$-derivative of $\theta \mapsto \mathbb{P}_\theta$ in θ_0. In the sequel, we choose w.l.o.g. $\theta_0 \equiv 0$.

Theorem 8.4.1 (§18 in Hewitt and Stromberg (1975)). *Under the assumptions of Definition 8.4.1 let $\theta_0 = 0$ and let densities be given by $f_\theta(y) \stackrel{\text{def}}{=} d\mathbb{P}_\theta/(d\mu)(y)$. Assume that there exists an open neighborhood $\mathcal{U}$ of 0 such that for μ-almost all y the mapping $\mathcal{U} \ni \theta \mapsto f_\theta(y)$ is absolutely continuous, i.e., it exists an integrable function $\tau \mapsto \dot{f}(y, \tau)$ on $\mathcal{U}$ with*

$$\int_{\theta_1}^{\theta_2} \dot{f}(y, \tau) d\tau = f_{\theta_2}(y) - f_{\theta_1}(y), \quad \theta_1 < \theta_2,$$

and assume that $\frac{\partial}{\partial \theta} f_\theta(y)|_{\theta=0} = \dot{f}(y,0)$ μ*-almost everywhere. Furthermore, assume that for* $\theta \in \mathcal{U}$ *the function* $y \mapsto \dot{f}(y,\theta)$ *is* μ*-integrable with*

$$\int \left|\dot{f}(y,\theta)\right| d\mu(y) \to \int \left|\dot{f}(y,0)\right| d\mu(y), \quad \theta \to 0.$$

Then, $\theta \mapsto \mathbb{P}_\theta$ *is differentiable in the mean in* 0 *with* $g = \dot{f}(\cdot,0)$.

Theorem 8.4.2. *Under the assumptions of Definition 8.4.1 assume that the densities* $\theta \mapsto f_\theta$ *are differentiable in the mean in* 0 *with* $L_1(\mu)$ *derivative* g. *Then,*

$$\theta^{-1}\log(f_\theta(y)/f_0(y)) = \theta^{-1}(\log f_\theta(y) - \log f_0(y))$$

converges for $\theta \to 0$ *to* $\dot{L}(y)$ *(say) in* $\mathbb{P}_0$ *probability. We call* $\dot{L}$ *the derivative of the (logarithmic) likelihood ratio or* score function. *It holds*

$$\dot{L}(y) = g(y)/f_0(y), \quad \int \dot{L}\, d\mathbb{P}_0 = 0, \{f_0 = 0\} \subseteq \{g = 0\} \quad \mathbb{P}_0\text{-almost surely.}$$

Proof. $\theta^{-1}(f_\theta/f_0 - 1) \to g/f_0$ converges in $L_1(\mathbb{P}_0)$ and, consequently, in $\mathbb{P}_0$ probability. The chain rule yields $\dot{L}(y) = g(y)/f_0(y)$. Noting that $\int (f_\theta - f_0) d\mu = 0$ we conclude

$$\int \dot{L}\, d\mathbb{P}_0 = \int g\, d\mu = 0.$$

Example 8.4.1. (a) Location parameter model:
Let $Y = \theta + X, \theta \geq 0$, and assume that X has a density f which is absolutely continuous with respect to the Lebesgue measure and does not dependent on θ. Then, the densities $\theta \mapsto f(y-\theta)$ of Y under θ are differentiable in the mean in zero with score function $\dot{L}(y) = -f'(y)/f(y)$ (differentiation with respect to y).

(b) Scale parameter model:
Let $Y = \exp(\theta)X$ and assume again that X has density f with the properties stated in part (a). Moreover, assume that $\int |xf'(x)|\, dx < \infty$. Then, the densities $\theta \mapsto \exp(-\theta) f(y \exp(-\theta))$ of Y under θ are differentiable in the mean in zero with score function $\dot{L}(y) = -(1 + yf'(y)/f(y))$.

Lemma 8.4.1. *Assume that the family* $\theta \mapsto \mathbb{P}_\theta$ *is differentiable in the mean with score function* $\dot{L}$ *in* $\theta_0 = 0$ *and that* c_i, $1 \leq i \leq n$, *are real constants. Then, also* $\theta \mapsto \bigotimes_{i=1}^n \mathbb{P}_{c_i \theta}$ *is differentiable in the mean in zero, and has score function*

$$(y_1, \ldots, y_n)^\top \mapsto \sum_{i=1}^n c_i \dot{L}(y_i).$$

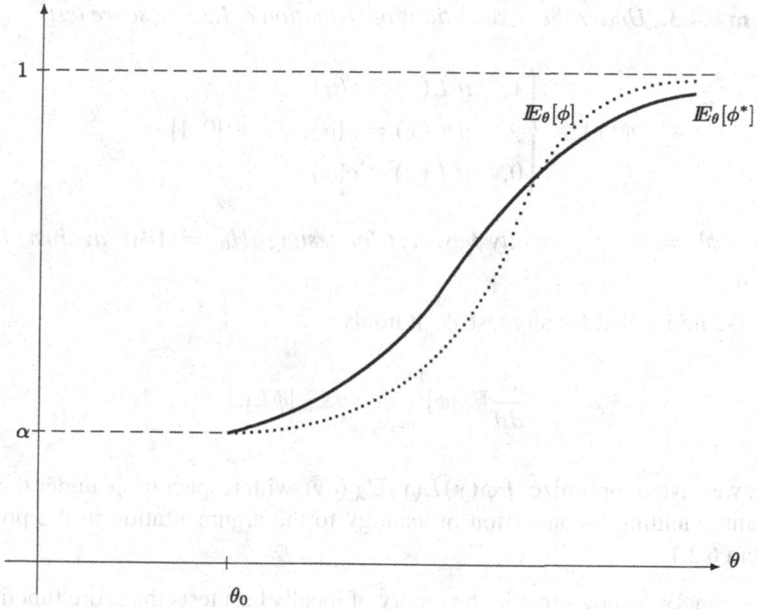

Fig. 8.1 Locally best test ϕ^* with expectation α under θ_0

Exercise 8.4.1. Prove Lemma 8.4.1.

Definition 8.4.2 (Score Test). Let $\theta \mapsto \mathbb{P}_\theta$ be differentiable in the mean in θ_0 with score function $\dot{L}$. Then, every test ϕ of the form

$$\phi(y) = \begin{cases} 1, & \text{if } \dot{L}(y) > c, \\ \gamma, & \text{if } \dot{L}(y) = c, \\ 0, & \text{if } \dot{L}(y) < c, \end{cases}$$

is called a score test. In this, $\gamma \in [0, 1]$ denotes a randomization constant.

Definition 8.4.3 (Locally Best Test). Let $(\mathbb{P}_\theta)_{\theta \in \Theta}$ with $\Theta \subseteq \mathbb{R}$ denote a family which is differentiable in the mean in $\theta_0 \in \mathring{\Theta}$. A test ϕ^* with $\mathbb{E}_{\theta_0}[\phi^*] = \alpha$ is called locally best test among all tests with expectation α under θ_0 for the test problem $H_0 = \{\theta_0\}$ versus $H_1 = \{\theta > \theta_0\}$ if

$$\frac{d}{d\theta}\mathbb{E}_\theta[\phi^*]\bigg|_{\theta=\theta_0} \geq \frac{d}{d\theta}\mathbb{E}_\theta[\phi]\bigg|_{\theta=\theta_0}$$

for all tests ϕ with $\mathbb{E}_{\theta_0}[\phi] = \alpha$.

Figure 8.1 illustrates the situation considered in Definition 8.4.3 graphically.

Theorem 8.4.3. *Under the assumptions of Definition 8.4.3, the score test*

$$\phi(y) = \begin{cases} 1, & \text{if } \dot{L}(y) > c(\alpha) \\ \gamma, & \text{if } \dot{L}(y) = c(\alpha), \quad \gamma \in [0,1] \\ 0, & \text{if } \dot{L}(y) < c(\alpha) \end{cases}$$

with $\mathbb{E}_{\theta_0}[\phi] = \alpha$ *is a locally best test for testing* $H_0 = \{\theta_0\}$ *against* $H_1 = \{\theta > \theta_0\}$.

Proof. We notice that for any test ϕ, it holds

$$\frac{d}{d\theta}\mathbb{E}_\theta[\phi]\bigg|_{\theta=\theta_0} = \mathbb{E}_{\theta_0}[\phi \dot{L}].$$

Hence, we have to optimize $\int \phi(y)\dot{L}(y)\mathbb{P}_{\theta_0}(dy)$ with respect to ϕ under the level constraint, yielding the assertion in analogy to the argumentation in the proof of Theorem 6.2.1.

Theorem 8.4.3 shows that in the theory of locally best tests the score function $\dot{L}$ takes the role that the likelihood ratio has in the LR theory. Notice that, for an i.i.d. sample $\boldsymbol{Y} = (Y_1, \ldots, Y_n)^\top$, the joint product measure $(P_\theta)^{\otimes n}$ has score function $(y_1, \ldots, y_n) \mapsto \sum_{i=1}^n \dot{L}(y_i)$ according to Lemma 8.4.1 and Theorem 8.4.3 can be applied to test $H_0 = \{\theta_0\}$ against $H_1 = \{\theta > \theta_0\}$ based on $\boldsymbol{Y}$.

Moreover, for k-sample problems with $k \geq 2$ groups and n jointly independent observations, Lemma 8.4.1 can be utilized to test the homogeneity hypothesis

$$H_0 = \left\{\mathbb{P}^{Y_1} = \mathbb{P}^{Y_2} = \ldots = \mathbb{P}^{Y_n} : \mathbb{P}^{Y_1} \text{ continuous}\right\}. \tag{8.10}$$

To this end, one considers parametric families $\theta \mapsto \mathbb{P}_{n,\theta}$ which belong to H_0 only in case of $\theta = 0$, i.e., $\mathbb{P}_{n,0} \in H_0$. For $\theta \neq 0$, $\mathbb{P}_{n,\theta}$ is a product measure with non-identical factors.

Example 8.4.2. (a) Regression model for a location parameter:
Let $Y_i = c_i\theta + X_i$, $1 \leq i \leq n$, where $\theta \geq 0$. In this, assume that the X_i are i.i.d. with Lebesgue density f which is independent of θ. Now, for a two-sample problem with n_1 observations in the first group and $n_2 = n - n_1$ observations in the second group, we set $c_1 = c_2 = \cdots = c_{n_1} = 1$ and $c_i = 0$ for all $n_1 + 1 \leq i \leq n$. Under alternatives, the observations in the first group are shifted by $\theta > 0$.

(b) Regression model for a scale parameter:
Let $c_i, 1 \leq i \leq n$, denote real regression coefficients and consider the model $Y_i = \exp(c_i\theta)X_i$, $1 \leq i \leq n$, $\theta \in \mathbb{R}$, where we assume again that the X_i are iid with Lebesgue density f which is independent of θ. Then, it holds

$$\frac{d\mathbb{P}_{n,\theta}}{d\lambda^n}(y) = \prod_{i=1}^{n} \exp(-c_i\theta) f(y_i \exp(-c_i\theta)).$$

Under $\theta_0 = 0$, the product measure $\mathbb{P}_{n,0}$ belongs to H_0, while under alternatives it does not.

8.4.2 Rank Tests

In this section, we will consider the case that only the ranks of the observations are trustworthy (or available). Theorem 8.4.4 will be utilized to define resulting rank tests based on parametric families $\mathbb{P}_{n,\theta}$ as considered in Example 8.4.2. It turns out that the score test based on ranks has a very simple structure.

Theorem 8.4.4. *Let $\theta \mapsto \mathbb{P}_\theta$ denote a parametric family which is differentiable in the mean in $\theta_0 = 0$ with respect to some reference measure μ, $L_1(\mu)$-differentiable for short, with score function $\dot{L}$. Furthermore, let $S : \mathcal{Y} \to \mathcal{S}$ measurable. Then, $\theta \mapsto \mathbb{P}_\theta^S$ is $L_1(\mu^S)$-differentiable with score function $s \mapsto \mathbb{E}_0[\dot{L} \mid S = s]$.*

Proof. First, we show that the $L_1(\mu^S)$-derivative of $\theta \mapsto \mathbb{P}_\theta^S$ is given by $s \mapsto \mathbb{E}_\mu[g \mid S = s]$, where g is the $L_1(\mu)$-derivative of $\theta \mapsto \mathbb{P}_\theta$. To this end, notice that

$$\frac{d\mathbb{P}_\theta^S}{d\mu^S}(s) = \mathbb{E}_\mu[f_\theta \mid S = s], \quad \text{where } f_\theta = \frac{d\mathbb{P}_\theta}{d\mu}.$$

Linearity of $\mathbb{E}_\mu[\cdot \mid S]$ and transformation of measures leads to

$$\int \left| \theta^{-1}\left(\frac{d\mathbb{P}_\theta^S}{d\mu^S} - \frac{d\mathbb{P}_0^S}{d\mu^S}\right) - \mathbb{E}_\mu[g \mid S = s] \right| d\mu^S(y)$$
$$= \int |\mathbb{E}_\mu[\theta^{-1}(f_\theta - f_0) - g \mid S]| \, d\mu.$$

Applying Jensen's inequality and Vitali's theorem, we conclude that $s \mapsto \mathbb{E}_\mu[g \mid S = s]$ is $L_1(\mu^S)$-derivative of $\theta \mapsto \mathbb{P}_\theta^S$. Now, the chain rule yields that the score function of $\mathbb{P}_\theta^S$ in zero is given by

$$s \mapsto \mathbb{E}_\mu[g \mid S = s]\{\mathbb{E}_\mu[\frac{d\mathbb{P}_0}{d\mu} \mid S = s]\}^{-1}$$

and the assertion follows by substituting $g = \dot{L}\, d\mathbb{P}_0/(d\mu)$ and verifying that

$$\mathbb{E}_0[\dot{L} \mid S]\, \mathbb{E}_\mu[\frac{d\,\mathbb{P}_0}{d\mu} \mid S] = \mathbb{E}_\mu[\dot{L}\frac{d\,\mathbb{P}_0}{d\mu} \mid S] \quad \mu\text{-almost surely.}$$

For applying the latter theorem to rank statistics, we need to gather some basic facts about ranks and order statistics.

Definition 8.4.4. Let $y = (y_1, \ldots, y_n)$ be a point in $\mathbb{R}^n$. Assume that the y_i are pairwise distinct and denote their ordered values by $y_{1:n} < y_{2:n} < \ldots < y_{n:n}$.

(a) For $1 \le i \le n$, the integer $r_i \equiv r_i(y) \stackrel{\text{def}}{=} \#\{j \in \{1, \ldots, n\} \colon y_j \le y_i\}$ is called the rank of y_i (in y). The vector $r(y) \stackrel{\text{def}}{=} (r_1(y), \ldots, r_n(y)) \in \mathcal{S}_n$ is called rank vector of y.
(b) The inverse permutation $d(y) = (d_1(y), \ldots, d_n(y)) \stackrel{\text{def}}{=} [r(y)]^{-1}$ is called the vector of antiranks of y, and the integer $d_i(y)$ is called antirank of i (the index that corresponds to the i-th smallest observation in y).

Now, let $Y_1, \ldots, Y_n$ with $Y_i : \mathcal{Y}_i \to \mathbb{R}$ be stochastically independent, continuously distributed random variables and denote the joint distribution of $(Y_1, \ldots, Y_n)$ by $\mathbb{P}$.

(c) Because of $\mathbb{P}(\bigcup_{i \ne j}\{Y_i = Y_j\}) = 0$ the following objects are $\mathbb{P}$-almost surely uniquely defined: $Y_{i:n}$ is called i-th order statistic of $\boldsymbol{Y} = (Y_1, \ldots, Y_n)^\top$, $R_i(\boldsymbol{Y}) \stackrel{\text{def}}{=} n\hat{F}_n(Y_i) = r_i(Y_1, \ldots, Y_n)$ is called rank of Y_i, $R(\boldsymbol{Y}) \stackrel{\text{def}}{=} (R_1(\boldsymbol{Y}), \ldots, R_n(\boldsymbol{Y}))^\top$ is called vector of rank statistics of $\boldsymbol{Y}$, $D_i(\boldsymbol{Y}) \stackrel{\text{def}}{=} d_i(Y_1, \ldots, Y_n)$ is called antirank of i with respect to $\boldsymbol{Y}$ and $D(\boldsymbol{Y}) \stackrel{\text{def}}{=} d(\boldsymbol{Y})$ is called vector of antiranks of $\boldsymbol{Y}$.

Lemma 8.4.2. *Under the assumptions of Definition 8.4.4, it holds*

(a) $i = r_{d_i} = d_{r_i}, \quad y_i = y_{r_i:n}, \quad y_{i:n} = y_{d_i}$.
(b) If $Y_1, \ldots, Y_n$ *are exchangeable random variables, then* $R(\boldsymbol{Y})$ *is uniformly distributed on* $\mathcal{S}_n$*, i.e.,* $\mathbb{P}(R(\boldsymbol{Y}) = \sigma) = 1/n!$ *for all permutations* $\sigma = (r_1, \ldots, r_n) \in \mathcal{S}_n$.
(c) If $U_1, \ldots, U_n$ *are i.i.d. with* $U_1 \sim \mathit{UNI}[0,1]$*, and* $Y_i = F^{-1}(U_i)$, $1 \le i \le n$*, for some distribution function* F*, then it holds* $Y_{i:n} = F^{-1}(U_{i:n})$*. If* F *is continuous, then it holds* $R(\boldsymbol{Y}) = R(U_1, \ldots, U_n)$.
(d) If $(Y_1, \ldots, Y_n)$ *are i.i.d. with c.d.f.* F *of* Y_1*, then we have*
   (i) $\mathbb{P}(Y_{i:n} \le y) = \sum_{j=i}^n \binom{n}{j} F(y)^j (1 - F(y))^{n-j}$.
   (ii) $\frac{d\mathbb{P}^{Y_{i:n}}}{d\mathbb{P}^{Y_1}}(y) = n\binom{n-1}{i-1} F(y)^{i-1}(1 - F(y))^{n-i}$*. If* $\mathbb{P}^{Y_1}$ *has Lebesgue density* f*, then* $\mathbb{P}^{Y_{i:n}}$ *has Lebesgue density* $f_{i:n}$*, given by* $f_{i:n}(y) = n\binom{n-1}{i-1} F(y)^{i-1}(1 - F(y))^{n-i} f(y)$.
   (iii) Letting $\mu \stackrel{\text{def}}{=} \mathbb{P}^{Y_1}$, $(Y_{i:n})_{1 \le i \le n}$ *has the joint* μ^n*-density* $(y_1, \ldots, y_n) \mapsto n!\, \mathbb{I}_{\{y_1 < y_2 < \ldots < y_n\}}$*. If* μ *has Lebesgue density* f*, then* $(Y_{i:n})_{1 \le i \le n}$ *has* λ^n*-density* $(y_1, \ldots, y_n) \mapsto n! \prod_{i=1}^n f(y_i)\, \mathbb{I}_{\{y_1 < y_2 < \ldots < y_n\}}$.

Remark 8.4.1. Lemma 8.4.2(c) (quantile transformation) shows the special importance of the distribution of order statistics of i.i.d. UNI[0, 1]-distributed random variables $U_1, \ldots, U_n$. According to Lemma 8.4.2(d), the order statistic $U_{i:n}$ has a Beta $(i, n-i+1)$ distribution with $\mathbb{E}[U_{i:n}] = i/(n+1)$ and $\mathrm{Var}(U_{i:n}) = (i(n-i+1))/((n+1)^2(n+2))$. For computing the joint distribution function of $(U_{1:n}, \ldots, U_{n:n})$, efficient recursive algorithms exist, for instance Bolshev's recursion and Steck's recursion (see Shorack and Wellner (1986), p. 362 ff.).

Theorem 8.4.5. *Let $Y = (Y_1, \ldots, Y_n)^\top$ be a vector of real-valued i.i.d. random variables with continuous $\mu = \mathbb{P}^{Y_1}$.*

(a) The random vectors $R(Y)$ and $(Y_{i:n})_{1 \le i \le n}$ are stochastically independent.
(b) Let $T : \mathbb{R}^n \to \mathbb{R}$ denote a mapping such that the statistic $T(Y)$ is integrable. For any $\sigma = (r_1, \ldots, r_n) \in \mathcal{S}_n$, it holds

$$\mathbb{E}[T(Y) \,|\, R(Y) = \sigma] = \mathbb{E}[T((Y_{r_i:n})_{1 \le i \le n})].$$

Proof. For proving part (a), let $\sigma = (r_1, \ldots, r_n) \in \mathcal{S}_n$ and Borel sets A_i, $1 \le i \le n$, arbitrary but fixed and define $(d_1, \ldots, d_n) \stackrel{\text{def}}{=} \sigma^{-1}$. We note that $R(Y) = \sigma$ if and only if $Y_{d_1} < Y_{d_2} < \ldots < Y_{d_n}$ and that $Y_{d_i} = Y_{i:n} \in A_i$ if and only if $Y_i \in A_{r_i}$.

Define $B \stackrel{\text{def}}{=} \{y \in \mathbb{R}^n : y_1 < y_2 < \ldots < y_n\}$. Then we obtain that

$$\begin{aligned}
&\mathbb{P}\,(R(Y) = \sigma, \forall 1 \le i \le n : Y_{i:n} \in A_i) \\
&= \mathbb{P}\,(\forall 1 \le i \le n : Y_{d_i} \in A_i, (Y_{d_i})_{1 \le i \le n} \in B) \\
&= \int_{\times_{i=1}^n A_{r_i}} \mathbb{I}_B(y_{d_1}, \ldots, y_{d_n}) d\mu^n(y_1, \ldots, y_n) \\
&= \int_{\times_{i=1}^n A_{r_i}} \mathbb{I}_B(y_1, \ldots, y_n) d\mu^n(y_1, \ldots, y_n),
\end{aligned}$$

because μ^n is invariant under the transformation $(y_1, \ldots, y_n) \mapsto (y_{d_1}, \ldots, y_{d_n})$ due to exchangeability.

Summation over all $\sigma \in \mathcal{S}_n$ yields

$$\mathbb{P}\,(\forall 1 \le i \le n : Y_{i:n} \in A_i) = n! \int_{\times_{i=1}^n A_{r_i}} \mathbb{I}_B(y_1, \ldots, y_n) d\mu^n(y_1, \ldots, y_n).$$

Making use of Lemma 8.4.2(b), we conclude

$$\begin{aligned}
&\mathbb{P}\big(R(Y) = \sigma, \forall 1 \le i \le n : Y_{i:n} \in A_i\big) \\
&= \mathbb{P}\big(R(Y) = \sigma\big)\mathbb{P}\big(\forall 1 \le i \le n : Y_{i:n} \in A_i\big),
\end{aligned}$$

hence, the assertion of part (a). For proving part (b), we verify

$$\begin{aligned}\mathbb{E}[T(Y) \mid R(Y)=\sigma] &= \int_{\{R(Y)=\sigma\}} \frac{T(Y)}{\mathbb{P}(R(Y)=\sigma)}\,d\mathbb{P}\\ &= \mathbb{E}[T((Y_{r_i:n})_{1\le i\le n}) \mid R(Y)=\sigma]\\ &= \mathbb{E}[T((Y_{r_i:n})_{1\le i\le n})],\end{aligned}$$

where we used that $Y = (Y_{r_i:n})_{i=1}^n$ if $R(Y)=\sigma$ in the second line and part (a) in the third line.

Now we are ready to apply Theorem 8.4.4 to vectors of rank statistics.

Corollary 8.4.1. *Let* $(\mathbb{P}_\theta)_{\theta\in\Theta}$ *with* $\Theta \subseteq \mathbb{R}$ *denote an* $L_1(\mu)$*-differentiable family with score function* $\dot{L}$ *in* $\theta_0 = 0$. *Let* $Y = (Y_1,\ldots,Y_n)^\top$ *be a sample from* $\mathbb{P}_{n,\theta} = \bigotimes_{i=1}^n \mathbb{P}_{c_i\theta}$. *Then,* $\mathbb{P}_{n,\theta}^R$ *has score function*

$$\begin{aligned}\sigma = (r_1,\ldots,r_n) \mapsto \mathbb{E}_{n,0}\left[\sum_{i=1}^n c_i \dot{L}(Y_i) \,\Big|\, R(Y)=\sigma\right] &= \sum_{i=1}^n c_i \mathbb{E}_{n,0}[\dot{L}(Y_i) \mid R(Y)=\sigma]\\ &= \sum_{i=1}^n c_i \mathbb{E}_{n,0}[\dot{L}(Y_{r_i:n})] = \sum_{i=1}^n c_i a(r_i)\end{aligned}$$

with $\mathbb{E}_{n,0}$ *denoting the expectation with respect to* $\mathbb{P}_{n,0}$ *and scores* $a(i) \stackrel{\text{def}}{=} \mathbb{E}_{n,0}[\dot{L}(Y_{i:n})]$.

Remark 8.4.2. (a) The test statistic $T(Y) = \sum_{i=1}^n c_i a(R_i(Y))$ is called a linear rank statistic.

(b) The hypothesis H_0 from (8.10) leads under conditioning on $R(Y)$ to a simple null hypothesis on $\mathcal{S}_n$, namely, the discrete uniform distribution on $\mathcal{S}_n$, see Lemma 8.4.2(b). Therefore, the critical value $c(\alpha)$ for the rank test $\phi = \phi(R(Y))$, given by

$$\phi(y) = \begin{cases} 1, & \text{if } T(y) > c(\alpha),\\ \gamma, & \text{if } T(y) = c(\alpha),\\ 0, & \text{if } T(y) < c(\alpha),\end{cases}$$

can be computed by traversing all possible permutations $\sigma \in \mathcal{S}_n$ and thereby determining the discrete distribution of $T(Y)$ under H_0. For large n, we can approximate $c(\alpha)$ by only traversing $B < n!$ randomly chosen permutations $\sigma \in \mathcal{S}_n$.

(c) For the scores, it holds $\sum_{i=1}^{n} a(i) = 0$. If $\dot{L}$ is isotone, then it holds $a(1) \leq a(2) \leq \ldots \leq a(n)$.

(d) Due to the relation $Y_{i:n} \stackrel{\mathcal{D}}{=} F^{-1}(U_{i:n})$, the scores are often given in the form $a(i) = \mathbb{E}[\dot{L} \circ F^{-1}(U_{i:n})]$ and the function $\dot{L} \circ F^{-1}$ is called score-generating function. For large n, one can approximately work with $b(i) \stackrel{\text{def}}{=} \dot{L} \circ F^{-1}(\frac{i}{n+1})$ (since $\mathbb{E}[U_{i:n}] = 1/(n+1)$, see Remark 8.4.1) or with $\tilde{b}(i) = n \int_{(i-1)/n}^{i/n} \dot{L} \circ F^{-1}(u) du$ instead of $a(i)$.

In the case that the score function is isotone, rank tests can also be used to test for stochastically ordered distributions in two-sample problems.

Lemma 8.4.3 (Two-Sample Problems with Stochastically Ordered Distributions). *Assume that $a(1) \leq a(2) \leq \ldots \leq a(n)$, cf. Remark 8.4.2(c), and let ϕ denote a rank test at level α for H_0 from* (8.10), *i.e., $\mathbb{E}_{H_0}[\phi] = \alpha$. Assume that $Y_1, \ldots, Y_{n_1}$ are i.i.d. with c.d.f. F_1 of Y_1 and $Y_{n_1+1}, \ldots, Y_n$ i.i.d. with c.d.f. F_2 of Y_{n_1+1}.*

(a) If $F_1 \geq F_2$, then $\mathbb{E}[\phi] \leq \alpha$.
(b) If $F_1 \leq F_2$, then $\mathbb{E}[\phi] \geq \alpha$.

Proof. Lemma 4.4 in Janssen (1998).

In location parameter models as considered in Example 8.4.1(a), the score function is isotone if and only if the density f is unimodal. The following example discusses some specific instances of such densities and derives the corresponding rank tests.

Example 8.4.3 (Two-Sample Rank Tests in Location Parameter Models with "Stochastically Larger" Alternatives).

(i) *Fisher–Yates test*:
Let f denote the density of $\mathcal{N}(0,1)$. Then it holds $\dot{L}(y) = y$ and we obtain

$$T = \sum_{i=1}^{n_1} a(R_i) \quad \text{with} \quad a(i) = \mathbb{E}[Y_{i:n}].$$

In this, $Y_{i:n}$ denotes the i-th order statistic of i.i.d. random variables $Y_1, \ldots, Y_n$ with $Y_1 \sim \mathcal{N}(0,1)$.

(ii) *Van der Waerden test*:
Let f be as in part (i). The score-generating function is given by $u \mapsto \Phi^{-1}(u)$. Following Remark 8.4.2(e), $b(i) = \Phi^{-1}(i/(n+1))$ are approximate scores, leading to the test statistic

$$T = \sum_{i=1}^{n_1} \Phi^{-1}\left(\frac{R_i}{n+1}\right).$$

(iii) *Wilcoxon's rank sum test*:
Let f be the density of the standard logistic distribution, given by $f(y) = \exp(-y)(1+\exp(-y))^{-2}$ with corresponding cdf $F(y) = (1+\exp(-y))^{-1}$. The score-generating function is in this case given by $u \mapsto 2u-1$, leading to the scores

$$a(i) = \mathbb{E}[\dot{L} \circ F^{-1}(U_{i:n})] = \frac{2i}{n+1} - 1.$$

These scores are an affine transformation of the identity and therefore, the test can equivalently be carried out by means of the test statistic

$$T = \sum_{i=1}^{n_1} R_i(Y),$$

which is the sum of the ranks in the first group.

(iv) *Median test*:
The Lebesgue density of the Laplace distribution is given by $f(y) = \exp(-|y|)/2$, with induced score-generating function $u \mapsto \operatorname{sgn}(\ln(2u)) = \operatorname{sgn}(2u-1)$. Approximate scores are therefore given by

$$b(i) = \dot{L} \circ F^{-1}\left(\frac{i}{n+1}\right) = \begin{cases} 1, & \text{if } i > (n+1)/2, \\ 0, & \text{if } i = (n+1)/2, \\ -1, & \text{if } i < (n+1)/2. \end{cases}$$

We conclude this section with the Savage test (or log-rank test), an example for a scale parameter test, cf. Example 8.4.1(b).

Example 8.4.4. Under the scale parameter model considered in Example 8.4.1(b), assume that X is exponentially distributed with density $f(x) = \exp(-x)\,\mathbb{1}_{(0,\infty)}(x)$. Then we obtain for $y > 0$ the score function

$$\dot{L}(y) = -\left(1 + y\frac{f'(y)}{f(y)}\right) = y - 1.$$

Exercise 8.4.2. Show that for i.i.d. random variables $Y_1, \ldots, Y_n$ with $Y_1 \sim \operatorname{Exp}(1)$, it holds

$$\mathbb{E}[Y_{i:n}] = \sum_{j=1}^{i} \frac{1}{n+1-j}.$$

Making use of the latter result, exact scores are given by

$$a(i) = \sum_{j=1}^{i} \frac{1}{n+1-j} - 1.$$

Since X is almost surely positive, the model $Y = \exp(\theta)X$ can be transformed into the location parameter model $\log(Y) = \theta + \log(X)$. For $X \sim \text{Exp}(1)$, it holds that $\log(X)$ possesses a reflected Gumbel distribution, satisfying

$$\mathbb{P}(\log(X) \leq x) = 1 - \exp(-\exp(x)), \quad x > 0.$$

8.4.3 *Permutation Tests*

Permutation tests can be regarded as special instances of rank tests for k-sample problems.

Example 8.4.5 (Two-Sample Problem in Gaussian Location Parameter Model). Let $(Y_i)_{1 \leq i \leq n}$ denote real-valued, stochastically independent random variables, where $Y_1, \ldots, Y_{n_1}$ are i.i.d. with $Y_1 \sim F_1$ and $Y_{n_1+1}, \ldots, Y_n$ are i.i.d. with $Y_{n_1+1} \sim F_2$. Assume that the test problem of interest is given by

$$H_0\colon \{F_1 = F_2\} \qquad \text{versus} \qquad H_1\colon \{F_1 \neq F_2\}. \tag{8.11}$$

In the special case that F_1 and F_2 are Gaussian cdfs which only differ in their means, one would compare the empirical group means to carry out a test for problem (8.11). More specifically, we let $n_2 \stackrel{\text{def}}{=} n - n_1$ and define group means by $\overline{Y}_{n_1} \stackrel{\text{def}}{=} n_1^{-1} \sum_{i=1}^{n_1} Y_i$ and $\overline{Y}_{n_2} \stackrel{\text{def}}{=} n_2^{-1} \sum_{j=n_1+1}^{n} Y_j$, assuming that $0 < n_1 < n$. The test statistic of the resulting two-sample Z-test is then given by $\tilde{T} \stackrel{\text{def}}{=} \left|\overline{Y}_{n_1} - \overline{Y}_{n_2}\right|$ and the test for (8.11) can easily be calibrated by noticing that $\overline{Y}_{n_1} - \overline{Y}_{n_2}$ is again normally distributed under H_0.

However, in the case of general F_1 and F_2, exact distributional results for $\tilde{T}$ are difficult to obtain. Assuming that F_1 and F_2 are continuous, we consider more general statistics of the form

$$T = \sum_{i=1}^{n} c_i g(Y_i) = \sum_{i=1}^{n} c_{D_i(Y)} g(Y_{i:n}) \tag{8.12}$$

for a given function $g : \mathbb{R} \to \mathbb{R}$ and real numbers $(c_i)_{1 \leq i \leq n}$.

The representation of T on the right-hand side of (8.12) establishes the connection to rank tests. For example, $|T|$ equals $\tilde{T}$ if we choose $g = id$, $c_i = n_1^{-1}$ for $i \leq n_1$ and $c_i = -n_2^{-1}$ for $i > n_1$. Under H_0 from (8.11), the antiranks $D(Y) = (D_i(Y))_{1 \leq i \leq n}$ and the order statistics $(Y_{i:n})_{1 \leq i \leq n}$ are stochastically independent, see Theorem 8.4.5. Due to this property, the two-sample permutation test based on T can be carried out according to the following resampling scheme.

Example 8.4.6 (Resampling Scheme for a Two-Sample Permutation Test). The following resampling scheme is appropriate for a one-sided "stochastically larger" alternative. The two-sided case is obtained by obvious modifications.

(A) Consider the order statistics $(Y_{i:n})_{1\leq i\leq n}$ and regard $a(i) \stackrel{\text{def}}{=} g(Y_{i:n})$ as random scores.
(B) Denote by $\tilde{D} = (\tilde{D}_i)_{1\leq i\leq n}$ a random vector which is uniformly distributed on $\mathcal{S}_n$ and let $c = c(\alpha, (Y_{i:n})_{1\leq i\leq n})$ denote the $(1-\alpha)$-quantile of the discretely distributed random variable $\tilde{D} \mapsto \sum_{i=1}^n c_{\tilde{D}_i} a(i)$.
(C) The permutation test ϕ for testing (8.11) is then given by

$$\phi = \begin{cases} 1, & T > c, \\ \gamma, & T = c, \\ 0 & T < c, \end{cases}$$

where $\gamma \in [0,1]$ denotes a randomization constant.

Remark 8.4.3. If we choose $g = id$ and $(c_j)_{1\leq j\leq n}$ as in Example 8.4.5, leading to $|T| = \tilde{T}$, then the test ϕ from Example 8.4.6 is called *Pitman's permutation test*, see Pitman (1937).

The permutation test principle can be adapted to test the more general null hypothesis

$$H_0\colon Y_1, \ldots, Y_n \quad \text{are i.i.d.} \tag{8.13}$$

In the generalized form, the $Y_j\colon 1 \leq j \leq n$ are not even restricted to be real-valued. The modified resampling scheme is given as follows.

Example 8.4.7 (Modified Resampling Scheme for General Permutation Tests).

(A) Consider n random variates Y_j, $1 \leq j \leq n$ with values in some space $\mathcal{Y}$ and a real-valued test statistic $T = T(Y_1, \ldots, Y_n)$.
(B) In the remainder, consider permutations π with values in $\mathcal{S}_n$ which are independent of $Y_1, \ldots, Y_n$.
(C) Denote by Q_0 the uniform distribution on $\mathcal{S}_n$ and let $c = c(Y_1, \ldots, Y_n)$ denote the $(1-\alpha)$-quantile of $t \mapsto Q_0(\{\pi \in \mathcal{S}_n : T(Y_{\pi(1)}, \ldots, Y_{\pi(n)}) \leq t\})$.
(D) The modified permutation test $\tilde{\phi}$ for testing (8.13) is then given by

$$\tilde{\phi} = \begin{cases} 1, & T > c, \\ \gamma, & T = c, \\ 0, & T < c. \end{cases}$$

Theorem 8.4.6. *Under the respective assumptions, the permutation test ϕ defined in Example 8.4.6 and the modified permutation test $\tilde{\phi}$ defined in Example 8.4.7 are under the null hypothesis H_0 from* (8.11) *or* (8.13), *respectively, tests of* exact level *α for any fixed $n \in \mathbb{N}$.*

Proof. Conditional to the order statistics (Example 8.4.6) or to the data themselves (Example 8.4.7), the critical value c and the randomization constant γ are chosen such that

$$\mathbb{E}_{\mathcal{L}(\tilde{D})}[\varphi \mid Y = y] = \mathbb{E}_{Q_0}[\tilde{\varphi} \mid Y = y] = \alpha$$

holds true. Furthermore, the antiranks $D(Y)$ are under H_0 from (8.11) stochastically independent of the order statistics. Analogously, the random permutations π are chosen stochastically independent of $(Y_1, \ldots, Y_n)$ in the case of $\tilde{\phi}$. The result of the theorem follows by averaging with respect to the distribution of Y.

8.5 Historical Remarks and Further Reading

The *Kolmogorov–Smirnov test* goes back to Kolmogorov (1933) and Smirnov (1948). The origins of the ω^2 *test* can be traced back to Cramér (1928) and the German lecture notes by von Mises (1931). The limiting ω^2 distribution has been derived in the work by Smirnov (1937). A comprehensive resource for *(nonparametric) goodness-of-fit tests* is the book edited by D'Agostino and Stephens (1986).

The concept of *Bayes factors* goes back to Jeffreys (1935) and is treated comprehensively by Kass and Raftery (1995). *Bayesian approaches* to hypothesis testing are discussed in Sect. 4.3.3 of Berger (1985); see also the references therein.

Our treatment of *score tests* mainly follows (Janssen 1998). The theory of *rank tests* is developed in the textbook by Hájek and Šidák (1967). The classical reference for *permutation tests* is Pitman (1937). Recent textbook and monograph references on the subject are Good (2005), Edgington and Onghena (2007), and Pesarin and Salmaso (2010).

Appendix A
Deviation Probability for Quadratic Forms

A.1 Introduction

This chapter presents a number of deviation probability bounds for a quadratic form $\|\boldsymbol{\xi}\|^2$ or more generally $\|B\boldsymbol{\xi}\|^2$ of a random p vector $\boldsymbol{\xi}$ satisfying a general exponential moment condition. Such quadratic forms arise in many applications. Baraud (2010) lists some statistical tasks relying on such deviation bounds including hypothesis testing for linear models or linear model selection. We also refer to Massart (2007) for an extensive overview and numerous results on probability bounds and their applications in statistical model selection. Limit theorems for quadratic forms can be found e.g. in Götze and Tikhomirov (1999) and Horváth and Shao (1999). Some concentration bounds for U-statistics are available in Bretagnolle (1999), Giné et al. (2000), Houdré and Reynaud-Bouret (2003). Most of results assumes that the components of the vector $\boldsymbol{\xi}$ are independent and bounded.

Hsu et al. (2012) study the tail behavior of the quadratic form under the condition of sub-Gaussianity of the random vector $\boldsymbol{\xi}$ and show that the deviation probability are essentially the same as in the Gaussian case. However, the assumption that the vector $\boldsymbol{\xi}$ has finite exponential moments of arbitrary order is quite strict and is not fulfilled in many applications. A particular example is given by the Poisson and exponential cases. In the present work we only suppose that some exponential moments of $\boldsymbol{\xi}$ are finite. This makes the problem much more involved and requires new approaches and tools.

If $\boldsymbol{\xi}$ is standard normal then $\|\boldsymbol{\xi}\|^2$ is chi-squared with p degrees of freedom. We aim to extend this behavior to the case of a general vector $\boldsymbol{\xi}$ satisfying the following exponential moment condition:

$$\log \mathbb{E} \exp(\boldsymbol{\gamma}^\top \boldsymbol{\xi}) \le \|\boldsymbol{\gamma}\|^2/2, \qquad \boldsymbol{\gamma} \in \mathbb{R}^p, \|\boldsymbol{\gamma}\| \le \mathtt{g}. \tag{A.1}$$

Here $\mathtt{g}$ is a positive constant which appears to be very important in our results. Namely, it determines the frontier between the Gaussian and non-Gaussian type deviation bounds. Our first result shows that under (A.1) the deviation bounds for

V. Spokoiny and T. Dickhaus, *Basics of Modern Mathematical Statistics*,
Springer Texts in Statistics, DOI 10.1007/978-3-642-39909-1,

the quadratic form $\|\xi\|^2$ are essentially the same as in the Gaussian case, if the value $\mathfrak{g}^2$ exceeds $\mathtt{C}p$ for a fixed constant $\mathtt{C}$. Further we extend the result to the case of a more general form $\|B\xi\|^2$. An important advantage of the presented approach which makes it different from all the previous studies is that there is no any additional conditions on the structure or origin of the vector ξ. For instance, we do not assume that ξ is a sum of independent or weakly dependent random variables, or components of ξ are independent. The results are exact stated in a non-asymptotic fashion, all the constants are explicit and the leading terms are sharp.

As a motivating example, we consider a linear regression model $\boldsymbol{Y} = \Psi^\top \boldsymbol{\theta}^* + \boldsymbol{\varepsilon}$ in which $\boldsymbol{Y}$ is a n-vector of observations, $\boldsymbol{\varepsilon}$ is the vector of errors with zero mean, and Ψ is a $p \times n$ design matrix. The ordinary least square estimator $\tilde{\boldsymbol{\theta}}$ for the parameter vector $\boldsymbol{\theta}^* \in \mathbb{R}^p$ reads as

$$\tilde{\boldsymbol{\theta}} = \bigl(\Psi\Psi^\top\bigr)^{-1}\Psi\boldsymbol{Y}$$

and it can be viewed as the maximum likelihood estimator in a Gaussian linear model with a diagonal covariance matrix, that is, $\boldsymbol{Y} \sim \mathcal{N}(\Psi^\top\boldsymbol{\theta}, \sigma^2\boldsymbol{I}_n)$. Define the $p \times p$ matrix

$$D_0^2 \stackrel{\text{def}}{=} \Psi\Psi^\top,$$

Then

$$D_0(\tilde{\boldsymbol{\theta}} - \boldsymbol{\theta}^*) = D_0^{-1}\boldsymbol{\zeta}$$

with $\boldsymbol{\zeta} \stackrel{\text{def}}{=} \Psi\boldsymbol{\varepsilon}$. The likelihood ratio test statistic for this problem is exactly $\|D_0^{-1}\boldsymbol{\zeta}\|^2/2$. Similarly, the model selection procedure is based on comparing such quadratic forms for different matrices D_0; see e.g. Baraud (2010).

Now we indicate how this situation can be reduced to a bound for a vector ξ satisfying the condition (A.1). Suppose for simplicity that the entries ε_i of the error vector $\boldsymbol{\varepsilon}$ are independent and have exponential moments.

(e_1) *There exist some constants* ν_0 *and* $\mathfrak{g}_1 > 0$, *and for every* i *a constant* s_i *such that* $\mathbb{E}\bigl(\varepsilon_i/\mathsf{s}_i\bigr)^2 \le 1$ *and*

$$\log \mathbb{E}\exp\bigl(\lambda\varepsilon_i/\mathsf{s}_i\bigr) \le \nu_0^2\lambda^2/2, \qquad |\lambda| \le \mathfrak{g}_1. \tag{A.2}$$

Here $\mathfrak{g}_1$ is a fixed positive constant. One can show that if this condition is fulfilled for some $\mathfrak{g}_1 > 0$ and a constant $\nu_0 \ge 1$, then one can get a similar condition with ν_0 arbitrary close to one and $\mathfrak{g}_1$ slightly decreased. A natural candidate for s_i is σ_i where $\sigma_i^2 = \mathbb{E}\varepsilon_i^2$ is the variance of ε_i. Under (A.2), introduce a $p \times p$ matrix V_0 defined by

$$V_0^2 \stackrel{\text{def}}{=} \sum \mathsf{s}_i^2 \Psi_i \Psi_i^\top,$$

where $\Psi_1, \ldots, \Psi_n \in \mathbb{R}^p$ are the columns of the matrix Ψ. Define also

$$\xi = V_0^{-1}\Psi\varepsilon,$$

$$N^{-1/2} \stackrel{\text{def}}{=} \max_i \sup_{\gamma \in \mathbb{R}^p} \frac{\mathsf{s}_i |\Psi_i^\top \gamma|}{\|V_0\gamma\|}.$$

Simple calculation shows that for $\|\gamma\| \le \mathtt{g} = \mathtt{g}_1 N^{1/2}$

$$\log \mathbb{E}\exp(\gamma^\top \xi) \le \nu_0^2 \|\gamma\|^2/2, \qquad \gamma \in \mathbb{R}^p,\ \|\gamma\| \le \mathtt{g}.$$

We conclude that (A.1) is nearly fulfilled under (e_1) and moreover, the value $\mathtt{g}^2$ is proportional to the effective sample size N. The results below allow to get a nearly χ^2-behavior of the test statistic $\|\xi\|^2$ which is a finite sample version of the famous Wilks phenomenon; see e.g. Fan et al. (2001), Fan and Huang (2005), Boucheron and Massart (2011).

Section A.2 reminds the classical results about deviation probability of a Gaussian quadratic form. These results are presented only for comparison and to make the presentation selfcontained.

Section A.3 studies the probability of the form $\mathbb{P}\big(\|\xi\| > \mathtt{y}\big)$ under the condition

$$\log \mathbb{E}\exp(\gamma^\top \xi) \le \nu_0^2 \|\gamma\|^2/2, \qquad \gamma \in \mathbb{R}^p,\ \|\gamma\| \le \mathtt{g}.$$

The general case can be reduced to $\nu_0 = 1$ by rescaling ξ and $\mathtt{g}$:

$$\log \mathbb{E}\exp(\gamma^\top \xi/\nu_0) \le \|\gamma\|^2/2, \qquad \gamma \in \mathbb{R}^p,\ \|\gamma\| \le \nu_0 \mathtt{g}$$

that is, $\nu_0^{-1}\xi$ fulfills (A.1) with a slightly increased $\mathtt{g}$.

The obtained result is extended to the case of a general quadratic form in Sect. A.4. Some more extensions motivated by different statistical problems are given in Sects. A.6 and A.7. They include the bound with sup-norm constraint and the bound under Bernstein conditions. Among the statistical problems demanding such bounds is estimation of the regression model with Poissonian or bounded random noise. More examples can be found in Baraud (2010). All the proofs are collected in Sect. A.8.

A.2 Gaussian Case

Our benchmark will be a deviation bound for $\|\xi\|^2$ for a standard Gaussian vector ξ. The ultimate goal is to show that under (A.1) the norm of the vector ξ exhibits behavior expected for a Gaussian vector, at least in the region of moderate deviations. For the reason of comparison, we begin by stating the result for a Gaussian vector ξ. We use the notation $a \vee b$ for the maximum of a and b, while $a \wedge b = \min\{a, b\}$.

Theorem A.2.1. *Let ξ be a standard normal vector in $\mathbb{R}^p$. Then for any $u > 0$, it holds*

$$\mathbb{P}\bigl(\|\xi\|^2 > p + u\bigr) \le \exp\bigl\{-(p/2)\phi(u/p)\bigr]\}$$

with

$$\phi(t) \stackrel{\mathrm{def}}{=} t - \log(1+t).$$

Let $\phi^{-1}(\cdot)$ stand for the inverse of $\phi(\cdot)$. For any $\mathtt{x}$,

$$\mathbb{P}\bigl(\|\xi\|^2 > p + p\phi^{-1}(2\mathtt{x}/p)\bigr) \le \exp(-\mathtt{x}).$$

This particularly yields with $\varkappa = 6.6$

$$\mathbb{P}\bigl(\|\xi\|^2 > p + \sqrt{\varkappa \mathtt{x} p} \vee (\varkappa \mathtt{x})\bigr) \le \exp(-\mathtt{x}).$$

This is a simple version of a well known result and we present it only for comparison with the non-Gaussian case. The message of this result is that the squared norm of the Gaussian vector ξ concentrates around the value p and its deviation over the level $p + \sqrt{\mathtt{x}p}$ is exponentially small in $\mathtt{x}$.

A similar bound can be obtained for a norm of the vector $B\xi$ where B is some given deterministic matrix. For notational simplicity we assume that B is symmetric. Otherwise one should replace it with $(B^\top B)^{1/2}$.

Theorem A.2.2. *Let ξ be standard normal in $\mathbb{R}^p$. Then for every $\mathtt{x} > 0$ and any symmetric matrix B, it holds with $\mathtt{p} = \operatorname{tr}(B^2)$, $\mathtt{v}^2 = 2\operatorname{tr}(B^4)$, and $a^* = \|B^2\|_\infty$*

$$\mathbb{P}\bigl(\|B\xi\|^2 > \mathtt{p} + (2\mathtt{v}\mathtt{x}^{1/2}) \vee (6a^*\mathtt{x})\bigr) \le \exp(-\mathtt{x}).$$

Below we establish similar bounds for a non-Gaussian vector ξ obeying (A.1).

A.3 A Bound for the ℓ_2-Norm

This section presents a general exponential bound for the probability $\mathbb{P}\bigl(\|\xi\| > \mathtt{y}\bigr)$ under (A.1). The main result tells us that if $\mathtt{y}$ is not too large, namely if $\mathtt{y} \le \mathtt{y}_c$ with $\mathtt{y}_c^2 \asymp \mathtt{g}^2$, then the deviation probability is essentially the same as in the Gaussian case.

To describe the value $\mathtt{y}_c$, introduce the following notation. Given $\mathtt{g}$ and p, define the values $w_0 = \mathtt{g}p^{-1/2}$ and w_c by the equation

$$\frac{w_c(1+w_c)}{(1+w_c^2)^{1/2}} = w_0 = \mathtt{g}p^{-1/2}. \qquad \text{(A.3)}$$

It is easy to see that $w_0/\sqrt{2} \le w_c \le w_0$. Further define

$$\mu_c \overset{\text{def}}{=} w_c^2/(1+w_c^2)$$

$$\mathtt{y}_c \overset{\text{def}}{=} \sqrt{(1+w_c^2)p},$$

$$\mathtt{x}_c \overset{\text{def}}{=} 0.5p\bigl[w_c^2 - \log(1+w_c^2)\bigr]. \tag{A.4}$$

Note that for $\mathtt{g}^2 \ge p$, the quantities $\mathtt{y}_c$ and $\mathtt{x}_c$ can be evaluated as $\mathtt{y}_c^2 \ge w_c^2 p \ge \mathtt{g}^2/2$ and $\mathtt{x}_c \gtrsim p w_c^2/2 \ge \mathtt{g}^2/4$.

Theorem A.3.1. *Let $\xi \in \mathbb{R}^p$ fulfill* (A.1). *Then it holds for each* $\mathtt{x} \le \mathtt{x}_c$

$$\mathbb{P}\bigl(\|\xi\|^2 > p + \sqrt{\varkappa \mathtt{x} p} \vee (\varkappa \mathtt{x}),\ \|\xi\| \le \mathtt{y}_c\bigr) \le 2\exp(-\mathtt{x}),$$

where $\varkappa = 6.6$. *Moreover, for* $\mathtt{y} \ge \mathtt{y}_c$, *it holds with* $\mathtt{g}_c = \mathtt{g} - \sqrt{\mu_c p} = \mathtt{g} w_c/(1+w_c)$

$$\begin{aligned}\mathbb{P}\bigl(\|\xi\| > \mathtt{y}\bigr) &\le 8.4\exp\bigl\{-\mathtt{g}_c\mathtt{y}/2 - (p/2)\log(1-\mathtt{g}_c/\mathtt{y})\bigr\} \\ &\le 8.4\exp\bigl\{-\mathtt{x}_c - \mathtt{g}_c(\mathtt{y}-\mathtt{y}_c)/2\bigr\}.\end{aligned}$$

The statements of Theorem A.4.1 can be simplified under the assumption $\mathtt{g}^2 \ge p$.

Corollary A.3.1. *Let ξ fulfill* (A.1) *and* $\mathtt{g}^2 \ge p$. *Then it holds for* $\mathtt{x} \le \mathtt{x}_c$

$$\mathbb{P}\bigl(\|\xi\|^2 \ge \mathfrak{z}(\mathtt{x},p)\bigr) \le 2e^{-\mathtt{x}} + 8.4e^{-\mathtt{x}_c}, * \tag{A.5}$$

$$\mathfrak{z}(\mathtt{x},p) \overset{\text{def}}{=} \begin{cases} p + \sqrt{\varkappa \mathtt{x} p}, & \mathtt{x} \le p/\varkappa, \\ p + \varkappa \mathtt{x} & p/\varkappa < \mathtt{x} \le \mathtt{x}_c, \end{cases} \tag{A.6}$$

with $\varkappa = 6.6$. *For* $\mathtt{x} > \mathtt{x}_c$

$$\mathbb{P}\bigl(\|\xi\|^2 \ge \mathfrak{z}_c(\mathtt{x},p)\bigr) \le 8.4e^{-\mathtt{x}}, \qquad \mathfrak{z}_c(\mathtt{x},p) \overset{\text{def}}{=} \bigl|\mathtt{y}_c + 2(\mathtt{x}-\mathtt{x}_c)/\mathtt{g}_c\bigr|^2.$$

This result implicitly assumes that $p \le \varkappa \mathtt{x}_c$ which is fulfilled if $w_0^2 = \mathtt{g}^2/p \ge 1$:

$$\varkappa \mathtt{x}_c = 0.5\varkappa\bigl[w_0^2 - \log(1+w_0^2)\bigr]p \ge 3.3\bigl[1-\log(2)\bigr]p > p.$$

For $\mathtt{x} \le \mathtt{x}_c$, the function $\mathfrak{z}(\mathtt{x},p)$ mimics the quantile behavior of the chi-squared distribution χ_p^2 with p degrees of freedom. Moreover, increase of the value $\mathtt{g}$ yields a growth of the sub-Gaussian zone. In particular, for $\mathtt{g} = \infty$, a general quadratic form $\|\xi\|^2$ has under (A.1) the same tail behavior as in the Gaussian case.

Finally, in the large deviation zone $\mathtt{x} > \mathtt{x}_c$ the deviation probability decays as $e^{-c\mathtt{x}^{1/2}}$ for some fixed c. However, if the constant $\mathtt{g}$ in the condition (A.1) is sufficiently large relative to p, then $\mathtt{x}_c$ is large as well and the large deviation zone $\mathtt{x} > \mathtt{x}_c$ can be ignored at a small price of $8.4e^{-\mathtt{x}_c}$ and one can focus on the deviation bound described by (A.5) and (A.6).

A.4 A Bound for a Quadratic Form

Now we extend the result to more general bound for $\|B\xi\|^2 = \xi^\top B^2 \xi$ with a given matrix B and a vector ξ obeying the condition (A.1). Similarly to the Gaussian case we assume that B is symmetric. Define important characteristics of B

$$\mathtt{p} = \operatorname{tr}(B^2), \qquad \mathtt{v}^2 = 2\operatorname{tr}(B^4), \qquad \lambda_B \stackrel{\text{def}}{=} \|B^2\|_\infty \stackrel{\text{def}}{=} \lambda_{\max}(B^2).$$

For simplicity of formulation we suppose that $\lambda_B = 1$, otherwise one has to replace $\mathtt{p}$ and $\mathtt{v}^2$ with $\mathtt{p}/\lambda_B$ and $\mathtt{v}^2/\lambda_B$.

Let $\mathtt{g}$ be shown in (A.1). Define similarly to the ℓ_2-case w_c by the equation

$$\frac{w_c(1+w_c)}{(1+w_c^2)^{1/2}} = \mathtt{g}\mathtt{p}^{-1/2}.$$

Define also $\mu_c = w_c^2/(1+w_c^2) \wedge 2/3$. Note that $w_c^2 \ge 2$ implies $\mu_c = 2/3$. Further define

$$\mathtt{y}_c^2 = (1+w_c^2)\mathtt{p}, \qquad 2\mathtt{x}_c = \mu_c \mathtt{y}_c^2 + \log\det\{\boldsymbol{I}_p - \mu_c B^2\}. \tag{A.7}$$

Similarly to the case with $B = \boldsymbol{I}_p$, under the condition $\mathtt{g}^2 \ge \mathtt{p}$, one can bound $\mathtt{y}_c^2 \ge \mathtt{g}^2/2$ and $\mathtt{x}_c \gtrsim \mathtt{g}^2/4$.

Theorem A.4.1. *Let a random vector* ξ *in* $\mathbb{R}^p$ *fulfill* (A.1). *Then for each* $\mathtt{x} < \mathtt{x}_c$

$$\mathbb{P}\bigl(\|B\xi\|^2 > \mathtt{p} + (2\mathtt{v}\mathtt{x}^{1/2}) \vee (6\mathtt{x}),\ \|B\xi\| \le \mathtt{y}_c\bigr) \le 2\exp(-\mathtt{x}).$$

Moreover, for $\mathtt{y} \ge \mathtt{y}_c$, *with* $\mathtt{g}_c = \mathtt{g} - \sqrt{\mu_c \mathtt{p}} = \mathtt{g}w_c/(1+w_c)$, *it holds*

$$\mathbb{P}\bigl(\|B\xi\| > \mathtt{y}\bigr) \le 8.4\exp\bigl(-\mathtt{x}_c - \mathtt{g}_c(\mathtt{y}-\mathtt{y}_c)/2\bigr).$$

Now we describe the value $\mathfrak{z}(\mathtt{x}, B)$ ensuring a small value for the large deviation probability $\mathbb{P}\bigl(\|B\xi\|^2 > \mathfrak{z}(\mathtt{x}, B)\bigr)$. For ease of formulation, we suppose that $\mathtt{g}^2 \ge 2\mathtt{p}$ yielding $\mu_c^{-1} \le 3/2$. The other case can be easily adjusted.

Corollary A.4.1. *Let* ξ *fulfill* (A.1) *with* $\mathtt{g}^2 \ge 2\mathtt{p}$. *Then it holds for* $\mathtt{x} \le \mathtt{x}_c$ *with* $\mathtt{x}_c$ *from* (A.7)*:*

$$\mathbb{P}\bigl(\|B\xi\|^2 \ge \mathfrak{z}(\mathtt{x}, B)\bigr) \le 2e^{-\mathtt{x}} + 8.4e^{-\mathtt{x}_c},$$

$$\mathfrak{z}(\mathtt{x}, B) \stackrel{\text{def}}{=} \begin{cases} \mathtt{p} + 2\mathtt{v}\mathtt{x}^{1/2}, & \mathtt{x} \le \mathtt{v}/18, \\ \mathtt{p} + 6\mathtt{x} & \mathtt{v}/18 < \mathtt{x} \le \mathtt{x}_c. \end{cases} \tag{A.8}$$

For $\mathtt{x} > \mathtt{x}_c$

$$\mathbb{P}\bigl(\|B\xi\|^2 \ge \mathfrak{z}_c(\mathtt{x}, B)\bigr) \le 8.4e^{-\mathtt{x}}, \qquad \mathfrak{z}_c(\mathtt{x}, B) \stackrel{\text{def}}{=} \bigl|\mathtt{y}_c + 2(\mathtt{x} - \mathtt{x}_c)/\mathtt{g}_c\bigr|^2.$$

A.5 Rescaling and Regularity Condition

The result of Theorem A.4.1 can be extended to a more general situation when the condition (A.1) is fulfilled for a vector ζ rescaled by a matrix V_0. More precisely, let the random p-vector ζ fulfills for some $p \times p$ matrix V_0 the condition

$$\sup_{\gamma \in \mathbb{R}^p} \log \mathbb{E} \exp\Bigl(\lambda \frac{\gamma^\top \zeta}{\|V_0 \gamma\|}\Bigr) \le \nu_0^2 \lambda^2/2, \qquad |\lambda| \le \mathtt{g}, \tag{A.9}$$

with some constants $\mathtt{g} > 0$, $\nu_0 \ge 1$. Again, a simple change of variables reduces the case of an arbitrary $\nu_0 \ge 1$ to $\nu_0 = 1$. Our aim is to bound the squared norm $\|D_0^{-1}\zeta\|^2$ of a vector $D_0^{-1}\zeta$ for another $p \times p$ positive symmetric matrix D_0^2. Note that condition (A.9) implies (A.1) for the rescaled vector $\xi = V_0^{-1}\zeta$. This leads to bounding the quadratic form $\|D_0^{-1} V_0 \xi\|^2 = \|B\xi\|^2$ with $B^2 = D_0^{-1} V_0^2 D_0^{-1}$. It obviously holds

$$\mathtt{p} = \operatorname{tr}(B^2) = \operatorname{tr}(D_0^{-2} V_0^2).$$

Now we can apply the result of Corollary A.4.1.

Corollary A.5.1. *Let* ζ *fulfill* (A.9) *with some* V_0 *and* $\mathtt{g}$. *Given* D_0, *define* $B^2 = D_0^{-1} V_0^2 D_0^{-1}$, *and let* $\mathtt{g}^2 \ge 2\mathtt{p}$. *Then it holds for* $\mathtt{x} \le \mathtt{x}_c$ *with* $\mathtt{x}_c$ *from* (A.7):

$$\mathbb{P}\bigl(\|D_0^{-1}\zeta\|^2 \ge \mathfrak{z}(\mathtt{x}, B)\bigr) \le 2e^{-\mathtt{x}} + 8.4e^{-\mathtt{x}_c},$$

with $\mathfrak{z}(\mathtt{x}, B)$ *from* (A.8). *For* $\mathtt{x} > \mathtt{x}_c$

$$\mathbb{P}\bigl(\|D_0^{-1}\zeta\|^2 \ge \mathfrak{z}_c(\mathtt{x}, B)\bigr) \le 8.4e^{-\mathtt{x}}, \qquad \mathfrak{z}_c(\mathtt{x}, B) \stackrel{\text{def}}{=} \bigl|\mathtt{y}_c + 2(\mathtt{x} - \mathtt{x}_c)/\mathtt{g}_c\bigr|^2.$$

In the *regular* case with $D_0 \ge \mathfrak{a} V_0$ for some $\mathfrak{a} > 0$, one obtains $\|B\|_\infty \le \mathfrak{a}^{-1}$ and

$$\mathtt{v}^2 = 2\operatorname{tr}(B^4) \le 2\mathfrak{a}^{-2}\mathtt{p}.$$

A.6 A Chi-Squared Bound with Norm-Constraints

This section extends the results to the case when the bound (A.1) requires some other conditions than the ℓ_2-norm of the vector $\boldsymbol{\gamma}$. Namely, we suppose that

$$\log \mathbb{E} \exp(\boldsymbol{\gamma}^\top \boldsymbol{\xi}) \le \|\boldsymbol{\gamma}\|^2/2, \qquad \boldsymbol{\gamma} \in \mathbb{R}^p, \ \|\boldsymbol{\gamma}\|_\circ \le \mathtt{g}_\circ, \tag{A.10}$$

where $\|\cdot\|_\circ$ is a norm which differs from the usual Euclidean norm. Our driving example is given by the sup-norm case with $\|\boldsymbol{\gamma}\|_\circ \equiv \|\boldsymbol{\gamma}\|_\infty$. We are interested to check whether the previous results of Sect. A.3 still apply. The answer depends on how massive the set $\mathcal{A}(r) = \{\boldsymbol{\gamma} : \|\boldsymbol{\gamma}\|_\circ \le r\}$ is in terms of the standard Gaussian measure on $\mathbb{R}^p$. Recall that the quadratic norm $\|\boldsymbol{\varepsilon}\|^2$ of a standard Gaussian vector $\boldsymbol{\varepsilon}$ in $\mathbb{R}^p$ concentrates around p at least for p large. We need a similar concentration property for the norm $\|\cdot\|_\circ$. More precisely, we assume for a fixed r_* that

$$\mathbb{P}\bigl(\|\boldsymbol{\varepsilon}\|_\circ \le r_*\bigr) \ge 1/2, \qquad \boldsymbol{\varepsilon} \sim \mathcal{N}(0, \boldsymbol{I}_p). \tag{A.11}$$

This implies for any value $\mathrm{u}_\circ > 0$ and all $\boldsymbol{u} \in \mathbb{R}^p$ with $\|\boldsymbol{u}\|_\circ \le \mathrm{u}_\circ$ that

$$\mathbb{P}\bigl(\|\boldsymbol{\varepsilon} - \boldsymbol{u}\|_\circ \le r_* + \mathrm{u}_\circ\bigr) \ge 1/2, \qquad \boldsymbol{\varepsilon} \sim \mathcal{N}(0, \boldsymbol{I}_p).$$

For each $\mathfrak{z} > p$, consider

$$\mu(\mathfrak{z}) = (\mathfrak{z} - p)/\mathfrak{z}.$$

Given $\mathrm{u}_\circ$, denote by $\mathfrak{z}_\circ = \mathfrak{z}_\circ(\mathrm{u}_\circ)$ the root of the equation

$$\frac{\mathtt{g}_\circ}{\mu(\mathfrak{z}_\circ)} - \frac{r_*}{\mu^{1/2}(\mathfrak{z}_\circ)} = \mathrm{u}_\circ. \tag{A.12}$$

One can easily see that this value exists and unique if $\mathrm{u}_\circ \ge \mathtt{g}_\circ - r_*$ and it can be defined as the largest $\mathfrak{z}$ for which $\frac{\mathtt{g}_\circ}{\mu(\mathfrak{z})} - \frac{r_*}{\mu^{1/2}(\mathfrak{z})} \ge \mathrm{u}_\circ$. Let $\mu_\circ = \mu(\mathfrak{z}_\circ)$ be the corresponding μ-value. Define also $\mathrm{x}_\circ$ by

$$2\mathrm{x}_\circ = \mu_\circ \mathfrak{z}_\circ + p \log(1 - \mu_\circ).$$

If $\mathrm{u}_\circ < \mathtt{g}_\circ - r_*$, then set $\mathfrak{z}_\circ = \infty$, $\mathrm{x}_\circ = \infty$.

Theorem A.6.1. *Let a random vector* $\boldsymbol{\xi}$ *in* $\mathbb{R}^p$ *fulfill* (A.10). *Suppose* (A.11) *and let, given* $\mathrm{u}_\circ$, *the value* $\mathfrak{z}_\circ$ *be defined by* (A.12). *Then it holds for any* $u > 0$

$$\mathbb{P}\bigl(\|\boldsymbol{\xi}\|^2 > p + u, \|\boldsymbol{\xi}\|_\circ \le \mathrm{u}_\circ\bigr) \le 2 \exp\bigl\{-(p/2)\phi(u)]\bigr\}. \tag{A.13}$$

yielding for $\mathtt{x} \le \mathtt{x}_\circ$

$$\mathbb{P}\bigl(\|\boldsymbol{\xi}\|^2 > p + \sqrt{\varkappa \mathtt{x} p} \vee (\varkappa \mathtt{x}),\ \|\boldsymbol{\xi}\|_\circ \le \mathtt{u}_\circ\bigr) \le 2\exp(-\mathtt{x}), \tag{A.14}$$

where $\varkappa = 6.6$. *Moreover, for* $\mathfrak{z} \ge \mathfrak{z}_\circ$, *it holds*

$$\begin{aligned}\mathbb{P}\bigl(\|\boldsymbol{\xi}\|^2 > \mathfrak{z},\ \|\boldsymbol{\xi}\|_\circ \le \mathtt{u}_\circ\bigr) &\le 2\exp\bigl\{-\mu_\circ \mathfrak{z}/2 - (p/2)\log(1-\mu_\circ)\bigr\} \\ &= 2\exp\bigl\{-\mathtt{x}_\circ - \mathtt{g}_\circ(\mathfrak{z} - \mathfrak{z}_\circ)/2\bigr\}.\end{aligned}$$

It is easy to check that the result continues to hold for the norm of $\Pi\boldsymbol{\xi}$ for a given sub-projector Π in $\mathbb{R}^p$ satisfying $\Pi = \Pi^\top$, $\Pi^2 \le \Pi$. As above, denote $\mathtt{p} \stackrel{\text{def}}{=} \operatorname{tr}(\Pi^2)$, $\mathtt{v}^2 \stackrel{\text{def}}{=} 2\operatorname{tr}(\Pi^4)$. Let r_* be fixed to ensure

$$\mathbb{P}\bigl(\|\Pi\boldsymbol{\varepsilon}\|_\circ \le r_*\bigr) \ge 1/2, \qquad \boldsymbol{\varepsilon} \sim \mathcal{N}(0, \boldsymbol{I}_p).$$

The next result is stated for $\mathtt{g}_\circ \ge r_* + \mathtt{u}_\circ$, which simplifies the formulation.

Theorem A.6.2. *Let a random vector* $\boldsymbol{\xi}$ *in* $\mathbb{R}^p$ *fulfill* (A.10) *and* Π *follows* $\Pi = \Pi^\top$, $\Pi^2 \le \Pi$. *Let some* $\mathtt{u}_\circ$ *be fixed. Then for any* $\mu_\circ \le 2/3$ *with* $\mathtt{g}_\circ \mu_\circ^{-1} - r_* \mu_\circ^{-1/2} \ge \mathtt{u}_\circ$,

$$\mathbb{E}\exp\Bigl\{\frac{\mu_\circ}{2}\bigl(\|\Pi\boldsymbol{\xi}\|^2 - \mathtt{p}\bigr)\Bigr\}\mathbb{1}\bigl(\|\Pi^2\boldsymbol{\xi}\|_\circ \le \mathtt{u}_\circ\bigr) \le 2\exp(\mu_\circ^2 \mathtt{v}^2/4), \tag{A.15}$$

where $\mathtt{v}^2 = 2\operatorname{tr}(\Pi^4)$. *Moreover, if* $\mathtt{g}_\circ \ge r_* + \mathtt{u}_\circ$, *then for any* $\mathfrak{z} \ge 0$

$$\begin{aligned}&\mathbb{P}\bigl(\|\Pi\boldsymbol{\xi}\|^2 > \mathfrak{z},\ \|\Pi^2\boldsymbol{\xi}\|_\circ \le \mathtt{u}_\circ\bigr) \\ &\quad \le \mathbb{P}\bigl(\|\Pi\boldsymbol{\xi}\|^2 > \mathtt{p} + (2\mathtt{v}\mathtt{x}^{1/2}) \vee (6\mathtt{x}),\ \|\Pi^2\boldsymbol{\xi}\|_\circ \le \mathtt{u}_\circ\bigr) \le 2\exp(-\mathtt{x}).\end{aligned}$$

A.7 A Bound for the ℓ_2-Norm Under Bernstein Conditions

For comparison, we specify the results to the case considered recently in Baraud (2010). Let $\boldsymbol{\zeta}$ be a random vector in $\mathbb{R}^n$ whose components ζ_i are independent and satisfy the Bernstein type conditions: for all $|\lambda| < c^{-1}$

$$\log \mathbb{E} e^{\lambda \zeta_i} \le \frac{\lambda^2 \sigma^2}{1 - c|\lambda|}. \tag{A.16}$$

Denote $\boldsymbol{\xi} = \boldsymbol{\zeta}/(2\sigma)$ and consider $\|\boldsymbol{\gamma}\|_\circ = \|\boldsymbol{\gamma}\|_\infty$. Fix $\mathtt{g}_\circ = \sigma/c$. If $\|\boldsymbol{\gamma}\|_\circ \le \mathtt{g}_\circ$, then $1 - c\gamma_i/(2\sigma) \ge 1/2$ and

$$\log \mathbb{E}\exp(\boldsymbol{\gamma}^\top \boldsymbol{\xi}) \le \sum_i \log \mathbb{E}\exp\Bigl(\frac{\gamma_i \zeta_i}{2\sigma}\Bigr) \le \sum_i \frac{|\gamma_i/(2\sigma)|^2 \sigma^2}{1 - c\gamma_i/(2\sigma)} \le \|\boldsymbol{\gamma}\|^2/2.$$

Let also S be some linear subspace of $\mathbb{R}^n$ with dimension $\mathtt{p}$ and Π_S denote the projector on S. For applying the result of Theorem A.6.1, the value r_* has to be fixed. We use that the infinity norm $\|\boldsymbol{\varepsilon}\|_\infty$ concentrates around $\sqrt{2\log p}$.

Lemma A.7.1. *It holds for a standard normal vector* $\boldsymbol{\varepsilon} \in \mathbb{R}^p$ *with* $r_* = \sqrt{2\log p}$

$$\mathbb{P}\bigl(\|\boldsymbol{\varepsilon}\|_\circ \le r_*\bigr) \ge 1/2.$$

Indeed

$$\mathbb{P}\bigl(\|\boldsymbol{\varepsilon}\|_\circ > r_*\bigr) \le \mathbb{P}\bigl(\|\boldsymbol{\varepsilon}\|_\infty > \sqrt{2\log p}\bigr) \le p\mathbb{P}\bigl(|\varepsilon_1| > \sqrt{2\log p}\bigr) \le 1/2.$$

Now the general bound of Theorem A.6.1 is applied to bounding the norm of $\|\Pi_S \boldsymbol{\xi}\|$. For simplicity of formulation we assume that $\mathtt{g}_\circ \ge \mathtt{u}_\circ + r_*$.

Theorem A.7.1. *Let* S *be some linear subspace of* $\mathbb{R}^n$ *with dimension* $\mathtt{p}$. *Let* $\mathtt{g}_\circ \ge \mathtt{u}_\circ + r_*$. *If the coordinates* ζ_i *of* $\boldsymbol{\zeta}$ *are independent and satisfy* (A.16), *then for all* $\mathtt{x}$,

$$\mathbb{P}\bigl((4\sigma^2)^{-1}\|\Pi_S\boldsymbol{\zeta}\|^2 > \mathtt{p} + \sqrt{\varkappa \mathtt{x} \mathtt{p}} \vee (\varkappa \mathtt{x}),\ \|\Pi_S\boldsymbol{\zeta}\|_\infty \le 2\sigma \mathtt{u}_\circ\bigr) \le 2\exp(-\mathtt{x}),$$

The bound of Baraud (2010) reads

$$\mathbb{P}\biggl(\|\Pi_S\boldsymbol{\zeta}\|_2 > \bigl(3\sigma \vee \sqrt{6cu}\bigr)\sqrt{\mathtt{x}+3\mathtt{p}},\ \|\Pi_S\boldsymbol{\zeta}\|_\infty \le 2\sigma\mathtt{u}_\circ\biggr) \le e^{-\mathtt{x}}.$$

As expected, in the region $\mathtt{x} \le \mathtt{x}_c$ of Gaussian approximation, the bound of Baraud is not sharp and actually quite rough.

A.8 Proofs

Proof of Theorem A.2.1

The proof utilizes the following well known fact, which can be obtained by straightforward calculus : for $\mu < 1$

$$\log \mathbb{E}\exp\bigl(\mu\|\boldsymbol{\xi}\|^2/2\bigr) = -0.5p\log(1-\mu).$$

Now consider any $u > 0$. By the exponential Chebyshev inequality

$$\begin{aligned}\mathbb{P}\bigl(\|\boldsymbol{\xi}\|^2 > p+u\bigr) &\le \exp\bigl\{-\mu(p+u)/2\bigr\}\mathbb{E}\exp\bigl(\mu\|\boldsymbol{\xi}\|^2/2\bigr) \\ &= \exp\bigl\{-\mu(p+u)/2 - (p/2)\log(1-\mu)\bigr\}.\end{aligned} \tag{A.17}$$

It is easy to see that the value $\mu = u/(u+p)$ maximizes $\mu(p+u) + p\log(1-\mu)$ w.r.t. μ yielding

$$\mu(p+u) + p\log(1-\mu) = u - p\log(1+u/p).$$

Further we use that $x - \log(1+x) \ge a_0 x^2$ for $x \le 1$ and $x - \log(1+x) \ge a_0 x$ for $x > 1$ with $a_0 = 1 - \log(2) \ge 0.3$. This implies with $x = u/p$ for $u = \sqrt{\varkappa \mathtt{x} p}$ or $u = \varkappa \mathtt{x}$ and $\varkappa = 2/a_0 < 6.6$ that

$$\mathbb{P}\bigl(\|\xi\|^2 \ge p + \sqrt{\varkappa \mathtt{x} p} \vee (\varkappa \mathtt{x})\bigr) \le \exp(-\mathtt{x})$$

as required.

Proof of Theorem A.2.2

The matrix B^2 can be represented as $U^\top \operatorname{diag}(a_1, \ldots, a_p) U$ for an orthogonal matrix U. The vector $\tilde{\xi} = U\xi$ is also standard normal and $\|B\xi\|^2 = \tilde{\xi}^\top U B^2 U^\top \tilde{\xi}$. This means that one can reduce the situation to the case of a diagonal matrix $B^2 = \operatorname{diag}(a_1, \ldots, a_p)$. We can also assume without loss of generality that $a_1 \ge a_2 \ge \ldots \ge a_p$. The expressions for the quantities $\mathtt{p}$ and $\mathtt{v}^2$ simplifies to

$$\mathtt{p} = \operatorname{tr}(B^2) = a_1 + \ldots + a_p,$$
$$\mathtt{v}^2 = 2\operatorname{tr}(B^4) = 2(a_1^2 + \ldots + a_p^2).$$

Moreover, rescaling the matrix B^2 by a_1 reduces the situation to the case with $a_1 = 1$.

Lemma A.8.1. *It holds*

$$\mathbb{E}\|B\xi\|^2 = \operatorname{tr}(B^2), \qquad \operatorname{Var}\bigl(\|B\xi\|^2\bigr) = 2\operatorname{tr}(B^4).$$

Moreover, for $\mu < 1$

$$\mathbb{E}\exp\{\mu\|B\xi\|^2/2\} = \det(1 - \mu B^2)^{-1/2} = \prod_{i=1}^{p}(1 - \mu a_i)^{-1/2}. \tag{A.18}$$

Proof. If B^2 is diagonal, then $\|B\xi\|^2 = \sum_i a_i \xi_i^2$ and the summands $a_i \xi_i^2$ are independent. It remains to note that $\mathbb{E}(a_i \xi_i^2) = a_i$, $\operatorname{Var}(a_i \xi_i^2) = 2a_i^2$, and for $\mu a_i < 1$,

$$\mathbb{E}\exp\{\mu a_i \xi_i^2/2\} = (1 - \mu a_i)^{-1/2}$$

yielding (A.18).

Given u, fix $\mu < 1$. The exponential Markov inequality yields

$$\begin{aligned}\mathbb{P}\bigl(\|B\xi\|^2 > \mathtt{p} + u\bigr) &\le \exp\Bigl\{-\frac{\mu(\mathtt{p}+u)}{2}\Bigr\}\mathbb{E}\exp\Bigl(\frac{\mu\|B\xi\|^2}{2}\Bigr)\\ &\le \exp\Bigl\{-\frac{\mu u}{2} - \frac{1}{2}\sum_{i=1}^{p}[\mu a_i + \log(1-\mu a_i)]\Bigr\}.\end{aligned}$$

We start with the case when $\mathtt{x}^{1/2} \le \mathtt{v}/3$. Then $u = 2\mathtt{x}^{1/2}\mathtt{v}$ fulfills $u \le 2\mathtt{v}^2/3$. Define $\mu = u/\mathtt{v}^2 \le 2/3$ and use that $t + \log(1-t) \ge -t^2$ for $t \le 2/3$. This implies

$$\begin{aligned}&\mathbb{P}\bigl(\|B\xi\|^2 > \mathtt{p} + u\bigr)\\ &\quad\le \exp\Bigl\{-\frac{\mu u}{2} + \frac{1}{2}\sum_{i=1}^{p}\mu^2 a_i^2\Bigr\} = \exp\bigl(-u^2/(4\mathtt{v}^2)\bigr) = e^{-\mathtt{x}}.\end{aligned} \tag{A.19}$$

Next, let $\mathtt{x}^{1/2} > \mathtt{v}/3$. Set $\mu = 2/3$. It holds similarly to the above

$$\sum_{i=1}^{p}[\mu a_i + \log(1-\mu a_i)] \ge -\sum_{i=1}^{p}\mu^2 a_i^2 \ge -2\mathtt{v}^2/9 \ge -2\mathtt{x}.$$

Now, for $u = 6\mathtt{x}$ and $\mu u/2 = 2\mathtt{x}$, (A.19) implies

$$\mathbb{P}\bigl(\|B\xi\|^2 > \mathtt{p} + u\bigr) \le \exp\{-(2\mathtt{x} - \mathtt{x})\} = \exp(-\mathtt{x})$$

as required.

Proof of Theorem A.3.1

The main step of the proof is the following exponential bound.

Lemma A.8.2. *Suppose* (A.1). *For any* $\mu < 1$ *with* $\mathtt{g}^2 > p\mu$, *it holds*

$$\mathbb{E}\exp\Bigl(\frac{\mu\|\xi\|^2}{2}\Bigr)\mathbb{1}\Bigl(\|\xi\| \le \mathtt{g}/\mu - \sqrt{p/\mu}\Bigr) \le 2(1-\mu)^{-p/2}. \tag{A.20}$$

Proof. Let $\boldsymbol{\varepsilon}$ be a standard normal vector in $\mathbb{R}^p$ and $\boldsymbol{u} \in \mathbb{R}^p$. The bound $\mathbb{P}\bigl(\|\boldsymbol{\varepsilon}\|^2 > p\bigr) \le 1/2$ and the triangle inequality imply for any vector $\boldsymbol{u}$ and any $\mathtt{r}$ with $\mathtt{r} \ge \|\boldsymbol{u}\| + p^{1/2}$ that $\mathbb{P}\bigl(\|\boldsymbol{u} + \boldsymbol{\varepsilon}\| \le \mathtt{r}\bigr) \ge 1/2$. Let us fix some ξ with $\|\xi\| \le \mathtt{g}/\mu - \sqrt{p/\mu}$ and denote by $\mathbb{P}_\xi$ the conditional probability given ξ. The previous arguments yield:

$$\mathbb{P}_\xi\bigl(\|\boldsymbol{\varepsilon} + \mu^{1/2}\xi\| \le \mu^{-1/2}\mathtt{g}\bigr) \ge 0.5.$$

It holds with $c_p = (2\pi)^{-p/2}$

$$
\begin{aligned}
& c_p \int \exp\Bigl(\gamma^\top \xi - \frac{\|\gamma\|^2}{2\mu}\Bigr) \mathbb{1}(\|\gamma\| \le \mathtt{g}) d\gamma \\
&= c_p \exp(\mu \|\xi\|^2/2) \int \exp\Bigl(-\frac{1}{2}\|\mu^{-1/2}\gamma - \mu^{1/2}\xi\|^2\Bigr) \mathbb{1}(\mu^{-1/2}\|\gamma\| \le \mu^{-1/2}\mathtt{g}) d\gamma \\
&= \mu^{p/2} \exp(\mu\|\xi\|^2/2) \mathbb{P}_\xi\bigl(\|\varepsilon + \mu^{1/2}\xi\| \le \mu^{-1/2}\mathtt{g}\bigr) \\
&\ge 0.5\mu^{p/2} \exp(\mu\|\xi\|^2/2),
\end{aligned}
$$

because $\|\mu^{1/2}\xi\| + p^{1/2} \le \mu^{-1/2}\mathtt{g}$. This implies in view of $p < \mathtt{g}^2/\mu$ that

$$
\begin{aligned}
& \exp(\mu\|\xi\|^2/2) \mathbb{1}\bigl(\|\xi\|^2 \le \mathtt{g}/\mu - \sqrt{p/\mu}\bigr) \\
& \qquad \le 2\mu^{-p/2} c_p \int \exp\Bigl(\gamma^\top \xi - \frac{\|\gamma\|^2}{2\mu}\Bigr) \mathbb{1}(\|\gamma\| \le \mathtt{g}) d\gamma .
\end{aligned}
$$

Further, by (A.1)

$$
\begin{aligned}
& c_p \mathbb{E} \int \exp\Bigl(\gamma^\top \xi - \frac{1}{2\mu}\|\gamma\|^2\Bigr) \mathbb{1}(\|\gamma\| \le \mathtt{g}) d\gamma \\
& \qquad \le c_p \int \exp\Bigl(-\frac{\mu^{-1} - 1}{2}\|\gamma\|^2\Bigr) \mathbb{1}(\|\gamma\| \le \mathtt{g}) d\gamma \\
& \qquad \le c_p \int \exp\Bigl(-\frac{\mu^{-1} - 1}{2}\|\gamma\|^2\Bigr) d\gamma \\
& \qquad \le (\mu^{-1} - 1)^{-p/2}
\end{aligned}
$$

and (A.20) follows.

Due to this result, the scaled squared norm $\mu\|\xi\|^2/2$ after a proper truncation possesses the same exponential moments as in the Gaussian case. A straightforward implication is the probability bound $\mathbb{P}\bigl(\|\xi\|^2 > p + u\bigr)$ for moderate values u. Namely, given $u > 0$, define $\mu = u/(u + p)$. This value optimizes the inequality (A.17) in the Gaussian case. Now we can apply a similar bound under the constraints $\|\xi\| \le \mathtt{g}/\mu - \sqrt{p/\mu}$. Therefore, the bound is only meaningful if $\sqrt{u + p} \le \mathtt{g}/\mu - \sqrt{p/\mu}$ with $\mu = u/(u + p)$, or, with $w = \sqrt{u/p} \le w_c$; see (A.3).

The largest value u for which this constraint is still valid, is given by $p + u = \mathtt{y}_c^2$. Hence, (A.20) yields for $p + u \le \mathtt{y}_c^2$

$$
\begin{aligned}
& \mathbb{P}\bigl(\|\xi\|^2 > p + u, \|\xi\| \le \mathtt{y}_c\bigr) \\
& \qquad \le \exp\Bigl\{-\frac{\mu(p + u)}{2}\Bigr\} \mathbb{E} \exp\Bigl(\frac{\mu\|\xi\|^2}{2}\Bigr) \mathbb{1}\Bigl(\|\xi\| \le \mathtt{g}/\mu - \sqrt{p/\mu}\Bigr)
\end{aligned}
$$

$$\le 2\exp\{-0.5[\mu(p+u) + p\log(1-\mu)]\}$$
$$= 2\exp\{-0.5[u - p\log(1+u/p)]\}.$$

Similarly to the Gaussian case, this implies with $\varkappa = 6.6$ that

$$\mathbb{P}\bigl(\|\xi\| \ge p + \sqrt{\varkappa \mathtt{x} p} \vee (\varkappa \mathtt{x}), \|\xi\| \le \mathtt{y}_c\bigr) \le 2\exp(-\mathtt{x}).$$

The Gaussian case means that (A.1) holds with $\mathtt{g} = \infty$ yielding $\mathtt{y}_c = \infty$. In the non-Gaussian case with a finite $\mathtt{g}$, we have to accompany the moderate deviation bound with a large deviation bound $\mathbb{P}\bigl(\|\xi\| > \mathtt{y}\bigr)$ for $\mathtt{y} \ge \mathtt{y}_c$. This is done by combining the bound (A.20) with the standard slicing arguments.

Lemma A.8.3. *Let* $\mu_0 \le \mathtt{g}^2/p$. *Define* $\mathtt{y}_0 = \mathtt{g}/\mu_0 - \sqrt{p/\mu_0}$ *and* $\mathtt{g}_0 = \mu_0 \mathtt{y}_0 = \mathtt{g} - \sqrt{\mu_0 p}$. *It holds for* $\mathtt{y} \ge \mathtt{y}_0$

$$\mathbb{P}\bigl(\|\xi\| > \mathtt{y}\bigr) \le 8.4(1 - \mathtt{g}_0/\mathtt{y})^{-p/2} \exp\bigl(-\mathtt{g}_0 \mathtt{y}/2\bigr) \tag{A.21}$$
$$\le 8.4\exp\{-\mathtt{x}_0 - \mathtt{g}_0(\mathtt{y} - \mathtt{y}_0)/2\}. \tag{A.22}$$

with $\mathtt{x}_0$ *defined by*

$$2\mathtt{x}_0 = \mu_0 \mathtt{y}_0^2 + p\log(1-\mu_0) = \mathtt{g}^2/\mu_0 - p + p\log(1-\mu_0).$$

Proof. Consider the growing sequence $\mathtt{y}_k$ with $\mathtt{y}_1 = \mathtt{y}$ and $\mathtt{g}_0 \mathtt{y}_{k+1} = \mathtt{g}_0 \mathtt{y} + k$. Define also $\mu_k = \mathtt{g}_0/\mathtt{y}_k$. In particular, $\mu_k \le \mu_1 = \mathtt{g}_0/\mathtt{y}$. Obviously

$$\mathbb{P}\bigl(\|\xi\| > \mathtt{y}\bigr) = \sum_{k=1}^{\infty} \mathbb{P}\bigl(\|\xi\| > \mathtt{y}_k, \|\xi\| \le \mathtt{y}_{k+1}\bigr).$$

Now we try to evaluate every slicing probability in this expression. We use that

$$\mu_{k+1}\mathtt{y}_k^2 = \frac{(\mathtt{g}_0 \mathtt{y} + k - 1)^2}{\mathtt{g}_0 \mathtt{y} + k} \ge \mathtt{g}_0 \mathtt{y} + k - 2,$$

and also $\mathtt{g}/\mu_k - \sqrt{p/\mu_k} \ge \mathtt{y}_k$ because $\mathtt{g} - \mathtt{g}_0 = \sqrt{\mu_0 p} > \sqrt{\mu_k p}$ and

$$\mathtt{g}/\mu_k - \sqrt{p/\mu_k} - \mathtt{y}_k = \mu_k^{-1}(\mathtt{g} - \sqrt{\mu_k p} - \mathtt{g}_0) \ge 0.$$

Hence by (A.20)

$$\mathbb{P}\Bigl(\|\xi\| > \mathtt{y}\Bigr) = \sum_{k=1}^{\infty} \mathbb{P}\Bigl(\|\xi\| > \mathtt{y}_k, \|\xi\| \le \mathtt{y}_{k+1}\Bigr)$$

$$
\begin{aligned}
&\le \sum_{k=1}^{\infty} \exp\Bigl(\frac{\mu_{k+1}\mathtt{y}_k^2}{2}\Bigr)\mathbb{E}\exp\Bigl(\frac{\mu_{k+1}\|\xi\|^2}{2}\Bigr)\mathbb{1}(\|\xi\| \le \mathtt{y}_{k+1}) \\
&\le \sum_{k=1}^{\infty} 2(1-\mu_{k+1})^{-p/2}\exp\Bigl(-\frac{\mu_{k+1}\mathtt{y}_k^2}{2}\Bigr) \\
&\le 2(1-\mu_1)^{-p/2}\sum_{k=1}^{\infty}\exp\Bigl(-\frac{\mathtt{g}_0\mathtt{y}+k-2}{2}\Bigr) \\
&= 2e^{1/2}(1-e^{-1/2})^{-1}(1-\mu_1)^{-p/2}\exp(-\mathtt{g}_0\mathtt{y}/2) \\
&\le 8.4(1-\mu_1)^{-p/2}\exp(-\mathtt{g}_0\mathtt{y}/2)
\end{aligned}
$$

and the first assertion follows. For $\mathtt{y} = \mathtt{y}_0$, it holds

$$
\mathtt{g}_0\mathtt{y}_0 + p\log(1-\mu_0) = \mu_0\mathtt{y}_0^2 + p\log(1-\mu_0) = 2\mathtt{x}_0
$$

and (A.21) implies $\mathbb{P}\bigl(\|\xi\| > \mathtt{y}_0\bigr) \le 8.4\exp(-\mathtt{x}_0)$. Now observe that the function $f(\mathtt{y}) = \mathtt{g}_0\mathtt{y}/2 + (p/2)\log(1-\mathtt{g}_0/\mathtt{y})$ fulfills $f(\mathtt{y}_0) = \mathtt{x}_0$ and $f'(\mathtt{y}) \ge \mathtt{g}_0/2$ yielding $f(\mathtt{y}) \ge \mathtt{x}_0 + \mathtt{g}_0(\mathtt{y}-\mathtt{y}_0)/2$. This implies (A.22).

The statements of the theorem are obtained by applying the lemmas with $\mu_0 = \mu_c = \mathtt{w}_c^2/(1+\mathtt{w}_c^2)$. This also implies $\mathtt{y}_0 = \mathtt{y}_c$, $\mathtt{x}_0 = \mathtt{x}_c$, and $\mathtt{g}_0 = \mathtt{g}_c = \mathtt{g} - \sqrt{\mu_c p}$; cf. (A.4).

Proof of Theorem A.4.1

The main steps of the proof are similar to the proof of Theorem A.3.1.

Lemma A.8.4. *Suppose* (A.1). *For any* $\mu < 1$ *with* $\mathtt{g}^2/\mu \ge \mathtt{p}$, *it holds*

$$
\mathbb{E}\exp\bigl(\mu\|B\xi\|^2/2\bigr)\mathbb{1}\bigl(\|B^2\xi\| \le \mathtt{g}/\mu - \sqrt{\mathtt{p}/\mu}\bigr) \le 2\det(\boldsymbol{I}_p - \mu B^2)^{-1/2}. \quad \text{(A.23)}
$$

Proof. With $c_p(B) = (2\pi)^{-p/2}\det(B^{-1})$

$$
\begin{aligned}
&c_p(B)\int \exp\Bigl(\boldsymbol{\gamma}^\top\xi - \frac{1}{2\mu}\|B^{-1}\boldsymbol{\gamma}\|^2\Bigr)\mathbb{1}(\|\boldsymbol{\gamma}\| \le \mathtt{g})d\boldsymbol{\gamma} \\
&= c_p(B)\exp\Bigl(\frac{\mu\|B\xi\|^2}{2}\Bigr)\int \exp\Bigl(-\frac{1}{2}\bigl\|\mu^{1/2}B\xi - \mu^{-1/2}B^{-1}\boldsymbol{\gamma}\bigr\|^2\Bigr)\mathbb{1}(\|\boldsymbol{\gamma}\| \le \mathtt{g})d\boldsymbol{\gamma} \\
&= \mu^{p/2}\exp\Bigl(\frac{\mu\|B\xi\|^2}{2}\Bigr)\mathbb{P}_\xi\bigl(\|\mu^{-1/2}B\boldsymbol{\varepsilon} + B^2\xi\| \le \mathtt{g}/\mu\bigr),
\end{aligned}
$$

where $\boldsymbol{\varepsilon}$ denotes a standard normal vector in $\mathbb{R}^p$ and $\mathbb{P}_{\boldsymbol{\xi}}$ means the conditional probability given $\boldsymbol{\xi}$. Moreover, for any $\boldsymbol{u} \in \mathbb{R}^p$ and $\mathtt{r} \geq \mathtt{p}^{1/2} + \|\boldsymbol{u}\|$, it holds in view of $\mathbb{P}\bigl(\|B\boldsymbol{\varepsilon}\|^2 > \mathtt{p}\bigr) \leq 1/2$

$$\mathbb{P}\bigl(\|B\boldsymbol{\varepsilon} - \boldsymbol{u}\| \leq \mathtt{r}\bigr) \geq \mathbb{P}\bigl(\|B\boldsymbol{\varepsilon}\| \leq \sqrt{\mathtt{p}}\bigr) \geq 1/2.$$

This implies

$$\begin{aligned}&\exp\Bigl(\mu\|B\boldsymbol{\xi}\|^2/2\Bigr)\mathbb{1}\bigl(\|B^2\boldsymbol{\xi}\| \leq \mathtt{g}/\mu - \sqrt{\mathtt{p}/\mu}\bigr)\\&\quad \leq 2\mu^{-p/2}c_p(B)\int \exp\Bigl(\boldsymbol{\gamma}^\top\boldsymbol{\xi} - \frac{1}{2\mu}\|B^{-1}\boldsymbol{\gamma}\|^2\Bigr)\mathbb{1}(\|\boldsymbol{\gamma}\| \leq \mathtt{g})d\boldsymbol{\gamma}.\end{aligned}$$

Further, by (A.1)

$$\begin{aligned}&c_p(B)\mathbb{E}\int \exp\Bigl(\boldsymbol{\gamma}^\top\boldsymbol{\xi} - \frac{1}{2\mu}\|B^{-1}\boldsymbol{\gamma}\|^2\Bigr)\mathbb{1}(\|\boldsymbol{\gamma}\| \leq \mathtt{g})d\boldsymbol{\gamma}\\&\quad \leq c_p(B)\int \exp\Bigl(\frac{\|\boldsymbol{\gamma}\|^2}{2} - \frac{1}{2\mu}\|B^{-1}\boldsymbol{\gamma}\|^2\Bigr)d\boldsymbol{\gamma}\\&\quad \leq \det(B^{-1})\det(\mu^{-1}B^{-2} - \boldsymbol{I}_p)^{-1/2} = \mu^{p/2}\det(\boldsymbol{I}_p - \mu B^2)^{-1/2}\end{aligned}$$

and (A.23) follows.

Now we evaluate the probability $\mathbb{P}\bigl(\|B\boldsymbol{\xi}\| > \mathtt{y}\bigr)$ for moderate values of $\mathtt{y}$.

Lemma A.8.5. *Let $\mu_0 < 1 \wedge (\mathtt{g}^2/\mathtt{p})$. With $\mathtt{y}_0 = \mathtt{g}/\mu_0 - \sqrt{\mathtt{p}/\mu_0}$, it holds for any $u > 0$*

$$\begin{aligned}&\mathbb{P}\bigl(\|B\boldsymbol{\xi}\|^2 > \mathtt{p} + u, \|B^2\boldsymbol{\xi}\| \leq \mathtt{y}_0\bigr)\\&\qquad \leq 2\exp\bigl\{-0.5\mu_0(\mathtt{p} + u) - 0.5\log\det(\boldsymbol{I}_p - \mu_0 B^2)\bigr\}.\end{aligned} \tag{A.24}$$

In particular, if B^2 is diagonal, that is, $B^2 = \operatorname{diag}\bigl(a_1, \ldots, a_p\bigr)$, then

$$\begin{aligned}&\mathbb{P}\bigl(\|B\boldsymbol{\xi}\|^2 > \mathtt{p} + u, \|B^2\boldsymbol{\xi}\| \leq \mathtt{y}_0\bigr)\\&\qquad \leq 2\exp\Bigl\{-\frac{\mu_0 u}{2} - \frac{1}{2}\sum_{i=1}^{p}\bigl[\mu_0 a_i + \log\bigl(1 - \mu_0 a_i\bigr)\bigr]\Bigr\}.\end{aligned} \tag{A.25}$$

Proof. The exponential Chebyshev inequality and (A.23) imply

$$\begin{aligned}&\mathbb{P}\bigl(\|B\boldsymbol{\xi}\|^2 > \mathtt{p} + u, \|B^2\boldsymbol{\xi}\| \leq \mathtt{y}_0\bigr)\\&\quad \leq \exp\Bigl\{-\frac{\mu_0(\mathtt{p} + u)}{2}\Bigr\}\mathbb{E}\exp\Bigl(\frac{\mu_0\|B\boldsymbol{\xi}\|^2}{2}\Bigr)\mathbb{1}\bigl(\|B^2\boldsymbol{\xi}\| \leq \mathtt{g}/\mu_0 - \sqrt{\mathtt{p}/\mu_0}\bigr)\\&\quad \leq 2\exp\bigl\{-0.5\mu_0(\mathtt{p} + u) - 0.5\log\det(\boldsymbol{I}_p - \mu_0 B^2)\bigr\}.\end{aligned}$$

Moreover, the standard change-of-basis arguments allow us to reduce the problem to the case of a diagonal matrix $B^2 = \operatorname{diag}(a_1, \ldots, a_p)$ where $1 = a_1 \ge a_2 \ge \ldots \ge a_p > 0$. Note that $\mathtt{p} = a_1 + \ldots + a_p$. Then the claim (A.24) can be written in the form (A.25).

Now we evaluate a large deviation probability that $\|B\xi\| > \mathtt{y}$ for a large $\mathtt{y}$. Note that the condition $\|B^2\|_\infty \le 1$ implies $\|B^2\xi\| \le \|B\xi\|$. So, the bound (A.24) continues to hold when $\|B^2\xi\| \le \mathtt{y}_0$ is replaced by $\|B\xi\| \le \mathtt{y}_0$.

Lemma A.8.6. *Let* $\mu_0 < 1$ *and* $\mu_0 \mathtt{p} < \mathtt{g}^2$. *Define* $\mathtt{g}_0$ *by* $\mathtt{g}_0 = \mathtt{g} - \sqrt{\mu_0 \mathtt{p}}$. *For any* $\mathtt{y} \ge \mathtt{y}_0 \stackrel{\text{def}}{=} \mathtt{g}_0/\mu_0$, *it holds*

$$\begin{aligned} \mathbb{P}\bigl(\|B\xi\| > \mathtt{y}\bigr) &\le 8.4 \det\{\boldsymbol{I}_p - (\mathtt{g}_0/\mathtt{y})B^2\}^{-1/2} \exp(-\mathtt{g}_0\mathtt{y}/2). \\ &\le 8.4 \exp\bigl(-\mathtt{x}_0 - \mathtt{g}_0(\mathtt{y} - \mathtt{y}_0)/2\bigr), \end{aligned} \tag{A.26}$$

where $\mathtt{x}_0$ *is defined by*

$$2\mathtt{x}_0 = \mathtt{g}_0\mathtt{y}_0 + \log\det\{\boldsymbol{I}_p - (\mathtt{g}_0/\mathtt{y}_0)B^2\}.$$

Proof. The slicing arguments of Lemma A.8.3 apply here in the same manner. One has to replace $\|\xi\|$ by $\|B\xi\|$ and $(1-\mu_1)^{-p/2}$ by $\det\{\boldsymbol{I}_p - (\mathtt{g}_0/\mathtt{y})B^2\}^{-1/2}$. We omit the details. In particular, with $\mathtt{y} = \mathtt{y}_0 = \mathtt{g}_0/\mu$, this yields

$$\mathbb{P}\bigl(\|B\xi\| > \mathtt{y}_0\bigr) \le 8.4 \exp(-\mathtt{x}_0).$$

Moreover, for the function $f(\mathtt{y}) = \mathtt{g}_0\mathtt{y} + \log\det\{\boldsymbol{I}_p - (\mathtt{g}_0/\mathtt{y})B^2\}$, it holds $f'(\mathtt{y}) \ge \mathtt{g}_0$ and hence, $f(\mathtt{y}) \ge f(\mathtt{y}_0) + \mathtt{g}_0(\mathtt{y} - \mathtt{y}_0)$ for $\mathtt{y} > \mathtt{y}_0$. This implies (A.26).

One important feature of the results of Lemma A.8.5 and Lemma A.8.6 is that the value $\mu_0 < 1 \wedge (\mathtt{g}^2/\mathtt{p})$ can be selected arbitrarily. In particular, for $\mathtt{y} \ge \mathtt{y}_c$, Lemma A.8.6 with $\mu_0 = \mu_c$ yields the large deviation probability $\mathbb{P}\bigl(\|B\xi\| > \mathtt{y}\bigr)$. For bounding the probability $\mathbb{P}\bigl(\|B\xi\|^2 > \mathtt{p} + u,\ \|B\xi\| \le \mathtt{y}_c\bigr)$, we use the inequality $\log(1-t) \ge -t - t^2$ for $t \le 2/3$. It implies for $\mu \le 2/3$ that

$$\begin{aligned} &-\log \mathbb{P}\bigl(\|B\xi\|^2 > \mathtt{p} + u, \|B\xi\| \le \mathtt{y}_c\bigr) \\ &\quad \ge \mu(\mathtt{p} + u) + \sum_{i=1}^{p} \log\bigl(1 - \mu a_i\bigr) \\ &\quad \ge \mu(\mathtt{p} + u) - \sum_{i=1}^{p} (\mu a_i + \mu^2 a_i^2) \ge \mu u - \mu^2 \mathtt{v}^2/2. \end{aligned} \tag{A.27}$$

Now we distinguish between $\mu_c = 2/3$ and $\mu_c < 2/3$ starting with $\mu_c = 2/3$. The bound (A.27) with $\mu = 2/3$ and with $u = (2\mathtt{v}\mathtt{x}^{1/2}) \vee (6\mathtt{x})$ yields

$$\mathbb{P}\bigl(\|B\xi\|^2 > \mathtt{p} + u,\ \|B\xi\| \le \mathtt{y}_c\bigr) \le 2\exp(-\mathtt{x});$$

see the proof of Theorem A.2.2 for the Gaussian case.

Now consider $\mu_c < 2/3$. For $\mathtt{x}^{1/2} \le \mu_c \mathtt{v}/2$, use $u = 2\mathtt{v}\mathtt{x}^{1/2}$ and $\mu_0 = u/\mathtt{v}^2$. It holds $\mu_0 = u/\mathtt{v}^2 \le \mu_c$ and $u^2/(4\mathtt{v}^2) = \mathtt{x}$ yielding the desired bound by (A.27). For $\mathtt{x}^{1/2} > \mu_c \mathtt{v}/2$, we select again $\mu_0 = \mu_c$. It holds with $u = 4\mu_c^{-1}\mathtt{x}$ that $\mu_c u/2 - \mu_c^2 \mathtt{v}^2/4 \ge 2\mathtt{x} - \mathtt{x} = \mathtt{x}$. This completes the proof.

Proof of Theorem A.6.1

The arguments behind the result are the same as in the one-norm case of Theorem A.3.1. We only outline the main steps.

Lemma A.8.7. *Suppose* (A.10) *and* (A.11). *For any* $\mu < 1$ *with* $\mathtt{g}_\circ > \mu^{1/2} r_*$, *it holds*

$$\mathbb{E}\exp(\mu\|\xi\|^2/2)\mathbb{1}\bigl(\|\xi\|_\circ \le \mathtt{g}_\circ/\mu - r_*/\mu^{1/2}\bigr) \le 2(1-\mu)^{-p/2}. \tag{A.28}$$

Proof. Let $\boldsymbol{\varepsilon}$ be a standard normal vector in $\mathbb{R}^p$ and $\boldsymbol{u} \in \mathbb{R}^p$. Let us fix some ξ with $\mu^{1/2}\|\xi\|_\circ \le \mu^{-1/2}\mathtt{g}_\circ - r_*$ and denote by $\mathbb{P}_\xi$ the conditional probability given ξ. It holds by (A.11) with $c_p = (2\pi)^{-p/2}$

$$\begin{aligned}
&c_p \int \exp\Bigl(\boldsymbol{\gamma}^\top\xi - \frac{1}{2\mu}\|\boldsymbol{\gamma}\|^2\Bigr)\mathbb{1}(\|\boldsymbol{\gamma}\|_\circ \le \mathtt{g}_\circ)d\boldsymbol{\gamma}\\
&= c_p \exp(\mu\|\xi\|^2/2)\int \exp\Bigl(-\frac{1}{2}\bigl\|\mu^{1/2}\xi - \mu^{-1/2}\boldsymbol{\gamma}\bigr\|^2\Bigr)\mathbb{1}(\|\mu^{-1/2}\boldsymbol{\gamma}\|_\circ \le \mu^{-1/2}\mathtt{g}_\circ)d\boldsymbol{\gamma}\\
&= \mu^{p/2}\exp(\mu\|\xi\|^2/2)\mathbb{P}_\xi\bigl(\|\boldsymbol{\varepsilon} - \mu^{1/2}\xi\|_\circ \le \mu^{-1/2}\mathtt{g}_\circ\bigr)\\
&\ge 0.5\mu^{p/2}\exp(\mu\|\xi\|^2/2).
\end{aligned}$$

This implies

$$\begin{aligned}
&\exp\Bigl(\frac{\mu\|\xi\|^2}{2}\Bigr)\mathbb{1}\bigl(\|\xi\|_\circ \le \mathtt{g}_\circ/\mu - r_*/\mu^{1/2}\bigr)\\
&\qquad \le 2\mu^{-p/2}c_p\int \exp\Bigl(\boldsymbol{\gamma}^\top\xi - \frac{1}{2\mu}\|\boldsymbol{\gamma}\|^2\Bigr)\mathbb{1}(\|\boldsymbol{\gamma}\|_\circ \le \mathtt{g}_\circ)d\boldsymbol{\gamma}.
\end{aligned}$$

Further, by (A.10)

$$c_p\mathbb{E}\int \exp\Bigl(\boldsymbol{\gamma}^\top\xi - \frac{1}{2\mu}\|\boldsymbol{\gamma}\|^2\Bigr)\mathbb{1}(\|\boldsymbol{\gamma}\|_\circ \le \mathtt{g}_\circ)d\boldsymbol{\gamma}$$

$$\leq c_p \int \exp\Bigl(-\frac{\mu^{-1}-1}{2}\|\gamma\|^2\Bigr) d\gamma \leq (\mu^{-1}-1)^{-p/2}$$

and (A.28) follows.

As in the Gaussian case, (A.28) implies for $\mathfrak{z} > p$ with $\mu = \mu(\mathfrak{z}) = (\mathfrak{z} - p)/\mathfrak{z}$ the bounds (A.13) and (A.14). Note that the value $\mu(\mathfrak{z})$ clearly grows with $\mathfrak{z}$ from zero to one, while $\mathtt{g}_\circ/\mu(\mathfrak{z}) - r_*/\mu^{1/2}(\mathfrak{z})$ is strictly decreasing. The value $\mathfrak{z}_\circ$ is defined exactly as the point where $\mathtt{g}_\circ/\mu(\mathfrak{z}) - r_*/\mu^{1/2}(\mathfrak{z})$ crosses $\mathtt{u}_\circ$, so that $\mathtt{g}_\circ/\mu(\mathfrak{z}) - r_*/\mu^{1/2}(\mathfrak{z}) \geq \mathtt{u}_\circ$ for $\mathfrak{z} \leq \mathfrak{z}_\circ$.

For $\mathfrak{z} > \mathfrak{z}_\circ$, the choice $\mu = \mu(\mathtt{y})$ conflicts with $\mathtt{g}_\circ/\mu(\mathfrak{z}) - r_*/\mu^{1/2}(\mathfrak{z}) \geq \mathtt{u}_\circ$. So, we apply $\mu = \mu_\circ$ yielding by the Markov inequality

$$\mathbb{P}\bigl(\|\xi\|^2 > \mathfrak{z},\ \|\xi\|_\circ \leq \mathtt{u}_\circ\bigr) \leq 2\exp\bigl\{-\mu_\circ\mathfrak{z}/2 - (p/2)\log(1-\mu_\circ)\bigr\},$$

and the assertion follows.

Proof of Theorem A.6.2

Arguments from the proof of Lemmas A.8.4 and A.8.7 yield in view of $\mathtt{g}_\circ\mu_\circ^{-1} - r_*\mu_\circ^{-1/2} \geq \mathtt{u}_\circ$

$$\begin{aligned}
&\mathbb{E}\exp\bigl\{\mu_\circ\|\Pi\xi\|^2/2\bigr\}\mathbb{1}\bigl(\|\Pi^2\xi\|_\circ \leq \mathtt{u}_\circ\bigr) \\
&\leq \mathbb{E}\exp\bigl(\mu_\circ\|\Pi\xi\|^2/2\bigr)\mathbb{1}\bigl(\|\Pi^2\xi\|_\circ \leq \mathtt{g}_\circ/\mu_\circ - \mathtt{p}/\mu_\circ^{1/2}\bigr) \\
&\leq 2\det(\boldsymbol{I}_p - \mu_\circ\Pi^2)^{-1/2}.
\end{aligned}$$

The inequality $\log(1-t) \geq -t - t^2$ for $t \leq 2/3$ and symmetricity of the matrix Π imply

$$-\log\det(\boldsymbol{I}_p - \mu_\circ\Pi^2) \leq \mu_\circ\mathtt{p} + \mu_\circ^2\mathtt{v}^2/2$$

cf. (A.27); the assertion (A.15) follows.

References

Baraud, Y. (2010). A Bernstein-type inequality for suprema of random processes with applications to model selection in non-Gaussian regression. *Bernoulli, 16*(4), 1064–1085.

Bayes, T. (1763). An essay towards solving a problem in the doctrine of chances. *Philosophical Transactions of the Royal Society of London, 53*, 370–418.

Berger, J. O. (1985). *Statistical decision theory and Bayesian analysis* (2nd ed.). Springer series in statistics. New York: Springer.

Bernardo, J. M. and Smith, A. F. (1994). *Bayesian theory.* Chichester: Wiley.

Borovkov, A. A. (1998). *Mathematical statistics* (A. Moullagaliev. Trans. from the Russian). Amsterdam: Gordon & Breach.

Boucheron, S., & Massart, P. (2011). A high-dimensional Wilks phenomenon. *Probability Theory and Related Fields, 150*, 405–433. doi:10.1007/s00440-010-0278-7.

Bretagnolle, J. (1999). A new large deviation inequality for U-statistics of order 2. *ESAIM, Probability and Statistics, 3*, 151–162.

Chihara, T. (2011). *An introduction to orthogonal polynomials. Dover books on mathematics.* Mineola, New York: Dover Publications.

Cramér, H. (1928). On the composition of elementary errors. I. Mathematical deductions. II. Statistical applications. *Skandinavisk Aktuarietidskrift, 11*, 13–74, 141–180.

D'Agostino, R. B., & Stephens, M. A. (1986). *Goodness-of-fit techniques. Statistics: Textbooks and monographs* (Vol. 68). New York, Basel: Marcel Dekker

de Boor, C. (2001). *A practical guide to splines* (Vol. 27). Applied mathematical sciences. New York: Springer.

de Finetti, B. (1937). Foresight: Its logical laws, its subjective sources (in French). *Annales de l'Institut Henri Poincaré, 7*, 1–68.

Dempster, A., Laird, N., & Rubin, D. (1977). Maximum likelihood from incomplete data via the EM algorithm. Discussion. *Journal of the Royal Statistical Society, Series B, 39*, 1–38.

Diaconis, P., & Ylvisaker, D. (1979). Conjugate priors for exponential families. *Annals of Statistics, 7*, 269–281.

Donoho, D. L. (1995). Nonlinear solution of linear inverse problems by wavelet-vaguelette decomposition. *Applied and Computational Harmonic Analysis, 2* (2), 101–126.

Edgington, E. S., & Onghena, P. (2007). *Randomization tests. With CD-ROM* (4th ed.). Boca Raton, FL: Chapman & Hall/CRC.

Fan, J., & Gijbels, I. (1996). *Local polynomial modelling and its applications: Monographs on statistics and applied probability 66. Chapman & Hall/CRC monographs on statistics & applied probability*. London: Taylor & Francis.

Fan, J., & Huang, T. (2005). Profile likelihood inferences on semiparametric varying-coefficient partially linear models. *Bernoulli, 11*(6), 1031–1057.

V. Spokoiny and T. Dickhaus, *Basics of Modern Mathematical Statistics*, Springer Texts in Statistics, DOI 10.1007/978-3-642-39909-1,

Fan, J., Zhang, C., & Zhang, J. (2001). Generalized likelihood ratio statistics and Wilks phenomenon. *Annals of Statistics, 29*(1):153–193.

Fisher, R. A. (1934). Two new properties of mathematical likelihood. *Proceedings of Royal Society London, Series A, 144*, 285–307.

Fisher, R. A. (1956). *Statistical methods and scientific inference.* Edinburgh, London: Oliver & Boyd.

Galerkin, B. G. (1915). On electrical circuits for the approximate solution of the Laplace equation (in Russian). *Vestnik Inzh., 19*, 897–908.

Gauß, C. F. (1995). *Theory of the combination of observations least subject to errors. Part one, part two, supplement* (G. W. Stewart, Trans.). Philadelphia, PA: SIAM.

Gill, R. D., & Levit, B. Y. (1995). Applications of the van Trees inequality: A Bayesian Cramér-Rao bound. *Bernoulli, 1*(1–2), 59–79.

Giné, E., Latała, R., & Zinn, J. (2000). Exponential and moment inequalities for U-statistics. In E. Giné et al. (Ed.), *High dimensional probability II. 2nd international conference*, University of Washington, DC, USA, August 1–6, 1999. Progress in Probability (Vol. 47, pp. 13–38). Boston, MA: Birkhäuser.

Good, I., & Gaskins, R. (1971). Nonparametric roughness penalties for probability densities. *Biometrika, 58*, 255–277.

Good, P. (2005). *Permutation, parametric and bootstrap tests of hypotheses* (3rd ed.). New York, NY: Springer.

Götze, F., & Tikhomirov, A. (1999). Asymptotic distribution of quadratic forms. *Annals of Probability, 27*(2), 1072–1098.

Green, P. J., & Silverman, B. (1994). *Nonparametric regression and generalized linear models: a roughness penalty approach.* London: Chapman & Hall.

Hájek, J., & Šidák, Z. (1967). *Theory of rank tests.* New York, London: Academic Press; Prague: Academia, Publishing House of the Czechoslovak Academy of Sciences.

Hewitt, E., & Stromberg, K. (1975). *Real and abstract analysis. A modern treatment of the theory of functions of a real variable* (3rd printing). *Graduate texts in mathematics* (Vol. 25). New York, Heidelberg, Berlin: Springer.

Hoerl, A. E. (1962). Application of ridge analysis to regression problems. *Chemical Engineering Progress, 58*, 54–59.

Horváth, L., & Shao, Q.-M. (1999). Limit theorems for quadratic forms with applications to Whittle's estimate. *Annals of Applied Probability, 9*(1), 146–187.

Hotelling, H. (1931). The generalization of student's ratio. *Annals of Mathematical Statistics, 2*, 360–378.

Houdré, C., & Reynaud-Bouret, P. (2003). Exponential inequalities, with constants, for U-statistics of order two. In E. Giné et al. (Ed.), *Stochastic inequalities and applications.* Selected papers presented at the Euroconference on "Stochastic inequalities and their applications", Barcelona, June 18–22, 2002. Progress in Probability (Vol. 56, pp. 55–69). Basel: Birkhäuser

Hsu, D., Kakade, S. M., & Zhang, T. (2012). A tail inequality for quadratic forms of subgaussian random vectors. *Electronic Communications in Probability, 17*(52), 6.

Huber, P. (1967). The behavior of maximum likelihood estimates under nonstandard conditions. In *Proceedings of the Fifth Berkeley Symposium on Mathematical Statistics and Probability*, University of California 1965/66, (Vol. 1, pp. 221–233).

Ibragimov, I., & Khas'minskij, R. (1981). *Statistical estimation. Asymptotic theory* (S. Kotz. Trans. from the Russian). New York, Heidelberg, Berlin: Springer.

Janssen, A. (1998). *Zur Asymptotik nichtparametrischer Tests. Lecture Notes. Skripten zur Stochastik Nr. 29.* Münster: Gesellschaft zur Förderung der Mathematischen Statistik.

Jeffreys, H. (1935). Some tests of significance, treated by the theory of probability. *Proceedings of the Cambridge Philosophical Society, 31*, 203–222.

Jeffreys, H. (1957). *Scientific inference* (2nd ed.). Cambridge: Cambridge University Press.

Jeffreys, H. (1961). *Theory of probability* (3rd ed.). The international series of monographs on physics. Oxford: Clarendon Press.

Kass, R. E., & Raftery, A. E. (1995). Bayes factors. *Journal of the American Statistical Association, 90*(430), 773–795.

Kendall, M. (1971). Studies in the history of probability and statistics. XXVI: The work of Ernst Abbe. *Biometrika, 58*, 369–373.

Kolmogorov, A. (1933). Sulla determinazione empirica di una legge di distribuzione. *Giornale dell'istituto italiano degli attuari, 4*, 83–91.

Kullback, S. (1997). *Information theory and statistics*. *Dover books on mathematics*. Mineola: Dover Publications.

Lehmann, E., & Casella, G. (1998). *Theory of point estimation* (2nd ed.). New York, NY: Springer

Lehmann, E. L. (1999). *Elements of large-sample theory.* New York, NY: Springer.

Lehmann, E. L. (2011). *Fisher, Neyman, and the creation of classical statistics.* New York, NY: Springer.

Lehmann, E. L., & Romano, J. P. (2005). *Testing statistical hypotheses* (3rd ed.). New York, NY: Springer.

Markoff, A. (1912). *Wahrscheinlichkeitsrechnung. Nach der 2. Auflage des russischen Werkes übersetzt von H. Liebmann.* Leipzig, Berlin: B. G. Teubner.

Massart, P. (2007). *Concentration inequalities and model selection. Ecole d'Eté de Probabilités de Saint-Flour XXXIII-2003.* Lecture notes in mathematics. Berlin: Springer.

Murphy, S., & van der Vaart, A. (2000). On profile likelihood. *Journal of the American Statistical Association, 95*(450):449–485.

Neyman, J., & Pearson, E. S. (1928). On the use and interpretation of certain test criteria for purposes of statistical inference. I, II. *Biometrika, 20*A, 175–240, 263–294.

Neyman, J., & Pearson, E. S. (1933). On the problem of the most efficient tests of statistical hypotheses. *Philosophical Transactions of the Royal Society of London. Series A, Containing Papers of a Mathematical or Physical Character, 231*, 289–337.

Pearson, K. (1900). On the criterion, that a given system of deviations from the probable in the case of a correlated system of variables is such that it can be reasonably supposed to have arisen from random sampling. *Philosophical Magazine Series (5), 50*, 157–175.

Pesarin, F., & Salmaso, L. (2010). *Permutation tests for complex data: Theory, applications and software.* Wiley series in probability and statistics. Hoboken, NJ: Wiley.

Pitman, E. (1937). Significance tests which may be applied to samples from any populations. *Journal of the Royal Statistical Society, 4*(1), 119–130.

Polzehl, J., & Spokoiny, V. (2006). Propagation-separation approach for local likelihood estimation. *Probability Theory and Related Fields, 135*(3):335–362.

Raiffa, H., & Schlaifer, R. (1961). *Applied statistical decision theory.* Cambridge, MA: MIT Press. Republished in Wiley Classic Library Series (2000).

Robert, C. P. (2001). *The Bayesian choice. From decision-theoretic foundations to computational implementation* (2nd ed.). New York, NY: Springer.

Savage, L. J. (1954). *The foundations of statistics.* Wiley Publications in Statistics (Vol. XV, 294 pp.). New York: Wiley.

Scheffé, H. (1959). *The analysis of variance.* A Wiley publication in mathematical statistics. New York, London: Wiley, Chapman & Hall.

Searle, S. (1971). *Linear models.* Wiley series in probability and mathematical statistics. New York: Wiley.

Shorack, G. R., & Wellner, J. A. (1986). *Empirical processes with applications to statistics.* Wiley series in probability and mathematical statistics. New York, NY: Wiley.

Smirnov, N. (1937). On the distribution of Mises' ω^2-criterion (in Russian). *Matematicheskiĭ Sbornik Novaya Seriya. 2*(5), 973–993.

Smirnov, N. (1948). Table for estimating the goodness of fit of empirical distributions. *Annals of Mathematical Statistics, 19*(2), 279–281.

Stein, C. (1956). Inadmissibility of the usual estimator for the mean of a multivariate normal distribution. In *Proceedings of the third Berkeley symposium on mathematical statistics and probability* (Vol. 1, pp. 197–206).

Strasser, H. (1985). *Mathematical theory of statistics. Statistical experiments and asymptotic decision theory* (Vol. 7). De Gruyter studies in mathematics. Berlin, New York: de Gruyter.

Student. (1908). The probable error of a mean. *Biometrika, 6*, 1–25.

Szegö, G. (1939). *Orthogonal polynomials* (Vol. 23). American Mathematical Society colloquium publications. Providence: American Mathematical Society.

Tikhonov, A. (1963). Solution of incorrectly formulated problems and the regularization method. *Soviet Mathematics - Doklady, 5*, 1035–1038.

Van Trees, H. L. (1968). *Detection, estimation, and modulation theory. Part I* (Vol. XIV, 697 pp.). New York, London, Sydney: Wiley.

von Mises, R. (1931). *Vorlesungen aus dem Gebiete der angewandten Mathematik. Bd. 1. Wahrscheinlichkeitsrechnung und ihre Anwendung in der Statistik und theoretischen Physik*. Leipzig, Wien: Franz Deuticke.

Wahba, G. (1990). *Spline models for observational data*. Philadelphia, PA: Society for Industrial and Applied Mathematics.

Wald, A. (1943). Tests of statistical hypotheses concerning several parameters when the number of observations is large. *Transactions of the American Mathematical Society, 54*, 426–482.

Wasserman, L. (2006). *All of nonparametric statistics*. Springer texts in statistics. Springer.

White, H. (1982). Maximum likelihood estimation of misspecified models. *Econometrica, 50*, 1–25.

Wilks, S. (1938). The large-sample distribution of the likelihood ratio for testing composite hypotheses. *Annals of Mathematical Statistics, 9*, 60–62.

Witting, H. (1985). *Mathematische Statistik I: Parametrische Verfahren bei festem Stichprobenumfang*. Stuttgart: B. G. Teubner.

Witting, H., & Müller-Funk, U. (1995). *Mathematische Statistik II. Asymptotische Statistik: Parametrische Modelle und nichtparametrische Funktionale*. Stuttgart: B. G. Teubner.

Index

V. Spokoiny and T. Dickhaus, *Basics of Modern Mathematical Statistics*,
Springer Texts in Statistics, DOI 10.1007/978-3-642-39909-1,

Printed in the United States
By Bookmasters